Notation

Elementary
Linear Algebra

Stewart Venit
Wayne Bishop

both of
California State University, Los Angeles

Prindle, Weber & Schmidt
Boston, Massachusetts

PWS PUBLISHERS

Prindle. Weber & Schmidt · ❦ · Willard Grant Press · **wG** · Duxbury Press · ♠
Statler Office Building · 20 Providence Street · Boston. Massachusetts 02116

Library of Congress Cataloging in Publication Data
Venit, Stewart.
   Elementary linear algebra.

   Includes index.
   1. Algebras, Linear.   I.   Bishop, Wayne, joint
author.II. Title.
QA184.V46   512'.5   80-18251
ISBN 0–87150–300–X

ISBN 0-87150-300-X

Printed in the United States of America.
10 9 8 7 6 5 4   — 85 84 83

PWS Publishers is a division of Wadsworth, Inc.

Text designed by Dana Andrus with the production staff of Prindle,
Weber & Schmidt. Artwork by Phil Carver & Friends. Composition by
Holmes Photosetting. Printing and binding by Halliday Lithograph.
Cover design by Helen Walden.

Preface

This text is designed primarily for a sophomore level course for majors in mathematics, engineering, and the physical and social sciences. Knowledge of the calculus is not required, except for some optional material near the end of the text. However, the mental preparation given by a quarter or two of calculus would benefit most students.

For the most part, we have emphasized the computational aspects of linear algebra. One of our basic goals has been to prepare a very readable text that develops students' abilities to apply their knowledge as well as to quote it. We have also tried to avoid the pitfall of stating and proving elegant theorems without showing how these theorems can be used to solve concrete problems. Throughout the text, algorithmic procedures are carefully described to guide students in their thinking, and often the statement and/or proof of a result is essentially the algorithm itself.

Since linear algebra is usually a transition course bridging the heavily computational algebra and the calculus with the more theoretical upper division courses in mathematics and other disciplines, we have tried to keep a reasonable balance between algorithms and theoretical understanding. Within each chapter, the easier more computational material appears at the beginning, while the more theoretical is reserved for the end. In a short course stressing mechanics, the instructor can use this organization to minimize theory without interrupting the flow. Or, by emphasizing material on theory, a fairly rigorous course could be created.

To make a gradual transition from the computational to the theoretical, we have delayed the development of abstract vector spaces until much later than usual in a text of this sort. Linear independence and basis are introduced for $\mathbf{R}^m$ in chapter 4, to ease the way through a difficult part of the material. So by chapter 7, when the idea of an abstract vector space is introduced, it is the *only* totally new idea: by then students have already worked extensively with linear independence, span, subspace, basis, and dimension in the context of $\mathbf{R}^m$.

Several other features help make this text especially suitable for an introductory course. First, the determinant of a matrix is defined in terms of cofactor expansion rather than by the use of permutations.

Also, we avoid the use of summation notation and proofs by induction. In fact, the more difficult proofs are often postponed until the end of the section or at least until an illustrative example has been worked.

The book ends with two chapters on applications. These could be presented with relevant material earlier in the text. Several applications could be used on completing chapter 2; others require background material from chapters 6 and 7. The prerequisites for the applications are listed in the introductions to chapters 8 and 9, and are also given in the accompanying flowchart.

As a bonus, computational notes in the text and special computational exercises are included. The computational notes relate the topic under discussion to digital computers. The exercises are designed to be solved with the aid of library (canned) programs available with most modern computer systems. In case canned programs are not available, sample programs are included in the appendix. These programs can be entered once on the local system (even by someone without programming experience) and used by all students in the future. They are useful for checking arithmetic quickly. However, the computational exercises can be omitted without losing any of the theoretical information or the flow of the course.

Although the recommended chapter order is the one used in the text, some instructors may prefer to do the vector space chapter (7) immediately after the material on independence and basis in $\mathbf{R}^m$ (chapter 4). To do this, simply omit the few examples in chapter 7 that deal with linear transformations.

This text can be used for quarter- or semester-length courses and by both poorly prepared and mathematically advanced students. Some possible combinations are:

1. **Minimally prepared students, quarter course:**
   All of chapters 1 and 2,
   Chapter 3, sections 3.1 through 3.3 (computational aspects)
   Chapter 4, sections 4.1 through 4.3,
   Chapter 5, sections 5.1 and 5.2
2. **Average class, semester course:**
   All of chapter 1 (but quickly),
   All of chapters 2 and 3,
   Chapter 4, sections 4.1 through 4.3,
   Chapter 5, sections 5.1 through 5.3,
   Chapter 6, section 6.1,
   Chapter 7, sections 7.1 through 7.3,
   Chapters 8 and 9, as appropriate.
3. **Advanced class, semester course:**
   Chapter 1, assigned reading,

All of chapters 2 through 7, with some early sections as assigned reading,

Chapters 8 and 9, as appropriate.

Other modifications are possible, and the text can be adapted to meet many different situations. For example, if the material in sections 1.1 through 1.3 has been covered already in the calculus, the course can begin with section 1.4. The instructor is referred to the flowchart for a detailed outline of the dependence of the material.

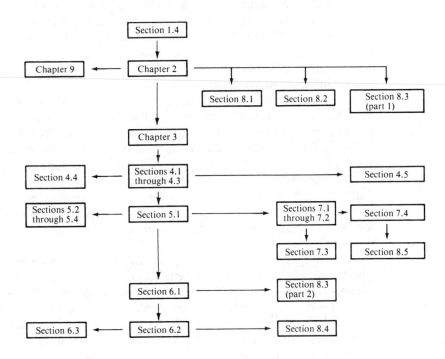

Interrelationship of Contents

Acknowledgments

We are grateful to the many people who helped us to bring this linear algebra project to fruition. Of special importance were those reviewers who stayed with the manuscript from its beginning to completion. They are: Gordon Brown, University of Colorado; Burton Fein, Oregon State University; David Kullman, Miami University (Ohio); Ancel Mewborn, University of North Carolina at Chapel Hill; Edgar Stout, University of Washington.

Others who reviewed the manuscript are: Jean Bevis, Georgia State University; Sam Councilman, California State University, Long Beach; Robert Devaney, Tufts University; Richard Porter, Northeastern University; Harris Shultz, California State University, Fullerton; Ray Spring, Ohio University; Harlan Stevens, Pennsylvania State University.

Special recognition should be given to Richard Chamberlain of California State University, Los Angeles who took great interest in our project. He made helpful suggestions on numerous occasions while class-testing the text and then reviewed the entire manuscript upon its completion.

We were fortunate to have as our typist, Katherine Numoto, who somehow managed to produce beautiful copy from our sometimes barely legible handwriting. We were also fortunate to deal with people at Prindle, Weber & Schmidt who were as interested as we were in producing a high quality text. These include John Kimmel and Barbara Schott who guided the manuscript to completion, and Helen Walden who supervised its production.

Finally we would like to thank our wives, Corinne and Judi, who offered us much encouragement and did not begrudge us the time it took to complete this project.

Contents

Geometry of R^m

In the physical sciences some entities, such as mass and pressure, can be adequately described by simply giving their magnitude. Others, such as force and velocity, require not only magnitude but direction as well. These latter entities are usually represented by *vectors*.

In sections 1.1 through 1.3 we will develop the properties of vectors in 2-space (the plane) and 3-space. Here it is easy to draw pictures of the situation and let intuition be our guide. Then in section 1.4 we will generalize our discussion to Euclidean m-space, **R**^{m}. The terminology and results given in this section will be used throughout the rest of the text.

1.1 Vectors in R^2 and R^3

Each point in a plane can be identified with a unique ordered pair of real numbers by use of a pair of perpendicular lines called *coordinate axes* (figure 1.1). In a similar manner each point in 3-space can be represented by a unique ordered triple of real numbers (figure 1.2). Throughout this text we will denote the set of real numbers by **R**, the set of ordered pairs of real numbers by **R**^{2}, and the set of ordered triples by **R**^{3}.

In addition to identifying a point in 2-space or 3-space, an ordered pair or triple will be used to identify the *directed line segment* from the

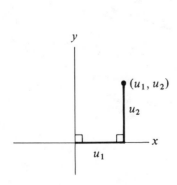

Figure 1.1

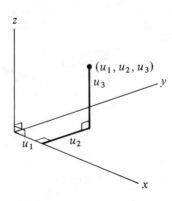

Figure 1.2

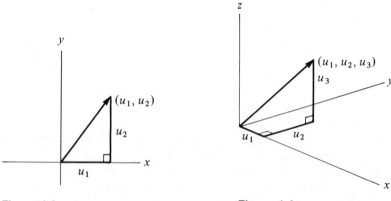

Figure 1.3   **Figure 1.4**

origin to the point indicated (figures 1.3 and 1.4). This interpretation of an ordered pair or triple is called a **vector** in $\mathbf{R}^2$ or $\mathbf{R}^3$. Vectors will be identified by bold-faced, lower-case, English letters. For example, $\mathbf{u} = (u_1, u_2, u_3)$ represents the vector in $\mathbf{R}^3$ with **components** u_1, u_2, and u_3. When we wish to speak of a *point* in $\mathbf{R}^2$ or $\mathbf{R}^3$, we use the same notation $\mathbf{u}$ but speak of "the point $\mathbf{u}$."

Two vectors $\mathbf{u}$ and $\mathbf{v}$ are **equal** if their *corresponding components* are equal; that is, $u_1 = v_1$ and $u_2 = v_2$ (and in $\mathbf{R}^3$ $u_3 = v_3$). The **zero vector** is the origin of the coordinate system, $\mathbf{0} = (0, 0)$ (or $(0, 0, 0)$). Thus the statement $\mathbf{u} \neq \mathbf{0}$ is equivalent to saying that at least one component of $\mathbf{u}$ is nonzero.

Two nonzero vectors $\mathbf{u}$ and $\mathbf{v}$ are *collinear* if the lines they determine are one and the same, in other words, if the *points* $\mathbf{u}$, $\mathbf{v}$, and $\mathbf{0}$ are collinear points (figure 1.5).

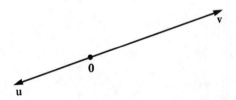

Figure 1.5

Length of Vector

Recall that the distance between any two points $\mathbf{u}$ and $\mathbf{v}$ in $\mathbf{R}^2$ can be given by the **distance formula,**

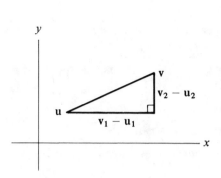

Figure 1.6

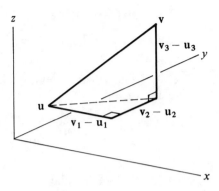

Figure 1.7

$$d(\mathbf{u}, \mathbf{v}) = \sqrt{(v_1 - u_1)^2 + (v_2 - u_2)^2}.$$

Similarly for $\mathbf{u}$ and $\mathbf{v}$ in $\mathbf{R}^3$

$$d(\mathbf{u}, \mathbf{v}) = \sqrt{(v_1 - u_1)^2 + (v_2 - u_2)^2 + (v_3 - u_3)^2}.$$

By considering the right triangles indicated in figures 1.6 and 1.7, we see that the distance formula is an application of the Pythagorean theorem. For example in figure 1.6, the hypotenuse of the right triangle is $d(\mathbf{u}, \mathbf{v})$, and the legs have lengths $v_1 - u_1$ and $v_2 - u_2$ (or more generally $|v_1 - u_1|$ and $|v_2 - u_2|$). Then

$$(d(\mathbf{u}, \mathbf{v}))^2 = (v_1 - u_1)^2 + (v_2 - u_2)^2,$$

and, by taking square roots, we obtain the first of the distance formulas. For the second formula this process is repeated twice.

Definition   The **length (norm, magnitude)** of a vector $\mathbf{u}$ is the distance from the origin to the point $\mathbf{u}$ and is denoted $\|\mathbf{u}\|$:

$$\|\mathbf{u}\| = d(\mathbf{0}, \mathbf{u}) = \begin{cases} \sqrt{u_1^2 + u_2^2}, & \text{if } \mathbf{u} \text{ is in } \mathbf{R}^2, \\ \sqrt{u_1^2 + u_2^2 + u_3^2}, & \text{if } \mathbf{u} \text{ is in } \mathbf{R}^3. \end{cases}$$

Note: If $\|\mathbf{u}\| = 0$, then $\mathbf{u} = \mathbf{0}$ (and conversely). If $\|\mathbf{u}\| = 1$, the vector $\mathbf{u}$ is called a **unit vector.**

Example 1.1

Find $\|\mathbf{u}\|$, where $\mathbf{u} = (2, -1, 3)$.

$$\|\mathbf{u}\| = \sqrt{2^2 + (-1)^2 + 3^2} = \sqrt{4 + 1 + 9} = \sqrt{14}.$$

Scalar Multiplication

An element of the set of real numbers $\mathbf{R}$ is called a **scalar**.

Definition Let c be a scalar and $\mathbf{u}$ a vector in $\mathbf{R}^2$ or $\mathbf{R}^3$. The **product** of c and $\mathbf{u}$ is the vector

$$c\mathbf{u} = \begin{cases} (cu_1, cu_2), & \text{if } \mathbf{u} \text{ is in } \mathbf{R}^2, \\ (cu_1, cu_2, cu_3), & \text{if } \mathbf{u} \text{ is in } \mathbf{R}^3. \end{cases}$$

In other words, $c\mathbf{u}$ is the vector obtained by multiplying each component of $\mathbf{u}$ by the scalar c.

Example 1.2

Find $c\mathbf{u}$ for each scalar c and vector $\mathbf{u}$.

a. $c = 3$, $\mathbf{u} = (1, -4, 4)$

$$c\mathbf{u} = (3, -12, 12).$$

b. $c = 0$, $\mathbf{u} = (2, 4)$

$$c\mathbf{u} = (0, 0) = \mathbf{0}.$$

Examples of the geometric interpretation of $c\mathbf{u}$ are given in figure 1.8 and will be justified by the next theorem.

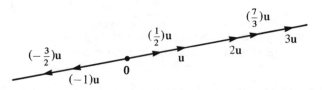

Figure 1.8

Theorem 1 Let $\mathbf{u}$ be a nonzero vector in $\mathbf{R}^2$ or $\mathbf{R}^3$, and let c be a scalar. Then $\mathbf{u}$ and $c\mathbf{u}$ are collinear and

a. if $c > 0$, then $\mathbf{u}$ and $c\mathbf{u}$ have the same direction,
b. if $c < 0$, then $\mathbf{u}$ and $c\mathbf{u}$ have opposite directions,
c. $\|c\mathbf{u}\| = |c|\,\|\mathbf{u}\|$.

Proof Collinearity of $\mathbf{u}$ and $c\mathbf{u}$ follows from 1a and 1b. The arguments for these statements can be based on similar triangles (similar tetrahedra in $\mathbf{R}^3$), since the lengths of corresponding legs are in equal ratios (figure 1.9). The direction is reversed for $c < 0$, since each component direction is reversed (figure 1.10).

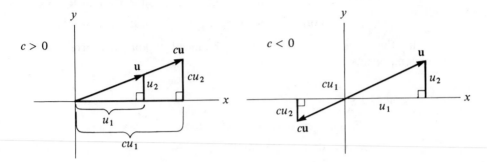

Figure 1.9   **Figure 1.10**

To verify 1c, just apply the distance formula. For example, if **u** is in $\mathbf{R}^2$,

$$\|c\mathbf{u}\| = \|(cu_1, cu_2)\| = \sqrt{(cu_1)^2 + (cu_2)^2}$$
$$= \sqrt{c^2(u_1^2 + u_2^2)} = \sqrt{c^2}\sqrt{u_1^2 + u_2^2} = |c|\,\|\mathbf{u}\|. \quad \blacksquare$$

Example 1.3

Find the *midpoint* of the vector **u**.

Since $(1/2)\mathbf{u}$ has the same direction as **u** and its length is $(1/2)\|\mathbf{u}\|$, the point $(1/2)\mathbf{u}$ is the midpoint.

Example 1.4

For a nonzero vector **u**, find a unit vector with the same direction.

We will show that the desired vector is $\dfrac{1}{\|\mathbf{u}\|}\mathbf{u}$. It has the same direction as **u** by theorem 1a and it has length 1, (it is a unit vector) because $\left\|\dfrac{1}{\|\mathbf{u}\|}\mathbf{u}\right\| = \dfrac{1}{\|\mathbf{u}\|}\|\mathbf{u}\|$ by theorem 1c.

Note: The scalar product result holds in reverse as well: if **u** and **v** are nonzero collinear vectors, then $\mathbf{v} = c\mathbf{u}$ for some scalar c. The argument can be based on similar triangles, where c is taken as the constant of proportionality (or its negative if **u** and **v** have opposite directions).

Vector Addition

Definition   Let **u** and **v** be vectors in **R**^{2} or **R**^{3}. The **sum** of **u** and **v** is the vector

$$\mathbf{u} + \mathbf{v} = \begin{cases} (u_1 + v_1,\ u_2 + v_2), & \text{if } \mathbf{u}, \mathbf{v} \text{ are in } \mathbf{R}^2, \\ (u_1 + v_1,\ u_2 + v_2,\ u_3 + v_3), & \text{if } \mathbf{u}, \mathbf{v} \text{ are in } \mathbf{R}^3. \end{cases}$$

In other words, **u** + **v** is the vector obtained by adding the corresponding components.

Example 1.5

Find the sum of each pair of vectors.

a. **u** = (2, 1, −1), **v** = (1, 2, 3)

$$\mathbf{u} + \mathbf{v} = (3, 3, 2).$$

b. **u** = (3, −1), **v** = (1, 3)

$$\mathbf{u} + \mathbf{v} = (4, 2).$$

We now look at the geometric interpretation of the sum of two vectors. The endpoint of **u** + **v** is located by the endpoint of the directed line segment with the same direction and length as **v**, but originating at the point **u** (figure 1.11). Another way to view **u** + **v** (for noncollinear vectors) is as the diagonal from **0** of the parallelogram determined by **u** and **v** (figure 1.12).

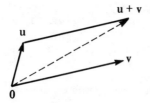

Figure 1.11

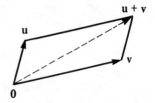

Figure 1.12

The above remarks are justified by the following theorem.

Theorem 2   For nonzero vectors **u** and **v** the directed segment from the point **u** to the point **u** + **v** is parallel and equal in length to the vector **v**.

Proof   We will show that the figure determined by the points **0**, **u**, **u** + **v**, and **v** is a parallelogram. In a plane it is sufficient to show that the two pairs of opposite sides are equal in length. By the distance formula

$$d(\mathbf{u}, \mathbf{u} + \mathbf{v}) = d((u_1, u_2), (u_1 + v_1, u_2 + v_2))$$
$$= \sqrt{((u_1 + v_1) - u_1)^2 + ((u_2 + v_2) - u_2)^2}$$
$$= \sqrt{v_1^2 + v_2^2} = \|\mathbf{v}\| = d(\mathbf{0}, \mathbf{v}).$$

Similarly since $d(\mathbf{v}, \mathbf{u} + \mathbf{v}) = d(\mathbf{0}, \mathbf{u})$, the figure is a parallelogram. In $\mathbf{R}^3$ these equations are as easy to check, but this is not quite enough to prove the assertion. It must also be confirmed that $\mathbf{u} + \mathbf{v}$ is *coplanar* with $\mathbf{0}$, $\mathbf{u}$, and $\mathbf{v}$. This detail will be verified in section 1.4, exercise 38. ∎

For each vector $\mathbf{u}$ the vector that has the same length as $\mathbf{u}$ but points in the opposite direction is called the **negative** of $\mathbf{u}$ and will be denoted $-\mathbf{u}$. By theorem 1, $(-1)\mathbf{u}$ has these properties, so $(-1)\mathbf{u} = -\mathbf{u}$. It follows from the definition of scalar multiplication that

$$-\mathbf{u} = \begin{cases} (-u_1, -u_2), & \text{if } \mathbf{u} \text{ is in } \mathbf{R}^2, \\ (-u_1, -u_2, -u_3), & \text{if } \mathbf{u} \text{ is in } \mathbf{R}^3. \end{cases}$$

Now we can define subtraction of vectors in terms of addition by "adding the negative." That is, we define the **difference** of vectors $\mathbf{u}$ and $\mathbf{v}$ to be

$$\mathbf{u} - \mathbf{v} = \mathbf{u} + (-\mathbf{v}).$$

The properties of vector addition and subtraction and scalar multiplication are given in theorem 3 and follow directly from the definitions of these operations. To demonstrate how the arguments can be constructed, the proofs of 3a, 3b, and 3f are included, and the others are left as exercises. In the proofs we assume the vectors are in $\mathbf{R}^2$ (for $\mathbf{R}^3$ just add a third component).

Theorem 3   Let $\mathbf{u}$, $\mathbf{v}$, and $\mathbf{w}$ be vectors in $\mathbf{R}^2$ or $\mathbf{R}^3$, and let c and d be scalars. Then

a. $\mathbf{u} + \mathbf{v} = \mathbf{v} + \mathbf{u}$ (vector addition is commutative),
b. $(\mathbf{u} + \mathbf{v}) + \mathbf{w} = \mathbf{u} + (\mathbf{v} + \mathbf{w})$ (vector addition is associative),
c. $\mathbf{u} + \mathbf{0} = \mathbf{u}$,
d. $\mathbf{u} + (-\mathbf{u}) = \mathbf{0}$,
e. $(cd)\mathbf{u} = c(d\mathbf{u})$,
f. $(c + d)\mathbf{u} = c\mathbf{u} + d\mathbf{u}$ ⎫
g. $c(\mathbf{u} + \mathbf{v}) = c\mathbf{u} + c\mathbf{v}$ ⎬   (distributive laws),
h. $1\mathbf{u} = \mathbf{u}$,
i. $(-1)\mathbf{u} = -\mathbf{u}$,
j. $0\mathbf{u} = \mathbf{0}$.

Proof of 3a $\mathbf{u} + \mathbf{v} = (u_1 + v_1, u_2 + v_2)$
$$= (v_1 + u_1, v_2 + u_2)$$
$$= \mathbf{v} + \mathbf{u}.$$

Proof of 3b
$$(\mathbf{u} + \mathbf{v}) + \mathbf{w} = (u_1 + v_1, u_2 + v_2) + (w_1, w_2)$$
$$= ((u_1 + v_1) + w_1, (u_2 + v_2) + w_2)$$
$$= (u_1 + (v_1 + w_1), u_2 + (v_2 + w_2))$$
$$= (u_1, u_2) + (v_1 + w_1, v_2 + w_2)$$
$$= \mathbf{u} + (\mathbf{v} + \mathbf{w}).$$

Proof of 3f
$$(c + d)\mathbf{u} = ((c + d)u_1, (c + d)u_2)$$
$$= (cu_1 + du_1, cu_2 + du_2)$$
$$= (cu_1, cu_2) + (du_1, du_2)$$
$$= c\mathbf{u} + d\mathbf{u}. \qquad \blacksquare$$

Translation

Subtraction of vectors is a particularly useful way of explaining directed line segments that do not begin at the origin. For points **p** and **q** in **R**^{2} or **R**^{3}, we will denote the directed line segment from **p** to **q** by the symbol $\overrightarrow{\mathbf{pq}}$. Two directed segments $\overrightarrow{\mathbf{pq}}$ and $\overrightarrow{\mathbf{rs}}$ are said to be **equivalent** if they have the same direction and length (figure 1.13). The next theorem relates equivalent directed segments to vector subtraction.

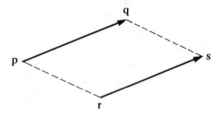

Figure 1.13

Theorem 4   Let **u** and **v** be distinct points in **R**^{2} or **R**^{3}. Then the vector **v** − **u** is equivalent to the directed line segment from **u** to **v**. In other words, $\overrightarrow{\mathbf{uv}}$ is parallel to the vector **v** − **u** and $d(\mathbf{u}, \mathbf{v}) = \| \mathbf{v} - \mathbf{u} \|$.

Proof   Viewing **u** as a vector, we add **u** and **v** − **u**,

$$\mathbf{u} + (\mathbf{v} - \mathbf{u}) = \mathbf{u} + (\mathbf{v} + (-\mathbf{u})) \qquad \text{(Definition of subtraction)}$$
$$= (\mathbf{u} + \mathbf{v}) + (-\mathbf{u}) \qquad \text{(theorem 3b)}$$
$$= (\mathbf{v} + \mathbf{u}) + (-\mathbf{u}) \qquad \text{(theorem 3a)}$$

$$= \mathbf{v} + (\mathbf{u} + (-\mathbf{u})) \qquad \text{(theorem 3b)}$$
$$= \mathbf{v} + \mathbf{0} \qquad \text{(theorem 3d)}$$
$$= \mathbf{v} \qquad \text{(theorem 3c).}$$

Since $\mathbf{v}$ is the sum of $\mathbf{u}$ and $\mathbf{v} - \mathbf{u}$, by theorem 2, the vector $\mathbf{v} - \mathbf{u}$ is parallel and equal in length, hence equivalent, to the directed line segment from $\mathbf{u}$ to $\mathbf{v}$ (figure 1.14). ■

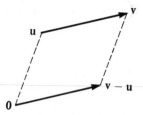

Figure 1.14

Note: The process of replacing the directed line segment $\overrightarrow{\mathbf{uv}}$ by the vector $\mathbf{v} - \mathbf{u}$ is often called *translating* $\mathbf{u}$ *to the origin.* In reverse, the process is called *translating to* $\mathbf{u}$.

Example 1.6

Let l_1 and l_2 be lines in $\mathbf{R}^3$. Suppose that l_1 is determined by the points $\mathbf{p} = (2, 2, 1)$ and $\mathbf{q} = (-1, 0, 2)$ and that l_2 is determined by the points $\mathbf{r} = (2, 1, 1)$ and $\mathbf{s} = (8, 5, -1)$. Are the lines parallel?

We translate the directed segments $\overrightarrow{\mathbf{pq}}$ and $\overrightarrow{\mathbf{rs}}$ to the origin and compare them,

$$\mathbf{q} - \mathbf{p} = (-3, -2, 1),$$
$$\mathbf{s} - \mathbf{r} = (6, 4, -2).$$

Since one of these is a scalar multiple of the other, $\mathbf{s} - \mathbf{r} = -2(\mathbf{q} - \mathbf{p})$, they determine the same line through the origin. But by theorem 4, $\mathbf{q} - \mathbf{p}$ is parallel to l_1 and $\mathbf{s} - \mathbf{r}$ is parallel to l_2. Thus l_1 and l_2 are parallel.

Alternate Notation

It is clear that a vector (a, b) in $\mathbf{R}^2$ may be written

$$(a, b) = a(1, 0) + b(0, 1).$$

Similarly (a, b, c) in $\mathbf{R}^3$ may be written

$$(a, b, c) = a(1, 0, 0) + b(0, 1, 0) + c(0, 0, 1).$$

In many engineering and physics books the symbols reserved for these special vectors are $\mathbf{i} = (1, 0, 0)$, $\mathbf{j} = (0, 1, 0)$, and $\mathbf{k} = (0, 0, 1)$. In **R**^{2} we have only $\mathbf{i}$ and $\mathbf{j}$ and drop the last component. Then the vector (a, b, c) is written as $a\mathbf{i} + b\mathbf{j} + c\mathbf{k}$.

Example 1.7

Express $4\mathbf{i} - 3\mathbf{k}$ as an ordered triple.

Since $\mathbf{j}$ is missing, its coefficient is 0, and the result is $4\mathbf{i} - 3\mathbf{k} = (4, 0, -3)$.

1.1 Exercises

In exercises 1 through 14 let the vectors be given as follows:

$$\mathbf{r} = (-1, 1), \qquad \mathbf{u} = (3, 1, 0),$$
$$\mathbf{s} = (2, 0), \qquad \mathbf{v} = (2, 0, 1),$$
$$\mathbf{t} = (3, -2), \qquad \mathbf{w} = (-1, -2, 3).$$

Perform each indicated operation.

1. $4\mathbf{w}$ 2. $\mathbf{s} + \mathbf{t}$

3. $\mathbf{u} + \mathbf{v}$ 4. $\mathbf{t} - \mathbf{r}$

5. $\mathbf{w} - \mathbf{v}$ 6. $3\mathbf{r} + 2\mathbf{s} - \mathbf{t}$

7. $\mathbf{u} - 5\mathbf{v} + 2\mathbf{w}$ 8. $\|\mathbf{t}\|$

9. $\|\mathbf{w}\|$ 10. $d(\mathbf{s}, \mathbf{t})$

11. $d(\mathbf{w}, \mathbf{v})$ 12. $\|\mathbf{w} - \mathbf{v}\|$

13. $\|\mathbf{w}\| - \|\mathbf{v}\|$ 14. $\|5(\mathbf{w} - \mathbf{v})\|$

In exercises 15 and 16 find the unit vector with the same direction as the given one.

15. $(3, -4)$ 16. $(2, 1, -3)$

In exercises 17 and 18 find the vector that is equivalent to the directed line segment from the first point to the second.

17. $(-1, 2)$, $(3, 4)$ 18. $(2, 1, 0)$, $(-1, 3, 4)$

19. Express $(2, 0, -3)$ in $\mathbf{i}, \mathbf{j}, \mathbf{k}$ form.

20. Express $3\mathbf{i} + 2\mathbf{j} - \mathbf{k}$ as an ordered triple.

21. Find the length of $2\mathbf{i} + 3\mathbf{j} - \mathbf{k}$.

22. Simplify $3(2\mathbf{i} + 7\mathbf{j} - \mathbf{k}) - 2(\mathbf{i} - 4\mathbf{j})$.

23. Prove the properties of theorem 3 that remain unproved. Assume the vectors are in **R**^{3}.

24. Prove that the midpoint of the line segment determined by the points $\mathbf{p}$ and $\mathbf{q}$

is $(1/2)\mathbf{p} + (1/2)\mathbf{q}$. (*Hint:* this can be done directly by the distance formula, but using theorem 4 is easier.)

25. Prove directly from the distance formula that $d(\mathbf{p}, \mathbf{q}) = \|\mathbf{p} - \mathbf{q}\|$.

26. Sketch a picture of the set of endpoints of all unit vectors in $\mathbf{R}^2$. What is the figure? What about in $\mathbf{R}^3$?

27. Describe the geometric figure consisting of the set of all solutions $\mathbf{x}$ in $\mathbf{R}^2$ to the equation $\|\mathbf{x} - \mathbf{p}\| = r$ for $\mathbf{p}$ in $\mathbf{R}^2$ and r in $\mathbf{R}$. What about in $\mathbf{R}^3$?

28. Let $\mathbf{p}$ and $\mathbf{q}$ be distinct points and t a real number. Let $\mathbf{x}(t) = (1 - t)\mathbf{p} + t\mathbf{q}$. Use theorem 4 to prove that

$$\frac{d(\mathbf{p}, \mathbf{x}(t))}{d(\mathbf{p}, \mathbf{q})} = |t|.$$

29. Use exercise 28 to find a point one-third of the distance from $\mathbf{p}$ to $\mathbf{q}$.

30. Let l_1 be the line determined by $(2, 1, 3)$ and $(1, 2, -1)$ and similarly l_2 by $(0, 2, 3)$ and $(-1, 1, 2)$. Is l_1 parallel to l_2?

31. Let l_1 be the line determined by $(3, 1, 2)$ and $(4, 3, 1)$ and similarly l_2 by $(1, 3, -3)$ and $(-1, -1, -1)$. Is l_1 parallel to l_2?

32. Let $\mathbf{p}, \mathbf{q}$, and $\mathbf{r}$ be the three vertices of a triangle. Using theorem 4, prove that the line segment that joins the midpoints of two of the sides is parallel to and half of the length of the third side.

1.2   Dot and Cross Products

In this section we will define and develop the properties of two types of vector products: the *dot product*, which yields a scalar and is of general interest and the *cross product*, which yields another vector and is useful only in $\mathbf{R}^3$.

Dot Product

Since two nonzero vectors are assumed to have a common initial point, $\mathbf{0}$, they form an angle. The following operation will be useful in describing this angle.

Definition   Let $\mathbf{u}$ and $\mathbf{v}$ be vectors in $\mathbf{R}^2$ or $\mathbf{R}^3$. The **dot product** (or **standard inner product**) of $\mathbf{u}$ and $\mathbf{v}$ is denoted by $\mathbf{u} \cdot \mathbf{v}$ and is defined as follows:

$$\mathbf{u} \cdot \mathbf{v} = \begin{cases} u_1 v_1 + u_2 v_2, & \text{if } \mathbf{u}, \mathbf{v} \text{ are in } \mathbf{R}^2, \\ u_1 v_1 + u_2 v_2 + u_3 v_3, & \text{if } \mathbf{u}, \mathbf{v} \text{ are in } \mathbf{R}^3. \end{cases}$$

In words, $\mathbf{u} \cdot \mathbf{v}$ is the sum of the products of corresponding components of $\mathbf{u}$ and $\mathbf{v}$.

Example 1.8

Compute the dot product of each pair of vectors.

a. $\mathbf{u} = (1, 2, 3)$, $\mathbf{v} = (-2, 4, 1)$

$\mathbf{u} \cdot \mathbf{v} = (1, 2, 3) \cdot (-2, 4, 1) = (1)(-2) + (2)(4) + (3)(1) = 9.$

b. $\mathbf{u} = (1, 2)$, $\mathbf{v} = (3, -4)$

$\mathbf{u} \cdot \mathbf{v} = (1, 2) \cdot (3, -4) = (1)(3) + (2)(-4) = -5.$

c. $\mathbf{u} = (2, 1, -3)$, $\mathbf{v} = (1, 1, 1)$

$\mathbf{u} \cdot \mathbf{v} = (2, 1, -3) \cdot (1, 1, 1) = (2)(1) + (1)(1) + (-3)(1) = 0.$

Notice that $\mathbf{u} \cdot \mathbf{v}$ is a *scalar* rather than a vector. For this reason dot product is often called **scalar product.** The algebraic properties of dot product are included in the next theorem.

Theorem 1   Let $\mathbf{u}$, $\mathbf{v}$, and $\mathbf{w}$ be vectors in $\mathbf{R}^2$ or $\mathbf{R}^3$, and let c be a scalar. Then

a. $\mathbf{u} \cdot \mathbf{v} = \mathbf{v} \cdot \mathbf{u}$ (dot product is commutative),
b. $c(\mathbf{u} \cdot \mathbf{v}) = (c\mathbf{u}) \cdot \mathbf{v} = \mathbf{u} \cdot (c\mathbf{v})$ (scalars factor out),
c. $\mathbf{u} \cdot (\mathbf{v} + \mathbf{w}) = \mathbf{u} \cdot \mathbf{v} + \mathbf{u} \cdot \mathbf{w}$ (distributive law),
d. $\mathbf{u} \cdot \mathbf{0} = 0$,
e. $\mathbf{u} \cdot \mathbf{u} = \|\mathbf{u}\|^2.$

Proof   Assume the vectors are in $\mathbf{R}^2$; the case of $\mathbf{R}^3$ is entirely analogous.

Proof of 1a
$$\begin{aligned} \mathbf{u} \cdot \mathbf{v} &= u_1 v_1 + u_2 v_2 \\ &= v_1 u_1 + v_2 u_2 \\ &= \mathbf{v} \cdot \mathbf{u}. \end{aligned}$$

Proof of 1b $c(\mathbf{u} \cdot \mathbf{v}) = c(u_1 v_1 + u_2 v_2)$

$$= c u_1 v_1 + c u_2 v_2 = \begin{cases} (c u_1) v_1 + (c u_2) v_2 = (c\mathbf{u}) \cdot \mathbf{v}, \\ u_1 (c v_1) + u_2 (c v_2) = \mathbf{u} \cdot (c\mathbf{v}). \end{cases}$$

Proof of 1e
$$\begin{aligned} \mathbf{u} \cdot \mathbf{u} &= u_1 u_1 + u_2 u_2 \\ &= u_1^2 + u_2^2 \\ &= \|\mathbf{u}\|^2. \end{aligned}$$

The proofs of 1c and 1d are left as exercises.   ■

The geometric interpretation of $\mathbf{u} \cdot \mathbf{v}$ is particularly interesting. If $\mathbf{u}$ and $\mathbf{v}$ are nonzero vectors, they form an *angle*, since they both originate at the origin $\mathbf{0}$ (figure 1.15). We will use *radian* measure for the angle (although degree measure could be used as well), and we will use the smallest nonnegative number possible. In other words, if θ is the measure of the angle, we have $0 \leq \theta \leq \pi$. The next theorem describes the relationship between this angle and $\mathbf{u} \cdot \mathbf{v}$.

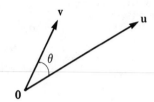

Figure 1.15

Theorem 2   Let $\mathbf{u}$ and $\mathbf{v}$ be vectors in $\mathbf{R}^2$ or $\mathbf{R}^3$, and let θ be the angle between $\mathbf{u}$ and $\mathbf{v}$. Then

$$\mathbf{u} \cdot \mathbf{v} = \|\mathbf{u}\| \, \|\mathbf{v}\| \cos \theta,$$

or equivalently for nonzero $\mathbf{u}$ and $\mathbf{v}$,

$$\cos \theta = \frac{\mathbf{u} \cdot \mathbf{v}}{\|\mathbf{u}\| \, \|\mathbf{v}\|}.$$

Proof   Consider the triangle determined by $\mathbf{u}$ and $\mathbf{v}$ (figure 1.16). Let $a = \|\mathbf{u}\|$, $b = \|\mathbf{v}\|$, and $c = \|\mathbf{v} - \mathbf{u}\|$. Recall from section 1.1 that $\mathbf{v} - \mathbf{u}$ is equivalent to the directed line segment $\overrightarrow{\mathbf{uv}}$, and thus c is the length of the third side of the triangle. To this triangle we apply the law of cosines, $c^2 = a^2 + b^2 - 2ab \cos \theta$. Substituting for a, b, and c, we have

$$\|\mathbf{v} - \mathbf{u}\|^2 = \|\mathbf{u}\|^2 + \|\mathbf{v}\|^2 - 2\|\mathbf{u}\| \, \|\mathbf{v}\| \cos \theta. \tag{1}$$

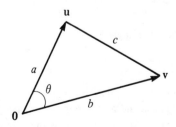

Figure 1.16

We need a more convenient form for $\|\mathbf{v} - \mathbf{u}\|^2$.

$$\|\mathbf{v} - \mathbf{u}\|^2 = (\mathbf{v} - \mathbf{u}) \cdot (\mathbf{v} - \mathbf{u}) \qquad \text{(theorem 1e)}$$
$$= (\mathbf{v} - \mathbf{u}) \cdot \mathbf{v} - (\mathbf{v} - \mathbf{u}) \cdot \mathbf{u} \qquad \text{(theorem 1c)}$$
$$= \mathbf{v} \cdot \mathbf{v} - \mathbf{u} \cdot \mathbf{v} - \mathbf{v} \cdot \mathbf{u} + \mathbf{u} \cdot \mathbf{u} \qquad \text{(theorem 1c)}$$
$$= \mathbf{v} \cdot \mathbf{v} - \mathbf{u} \cdot \mathbf{v} - \mathbf{u} \cdot \mathbf{v} + \mathbf{u} \cdot \mathbf{u} \qquad \text{(theorem 1a)}$$
$$= \|\mathbf{v}\|^2 - 2(\mathbf{u} \cdot \mathbf{v}) + \|\mathbf{u}\|^2 \qquad \text{(theorem 1e)}.$$

Returning now to equation (1),

$$\|\mathbf{v}\|^2 - 2(\mathbf{u} \cdot \mathbf{v}) + \|\mathbf{u}\|^2 = \|\mathbf{u}\|^2 + \|\mathbf{v}\|^2 - 2\|\mathbf{u}\|\, \|\mathbf{v}\| \cos\theta,$$
$$-2(\mathbf{u} \cdot \mathbf{v}) = -2\|\mathbf{u}\|\, \|\mathbf{v}\| \cos\theta,$$
$$\mathbf{u} \cdot \mathbf{v} = \|\mathbf{u}\|\, \|\mathbf{v}\| \cos\theta. \qquad \blacksquare$$

Example 1.9

Find the cosine of the angle between each pair of vectors.

a. $\mathbf{u} = (1, 2, 3)$, $\mathbf{v} = (-2, 4, 1)$

$$\cos\theta = \frac{\mathbf{u} \cdot \mathbf{v}}{\|\mathbf{u}\|\, \|\mathbf{v}\|}$$

$$= \frac{(1)(-2) + (2)(4) + (3)(1)}{\sqrt{1^2 + 2^2 + 3^2}\,\sqrt{(-2)^2 + 4^2 + 1^2}}$$

$$= \frac{9}{\sqrt{14}\,\sqrt{21}}$$

$$= \frac{3\sqrt{6}}{14}.$$

(Using tables or a hand calculator, θ is approximately 1.02 radians.)

b. $\mathbf{u} = (2, 1, -3)$, $\mathbf{v} = (1, 1, 1)$

$$\cos\theta = \frac{\mathbf{u} \cdot \mathbf{v}}{\|\mathbf{u}\|\, \|\mathbf{v}\|}$$

$$= \frac{0}{\|\mathbf{u}\|\, \|\mathbf{v}\|}$$

$$= 0$$

(Here the angle θ is $\pi/2$ radians, so θ is a right angle.)

Example 1.10

Find the cosine of the angle at **p** determined by the directed line segments $\overrightarrow{\mathbf{pq}}$ and $\overrightarrow{\mathbf{pr}}$, where $\mathbf{p} = (1, 2)$, $\mathbf{q} = (1 + \sqrt{3}, -1)$ and $\mathbf{r} = (1 - \sqrt{3}, 3)$.
We need vectors to apply theorem 2, so we translate $\overrightarrow{\mathbf{pq}}$ and $\overrightarrow{\mathbf{pr}}$ to the

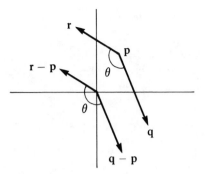

Figure 1.17

origin by subtracting **p** from **q** and **r**. This results in an angle at the origin determined by the vectors **q** − **p** and **r** − **p** (figure 1.17). Since these vectors are respectively equivalent (and hence parallel) to $\overrightarrow{\mathbf{pq}}$ and $\overrightarrow{\mathbf{pr}}$, the angle they form is the same, and we may use theorem 2,

$$
\begin{aligned}
\cos \theta &= \frac{(\mathbf{q} - \mathbf{p}) \cdot (\mathbf{r} - \mathbf{p})}{\|\mathbf{q} - \mathbf{p}\|\, \|\mathbf{r} - \mathbf{p}\|} \\
&= \frac{(\sqrt{3},\, -3) \cdot (-\sqrt{3},\, 1)}{\|(\sqrt{3},\, -3)\|\, \|(-\sqrt{3},\, 1)\|} \\
&= \frac{-6}{\sqrt{12}\,\sqrt{4}} \\
&= -\frac{\sqrt{3}}{2}.
\end{aligned}
$$

(This time the angle θ is $5\pi/6$ radians.)

We will investigate further the relationship between dot product and angle after introducing the following concept:

Definition   Vectors **u** and **v** in $\mathbf{R}^2$ or $\mathbf{R}^3$ are **orthogonal** if $\mathbf{u} \cdot \mathbf{v} = 0$.

For example, the vectors $\mathbf{u} = (2, 1, -3)$ and $\mathbf{v} = (1, 1, 1)$ of example 1.8 are orthogonal. In example 1.9 we saw that these same vectors form a right angle. The words *orthogonal* (from Greek), *perpendicular* (from Latin), and *normal* (also Latin) mean essentially the same thing, "meet at right angles." However, usage sometimes makes one word preferable to another. Two vectors are usually said to be *orthogonal*, two lines or planes are *perpendicular*, and a vector is *normal* to a plane.

Theorem 3   Let **u** and **v** be nonzero vectors in **R**^{2} or **R**^{3}, and let θ be the angle they form. Then θ is

> i. an acute angle if $\mathbf{u} \cdot \mathbf{v} > 0$,
> ii. a right angle if $\mathbf{u} \cdot \mathbf{v} = 0$,
> iii. an obtuse angle if $\mathbf{u} \cdot \mathbf{v} < 0$.

Proof   From theorem 2 we have the equation

$$\mathbf{u} \cdot \mathbf{v} = \|\mathbf{u}\| \, \|\mathbf{v}\| \cos \theta.$$

Since $\|\mathbf{u}\|$ and $\|\mathbf{v}\|$ are positive, it follows that $\mathbf{u} \cdot \mathbf{v}$ is zero, positive, or negative as $\cos \theta$ is zero, positive, or negative. But $\cos \theta > 0$ implies θ is an acute angle. Similarly $\cos \theta = 0$ implies θ is a right angle, and $\cos \theta < 0$ implies θ is obtuse.   ∎

   In example 1.9 we calculated $\cos \theta$ to be $3\sqrt{6}/14$ and gave the approximate measure of the angle. Without tables or a calculator θ is seen to be an acute angle by theorem 3. Similarly, the angle in example 1.10 must be obtuse, since $(\mathbf{q} - \mathbf{p}) \cdot (\mathbf{r} - \mathbf{p}) = -6$.

Cross Product

It may have seemed surprising that the dot product of two vectors is a scalar instead of a vector. There is in **R**^{3} another type of product of two vectors that produces a vector.

Definition   Let $\mathbf{u} = (u_1, u_2, u_3)$ and $\mathbf{v} = (v_1, v_2, v_3)$ be vectors in **R**^{3}. The **cross product** of **u** and **v** is the vector

$$\mathbf{u} \times \mathbf{v} = (u_2 v_3 - u_3 v_2, \; u_3 v_1 - u_1 v_3, \; u_1 v_2 - u_2 v_1).$$

   Fortunately there is a convenient way of remembering this formula. Given an array of numbers

$$\begin{bmatrix} a & b \\ c & d \end{bmatrix},$$

we define the **determinant** of the array as the difference of the diagonal products,

$$\det \begin{bmatrix} a & b \\ c & d \end{bmatrix} = ad - bc.$$

With this notation the expression for $\mathbf{u} \times \mathbf{v}$ becomes

$$\mathbf{u} \times \mathbf{v} = \left(\det \begin{bmatrix} u_2 & u_3 \\ v_2 & v_3 \end{bmatrix}, \ -\det \begin{bmatrix} u_1 & u_3 \\ v_1 & v_3 \end{bmatrix}, \ \det \begin{bmatrix} u_1 & u_2 \\ v_1 & v_2 \end{bmatrix} \right).$$

The arrays that appear in this expression are easy to remember by use of the following device: write the components of $\mathbf{u}$ above those of $\mathbf{v}$, forming the rectangular array

$$\begin{bmatrix} u_1 & u_2 & u_3 \\ v_1 & v_2 & v_3 \end{bmatrix}.$$

Then the square arrays that appear in the expression for $\mathbf{u} \times \mathbf{v}$ are obtained by successively deleting the first, second, and third columns of this array.

Example 1.11

Find $\mathbf{u} \times \mathbf{v}$ where $\mathbf{u} = (1, 2, -1)$ and $\mathbf{v} = (0, 2, 3)$.

Construct the array $\begin{bmatrix} 1 & 2 & -1 \\ 0 & 2 & 3 \end{bmatrix}$. Then

$$\mathbf{u} \times \mathbf{v} = \left(\det \begin{bmatrix} 2 & -1 \\ 2 & 3 \end{bmatrix}, \ -\det \begin{bmatrix} 1 & -1 \\ 0 & 3 \end{bmatrix}, \ \det \begin{bmatrix} 1 & 2 \\ 0 & 2 \end{bmatrix} \right)$$

$$= ((2)(3) - (-1)(2), \ -(1)(3) + (-1)(0), \ (1)(2) - (2)(0))$$

$$= (6 + 2, \ -3 - 0, \ 2 - 0)$$

$$= (8, -3, 2).$$

For our purposes the most important property of cross product is given by the following theorem.

Theorem 4   The vector $\mathbf{u} \times \mathbf{v}$ is orthogonal to both $\mathbf{u}$ and $\mathbf{v}$.

Proof   We compute $\mathbf{u} \cdot (\mathbf{u} \times \mathbf{v})$,

$$(u_1, u_2, u_3) \cdot (u_2 v_3 - u_3 v_2, \ u_3 v_1 - u_1 v_3, \ u_1 v_2 - u_2 v_1)$$

$$= u_1 u_2 v_3 - u_1 u_3 v_2 + u_2 u_3 v_1 - u_2 u_1 v_3 + u_3 u_1 v_2 - u_3 u_2 v_1$$

$$= 0.$$

Similarly $\mathbf{v} \cdot (\mathbf{u} \times \mathbf{v}) = 0$.   ∎

Example 1.12

Find a vector that is orthogonal to both $\mathbf{u} = (1, 2, -1)$ and $\mathbf{v} = (0, 2, 3)$.

These are the vectors of example 1.11, so by theorem 4, $\mathbf{u} \times \mathbf{v} = (8, -3, 2)$ is orthogonal to both $\mathbf{u}$ and $\mathbf{v}$.

Other properties of cross product are given in the following theorem.

Theorem 5   Let $\mathbf{u}, \mathbf{v}$ be vectors in $\mathbf{R}^3$ and c a scalar. Then

a. $\mathbf{u} \times \mathbf{v} = -(\mathbf{v} \times \mathbf{u})$,

b. $\mathbf{u} \times (\mathbf{v} + \mathbf{w}) = (\mathbf{u} \times \mathbf{v}) + (\mathbf{u} \times \mathbf{w})$ $\left.\right\}$ (distributive laws),

c. $(\mathbf{u} + \mathbf{v}) \times \mathbf{w} = (\mathbf{u} \times \mathbf{w}) + (\mathbf{v} \times \mathbf{w})$

d. $c(\mathbf{u} \times \mathbf{v}) = (c\mathbf{u}) \times \mathbf{v} = \mathbf{u} \times (c\mathbf{v})$ (scalars factor out),

e. $\mathbf{u} \times \mathbf{0} = \mathbf{0} \times \mathbf{u} = \mathbf{0}$,

f. $\mathbf{u} \times \mathbf{u} = \mathbf{0}$,

g. $\|\mathbf{u} \times \mathbf{v}\| = \|\mathbf{u}\| \, \|\mathbf{v}\| \sin \theta = \sqrt{\|\mathbf{u}\|^2 \, \|\mathbf{v}\|^2 - (\mathbf{u} \cdot \mathbf{v})^2}$,

where θ is the angle determined by $\mathbf{u}$ and $\mathbf{v}$.

Proof of 5a   Let $\mathbf{u} = (u_1, u_2, u_3)$ and $\mathbf{v} = (v_1, v_2, v_3)$, and simply compute both $\mathbf{u} \times \mathbf{v}$ and $\mathbf{v} \times \mathbf{u}$. The important fact here is the geometrical interpretation: $\mathbf{u} \times \mathbf{v}$ and $\mathbf{v} \times \mathbf{u}$ have the same length but point in opposite directions.

Proof of 5b   Let $\mathbf{u}$ and $\mathbf{v}$ be given as above, and $\mathbf{w} = (w_1, w_2, w_3)$. Then

$$\mathbf{u} \times (\mathbf{v} + \mathbf{w}) = (u_1, u_2, u_3) \times (v_1 + w_1, v_2 + w_2, v_3 + w_3)$$
$$= (u_2(v_3 + w_3) - u_3(v_2 + w_2), u_3(v_1 + w_1) - u_1(v_3 + w_3),$$
$$u_1(v_2 + w_2) - u_2(v_1 + w_1))$$
$$= (u_2 v_3 + u_2 w_3 - u_3 v_2 - u_3 w_2, u_3 v_1 + u_3 w_1 - u_1 v_3 - u_1 w_3,$$
$$u_1 v_2 + u_1 w_2 - u_2 v_1 - u_2 w_1)$$
$$= (u_2 v_3 - u_3 v_2, u_3 v_1 - u_1 v_3, u_1 v_2 - u_2 v_1)$$
$$+ (u_2 w_3 - u_3 w_2, u_3 w_1 - u_1 w_3, u_1 w_2 - u_2 w_1)$$
$$= \mathbf{u} \times \mathbf{v} + \mathbf{u} \times \mathbf{w}.$$

Proofs of 5c through 5f are straightforward computations and are left as exercises.

Proof of 5g   Notice the similarity of the statement 5g with the rule relating dot product and the cosine of the angle, $\|\mathbf{u}\| \, \|\mathbf{v}\| \cos \theta = \mathbf{u} \cdot \mathbf{v}$. Starting with this equation, we square both sides and proceed as follows:

$$\|\mathbf{u}\|^2 \|\mathbf{v}\|^2 \cos^2 \theta = (\mathbf{u} \cdot \mathbf{v})^2,$$
$$\|\mathbf{u}\|^2 \|\mathbf{v}\|^2 (1 - \sin^2 \theta) = (\mathbf{u} \cdot \mathbf{v})^2,$$
$$\|\mathbf{u}\|^2 \|\mathbf{v}\|^2 \sin^2 \theta = \|\mathbf{u}\|^2 \|\mathbf{v}\|^2 - (\mathbf{u} \cdot \mathbf{v})^2.$$

Taking square roots,

$$\|\mathbf{u}\| \, \|\mathbf{v}\| \sin \theta = \sqrt{\|\mathbf{u}\|^2 \|\mathbf{v}\|^2 - (\mathbf{u} \cdot \mathbf{v})^2}.$$

It is a straightforward but tedious computation to verify that $\|\mathbf{u} \times \mathbf{v}\| = \sqrt{\|\mathbf{u}\|^2 \|\mathbf{v}\|^2 - (\mathbf{u} \cdot \mathbf{v})^2}$. Simply expand each expression and verify that they are the same. The details are left as an exercise. ∎

Note: Although cross product is distributive over addition (theorem 5b and 5c), it is *not* commutative (theorem 5a). Moreover it is easily shown that the cross product is also not associative (see exercise 21).

The geometrical interpretation of theorem 5g is that the length of $\mathbf{u} \times \mathbf{v}$ is the area of the parallelogram determined by $\mathbf{u}$ and $\mathbf{v}$ (figure 1.18).

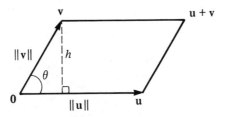

Figure 1.18

To see why this is so, let h be the height of the parallelogram from the point $\mathbf{v}$ to the side determined by $\mathbf{u}$. Then the area A of the parallelogram is given by $A = \|\mathbf{u}\| h$. However, $\sin \theta = h/\|\mathbf{v}\|$, or $h = \|\mathbf{v}\| \sin \theta$. Making the substitution, we conclude that $A = \|\mathbf{u}\| \, \|\mathbf{v}\| \sin \theta$. This fact together with theorem 4 gives a geometrical description of $\mathbf{u} \times \mathbf{v}$ as follows.

Theorem 6   For noncollinear vectors $\mathbf{u}$ and $\mathbf{v}$ in $\mathbf{R}^3$, the vector $\mathbf{u} \times \mathbf{v}$ is perpendicular to the plane determined by the vectors $\mathbf{u}$ and $\mathbf{v}$ and has length equal to the area of the parallelogram that they determine (figure 1.19).

Example 1.13

Find the area of the triangle determined by the three points $\mathbf{p} = (1, 2, 3)$, $\mathbf{q} = (-3, 2, 1)$, and $\mathbf{r} = (2, 4, 5)$.

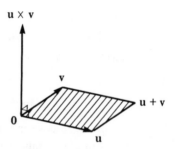

Figure 1.19

We translate $\overrightarrow{pq}$ and $\overrightarrow{pr}$ to the origin to obtain the vectors $\mathbf{u} = \mathbf{q} - \mathbf{p}$ and $\mathbf{v} = \mathbf{r} - \mathbf{p}$. By the preceding remarks $\|\mathbf{u} \times \mathbf{v}\|$ is the area of the parallelogram determined by $\mathbf{u}$ and $\mathbf{v}$. Thus the area of the triangle determined by $\mathbf{u}$ and $\mathbf{v}$ is $(1/2)\|\mathbf{u} \times \mathbf{v}\|$ (figure 1.20). Since $\mathbf{u} = (-4, 0, -2)$, and $\mathbf{v} = (1, 2, 2)$, we have

$$\frac{1}{2}\|\mathbf{u} \times \mathbf{v}\| = \frac{1}{2}\left\|\left(\det\begin{bmatrix} 0 & -2 \\ 2 & 2 \end{bmatrix}, -\det\begin{bmatrix} -4 & -2 \\ 1 & 2 \end{bmatrix}, \det\begin{bmatrix} -4 & 0 \\ 1 & 2 \end{bmatrix}\right)\right\|$$

$$= \frac{1}{2}\|(4, 6, -8)\| = \|(2, 3, -4)\| = \sqrt{29}.$$

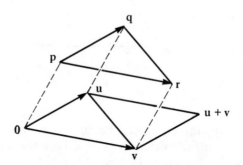

Figure 1.20

1.2 Exercises

In exercises 1 through 8 let the vectors be given as follows:

$$\mathbf{s} = (-1, 2), \qquad\qquad \mathbf{u} = (3, -1, 4),$$
$$\mathbf{t} = (2, 3), \qquad\qquad \mathbf{v} = (-1, -3, 1),$$
$$\mathbf{w} = (-1, 1, 2).$$

Perform each indicated operation.

1. $\mathbf{s} \cdot \mathbf{t}$

2. $\mathbf{s} \cdot (3\mathbf{t})$

3. $\mathbf{u} \cdot \mathbf{v}$

4. $(3\mathbf{v}) \cdot (-2\mathbf{u})$

5. $\mathbf{u} \times \mathbf{v}$

6. $(\mathbf{u} \times \mathbf{v}) \cdot \mathbf{w}$

7. $\mathbf{v} \times (2\mathbf{u})$

8. $(2\mathbf{u} + \mathbf{v}) \times (3\mathbf{w})$

In exercises 9 and 10 determine if the angle formed by the two vectors is acute, right, or obtuse.

9. $(-2, 1)$, $(5, 2)$

10. $(-1, 2, 3)$, $(2, 0, 4)$

In exercises 11 and 12 determine the cosine of the angle formed by the two vectors.

11. $(-2, 1)$, $(5, 2)$

12. $(-1, 2, 3)$, $(2, 0, 4)$

In exercises 13 and 14 find a vector that is orthogonal to both vectors.

13. $(1, 1, 3)$, $(2, 1, -1)$

14. $(-1, 2, 3)$, $(2, 0, 4)$

15. Find the area of the triangle with vertices at $(2, 1, 3)$, $(1, 0, 2)$, and $(-1, 1, 2)$.

16. Find the area of the parallelogram determined by the vectors $\mathbf{u} = (-1, 1, 0)$ and $\mathbf{v} = (2, 3, -1)$.

17. Prove that for any vector $\mathbf{u} = (a, b)$ in $\mathbf{R}^2$ the vector $\mathbf{r} = (-b, a)$ is orthogonal to $\mathbf{u}$.

The following expressions are undefined. Explain why.

18. $\mathbf{u} \times (\mathbf{v} \cdot \mathbf{w})$

19. $\mathbf{u} \cdot (\mathbf{v} \cdot \mathbf{w})$

20. $\|\mathbf{u} \cdot \mathbf{v}\|$

21. Show that cross product is not associative. That is, it is not always true that $(\mathbf{u} \times \mathbf{v}) \times \mathbf{w} = \mathbf{u} \times (\mathbf{v} \times \mathbf{w})$. (*Hint:* try $(\mathbf{i} \times \mathbf{i}) \times \mathbf{j}$ and $\mathbf{i} \times (\mathbf{i} \times \mathbf{j})$, where $\mathbf{i} = (1, 0, 0)$ and $\mathbf{j} = (0, 1, 0)$.)

22. Use theorem 5g to prove that $\mathbf{u}$ and $\mathbf{v}$ are collinear if and only if $\mathbf{u} \times \mathbf{v} = \mathbf{0}$.

23. Prove the properties in theorem 1 that remain unproved.

24. Use theorem 1e to prove the *parallelogram law*,

$$\|\mathbf{u} + \mathbf{v}\|^2 + \|\mathbf{u} - \mathbf{v}\|^2 = 2\|\mathbf{u}\|^2 + 2\|\mathbf{v}\|^2.$$

Interpret this result geometrically to conclude "The sum of the squares of the sides of a parallelogram is equal to the sum of the squares of the diagonals."

25. Use theorem 2 and the trigonometric identity, $\cos(\pi - \theta) = -\cos\theta$, to conclude that the angles formed by $\mathbf{u}$ and $\mathbf{v}$ and by $-\mathbf{u}$ and $\mathbf{v}$ are supplementary (add to π radians).

26. Prove the properties of theorem 5 that remain unproved.

27. Let $\mathbf{u} = (u_1, u_2, u_3)$ be a unit vector in $\mathbf{R}^3$ in the direction of $\mathbf{v}$, that is, $\mathbf{u} = (1/\|\mathbf{v}\|)\mathbf{v}$. Use theorem 2 to prove that u_1, u_2, and u_3 are the cosines of the angles formed by $\mathbf{v}$ with $(1, 0, 0)$, $(0, 1, 0)$, and $(0, 0, 1)$, respectively. The components of $\mathbf{u}$ are called the *direction cosines* of $\mathbf{v}$.

28. Find the direction cosines of the vector $(1, 2, -2)$.

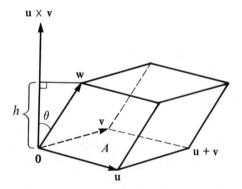

Figure 1.21

29. Use figure 1.21 to show that the *volume of the parallelepiped* determined by the vectors **u**, **v**, and **w** in **R**^{3} is given by

$$V = |(\mathbf{u} \times \mathbf{v}) \cdot \mathbf{w}|.$$

(Recall that $V = Ah$, the area of the base times the perpendicular height.)

30. Use the result of exercise 29 to find the volume of the parallelepiped determined by the vectors **u** = (2, 1, −1), **v** = (3, −1, 0), and **w** = (1, 1, −1).

1.3   Lines and Planes

In this section we will show how vector equations and functions can be used to describe lines in **R**^{2} and lines and planes in **R**^{3}.

Point-Normal Form for a Plane

It is intuitively clear that the set of all directed line segments in **R**^{3} from a given point and perpendicular to a given directed line segment is a plane (figure 1.22). We will use this idea to develop an algebraic description of a plane.

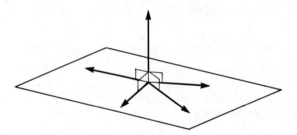

Figure 1.22

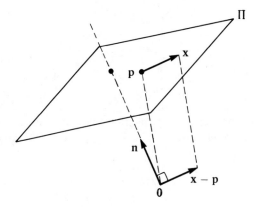

Figure 1.23

Let **p** be a point in $\mathbf{R}^3$, let **n** be a nonzero vector, and let Π denote the plane that contains the point **p** and is perpendicular to the line determined by **n** (figure 1.23). Let **x** be an arbitrary point in Π, and consider the directed segment $\overrightarrow{px}$. The vector $\mathbf{x} - \mathbf{p}$ is equivalent to $\overrightarrow{px}$ (theorem 4 of section 1.1) and therefore is orthogonal to **n**. In terms of dot product we have

$$\mathbf{n} \cdot (\mathbf{x} - \mathbf{p}) = 0. \tag{1}$$

In other words, any point **x** in the plane Π satisfies equation (1). Furthermore, if a point **y** is not in Π, then $\mathbf{p} - \mathbf{y}$ cannot be orthogonal to **n**. Thus equation (1) completely describes the plane Π. Equation (1) is called an equation in **point-normal form** for the plane, and **n** is called a normal for the plane. Notice that any nonzero scalar multiple of **n** and any other point in the plane yield an equally valid equation, so the point-normal form is by no means unique.

Example 1.14

Find a point-normal form for the plane that contains the point $\mathbf{p} = (2, 1, 3)$ with normal vector $\mathbf{n} = (1, 2, -2)$.

From equation (1) we have

$$\mathbf{n} \cdot (\mathbf{x} - \mathbf{p}) = 0, \quad \text{or} \quad (1, 2, -2) \cdot (\mathbf{x} - (2, 1, 3)) = 0.$$

Sometimes an expanded form of equation (1) is taken as the point-normal form. Specifically let $\mathbf{n} = (n_1, n_2, n_3)$, $\mathbf{p} = (p_1, p_2, p_3)$, and and $\mathbf{x} = (x, y, z)$. Then $\mathbf{n} \cdot (\mathbf{x} - \mathbf{p}) = 0$ becomes

$$(n_1, n_2, n_3) \cdot ((x, y, z) - (p_1, p_2, p_3)) = 0$$

$$(n_1, n_2, n_3) \cdot (x - p_1, y - p_2, z - p_3) = 0$$
$$n_1(x - p_1) + n_2(y - p_2) + n_3(z - p_3) = 0. \tag{2}$$

Equation (2) is equivalent to equation (1), and it often goes by the same name. For instance the equation of example 1.14 may be written

$$(x - 2) + 2(y - 1) - 2(z - 3) = 0.$$

The most efficient procedure for writing a "clean" equation for a plane through **p** with normal vector **n** results from expanding (1) in a slightly different manner,

$$\mathbf{n} \cdot (\mathbf{x} - \mathbf{p}) = 0$$
$$\mathbf{n} \cdot \mathbf{x} - \mathbf{n} \cdot \mathbf{p} = 0$$
$$\mathbf{n} \cdot \mathbf{x} = \mathbf{n} \cdot \mathbf{p}.$$

Letting $\mathbf{n} = (n_1, n_2, n_3)$, $\mathbf{x} = (x, y, z)$, and $\mathbf{n} \cdot \mathbf{p} = b$, we obtain an equation for the plane in **standard form:**

$$n_1 x + n_2 y + n_3 z = b. \tag{3}$$

Example 1.15

Find an equation in standard form for the plane through the given point with normal **n**.

a. $\mathbf{p} = (-1, 1, 2)$, $\mathbf{n} = (2, 3, 4)$

Since
$$(2, 3, 4) \cdot (-1, 1, 2) = 9,$$
equation (3) yields
$$2x + 3y + 4z = 9.$$

b. the origin, $\mathbf{n} = (2, -1, 3)$

Here $\mathbf{p} = \mathbf{0}$ and $\mathbf{n} \cdot \mathbf{p} = 0$, so we have simply
$$2x - y + 3z = 0.$$

Suppose we are given an equation for a plane in standard form

$$ax + by + cz = d, \quad (a, b, c) \neq \mathbf{0},$$

and we wish to transform it into point-normal form. The plane in question must be orthogonal to the vector $\mathbf{n} = (a, b, c)$, but we still need a point in the plane to get a point-normal form for it. One such point is easily found. If $a \neq 0$, then $(d/a, 0, 0)$ satisfies the equation. If $b \neq 0$, we can use $(0, d/b, 0)$, and similarly, if $c \neq 0$ we can use $(0, 0, d/c)$. Letting **p** be any of these, it is a simple matter to show that $\mathbf{n} \cdot (\mathbf{x} - \mathbf{p}) = 0$ reduces to the original equation $ax + by + cz = d$.

Example 1.16

Find an equation in point-normal form for the plane described by the equation $3x - y + 2z = 5$.

A normal vector is $(3, -1, 2)$, and the point $(0, -5, 0)$ satisfies the equation. Thus

$$(3, -1, 2) \cdot (\mathbf{x} - (0, -5, 0)) = 0$$

(or $3(x - 0) - 1(y + 5) + 2(z - 0) = 0$) is one such equation.

Plane Determined by Three Points

A plane Π in $\mathbf{R}^3$ is determined by three noncollinear points $\mathbf{p}$, $\mathbf{q}$, and $\mathbf{r}$. If three such points are specified, the cross product of two vectors can be used to establish a normal vector. Specifically the vector $\mathbf{q} - \mathbf{p}$ is equivalent to $\overrightarrow{\mathbf{pq}}$, and $\mathbf{r} - \mathbf{p}$ is equivalent to $\overrightarrow{\mathbf{pr}}$ (figure 1.24). Since $\mathbf{p}$, $\mathbf{q}$, and $\mathbf{r}$ are not collinear, $\mathbf{q} - \mathbf{p}$ and $\mathbf{r} - \mathbf{p}$ are noncollinear vectors and by theorem 4 of section 1.2, $\mathbf{n} = (\mathbf{q} - \mathbf{p}) \times (\mathbf{r} - \mathbf{p})$ is a nonzero vector orthogonal to both $\mathbf{q} - \mathbf{p}$ and $\mathbf{r} - \mathbf{p}$. Thus $\mathbf{n}$ is normal to the plane determined by $\mathbf{q} - \mathbf{p}$ and $\mathbf{r} - \mathbf{p}$, and hence to Π as well. Using this vector, $\mathbf{n}$, the equation $\mathbf{n} \cdot (\mathbf{x} - \mathbf{p}) = 0$ describes the plane Π.

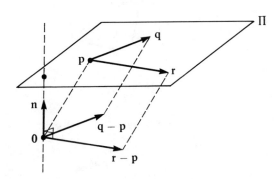

Figure 1.24

Example 1.17

Find an equation for the plane determined by the points $\mathbf{p} = (2, 1, 3)$, $\mathbf{q} = (4, 2, -1)$, and $\mathbf{r} = (6, 2, 2)$.

For a normal vector we have

$$\mathbf{n} = (\mathbf{q} - \mathbf{p}) \times (\mathbf{r} - \mathbf{p}) = (2, 1, -4) \times (4, 1, -1) = (3, -14, -2).$$

Thus a point-normal form for the plane is

$$(3, -14, -2) \cdot (\mathbf{x} - (2, 1, 3)) = 0,$$

or in standard form,

$$3x - 14y - 2z = -14.$$

Point-Normal Form for a Line in R²

Equation (1) is interesting in **R**^{2} as well. If **n** is a nonzero vector and **p** is a point in **R**^{2}, exactly the same logic as that which justified equation (1) gives the following result: the set of all solutions **x** in **R**^{2} to the vector equation

$$\mathbf{n} \cdot (\mathbf{x} - \mathbf{p}) = 0$$

is the line through **p**, which is perpendicular to the line determined by **n** (figure 1.25). This is called an equation in **point-normal** form for a line in **R**^{2}.

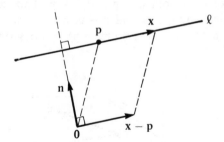

Figure 1.25

Letting $\mathbf{n} = (n_1, n_2)$, $\mathbf{p} = (p_1, p_2)$, and $\mathbf{x} = (x, y)$, we have the analog of equation (2),

$$n_1(x - p_1) + n_2(y - p_2) = 0. \tag{4}$$

The analog of equation (3) is called the **standard form** for the equation of a line. Letting $b = \mathbf{n} \cdot \mathbf{p}$, it has the form

$$n_1 x + n_2 y = b. \tag{5}$$

As before, if a line in **R**^{2} is given by an equation in standard form,

$$ax + by = c, \quad (a, b) \neq \mathbf{0},$$

then $\mathbf{n} = (a, b)$ is a normal vector to the line. For a point on the line we may use the y-intercept $(0, c/b)$ if $b \neq 0$ or the x-intercept $(c/a, 0)$ if $a \neq 0$.

Example 1.18

Find an equation in point-normal form for the line described by

$$3x - 2y = 5.$$

The vector $\mathbf{n} = (3, -2)$ is orthogonal to the line, and the x-intercept $(5/3, 0)$ is on the line, so a point-normal equation for the line is

$$(3, -2) \cdot (\mathbf{x} - (5/3, 0)) = 0$$

(or $3(x - 5/3) - 2(y - 0) = 0$).

Example 1.19

Find an equation for the line that contains the points $(-1, 1)$ and $(4, 3)$.

The directed segment from $(-1, 1)$ to $(4, 3)$ is parallel to the vector $(4 - (-1), 3 - 1) = (5, 2)$. Since the vector $(-2, 5)$ is orthogonal to $(5, 2)$, it is also orthogonal to the desired line. Now $(-2, 5) \cdot (-1, 1) = 7$, so by equation (5) we have $-2x + 5y = 7$.

Point-Parallel Form for a Line

Since the usual form for the equation of a line in $\mathbf{R}^2$, $ax + by = c$, becomes an equation of a plane in $\mathbf{R}^3$, it is natural to ask how we can describe a line in $\mathbf{R}^3$. Let $\mathbf{v}$ be a nonzero vector and $\mathbf{p}$ a point in $\mathbf{R}^3$. We wish to describe the line through $\mathbf{p}$ and parallel to $\mathbf{v}$ (figure 1.26).

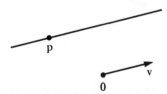

Figure 1.26

From section 1.1 we know that a point is on the line determined by the vector $\mathbf{v}$ if and only if it has the form $t\mathbf{v}$ for some scalar t. From the properties of vector addition, the directed segment from $\mathbf{p}$ to $\mathbf{p} + t\mathbf{v}$ is parallel and equal in length to the vector $t\mathbf{v}$. But then the point $\mathbf{p} + t\mathbf{v}$ must lie on the line determined by $\mathbf{p}$ and $\mathbf{p} + \mathbf{v}$ (figure 1.27).

The following equation is called the **point-parallel form** for the line through the point $\mathbf{p}$ and parallel to the vector $\mathbf{v}$,

$$\mathbf{x}(t) = \mathbf{p} + t\mathbf{v}. \tag{6}$$

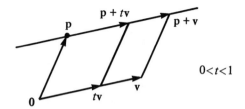

Figure 1.27

Notice that this equation is not an equation in the same sense as equations (1) through (5). It is instead a *function* where different points $\mathbf{x}(t)$ are obtained for different values of the variable t. For instance, $\mathbf{x}(0) = \mathbf{p}$ and $\mathbf{x}(1) = \mathbf{p} + \mathbf{v}$. Similarly $\mathbf{x}(1/2)$ is the midpoint of the segment from $\mathbf{p}$ to $\mathbf{p} + \mathbf{v}$ and $\mathbf{x}(2)$ and $\mathbf{x}(-1)$ are as pictured in figure 1.28.

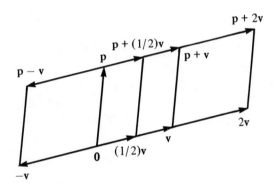

Figure 1.28

It is also important to notice that, although we were discussing lines in **R**^{3}, no special use of **R**^{3} was made in the discussion leading to equation (6). Therefore the interpretation of the equation holds in both **R**^{2} and **R**^{3}.

Example 1.20

Give a point-parallel form for the line through the point $\mathbf{p} = (2, 2, 0)$ and parallel to the vector $\mathbf{v} = (-3, -1, 2)$.

From equation (6) we have

$$\mathbf{x}(t) = (2, 2, 0) + t(-3, -1, 2).$$

Example 1.21

Give a point-parallel form for the line in $\mathbf{R}^2$ given by the equation $y = 2x - 1$.

We need to find a point on the given line and a vector parallel to it. The given line has the y-intercept $(0, -1)$, so let $\mathbf{p} = (0, -1)$. Moreover the line has slope $2 = 2/1$, so the vector $\mathbf{v} = (1, 2)$ is parallel to it. Thus we have

$$\mathbf{x}(t) = (0, -1) + t(1, 2).$$

It is sometimes convenient to write equation (6) in terms of its coordinates. To do this, let $\mathbf{p}$, $\mathbf{v}$, and $\mathbf{x} = \mathbf{x}(t)$ be points in $\mathbf{R}^3$ with $\mathbf{p} = (p_1, p_2, p_3)$, $\mathbf{v} = (v_1, v_2, v_3)$, and $\mathbf{x} = (x, y, z)$. Then equation (6) becomes

$$(x, y, z) = (p_1, p_2, p_3) + t(v_1, v_2, v_3)$$
$$= (p_1 + tv_1, p_2 + tv_2, p_3 + tv_3).$$

Equating coordinates we have

$$x = p_1 + tv_1,$$
$$y = p_2 + tv_2, \qquad (7)$$
$$z = p_3 + tv_3.$$

These equations are called **parametric equations** for the line and taken together are equivalent to the original equation (6). Of course there will only be equations for x and y in the case of $\mathbf{R}^2$.

Example 1.22

Find parametric equations for the line through the point $(2, 2, 0)$ and parallel to the vector $(-3, -1, 2)$.

From example 1.20 we have the point-parallel form for the line,

$$\mathbf{x}(t) = (2, 2, 0) + t(-3, -1, 2).$$

Equating coordinates, we have the parametric equations,

$$x = 2 - 3t,$$
$$y = 2 - t,$$
$$z = 2t.$$

Example 1.23

Find the point of intersection of the line $\mathbf{x}(t) = (2, 1, 3) + t(2, -2, 1)$ with the plane $x + 2y - z = 7$.

We substitute the parametric equations for the line,

$$x = 2 + 2t,$$
$$y = 1 - 2t,$$
$$z = 3 + t,$$

into the equation for the plane,

$$(2 + 2t) + 2(1 - 2t) - (3 + t) = 7.$$

Solving for t, we obtain $t = -2$. Thus the point is

$$\mathbf{x}(-2) = (2, 1, 3) + (-2)(2, -2, 1)$$
$$= (-2, 5, 1).$$

The *distance* from a point $\mathbf{p}$ to a plane Π in $\mathbf{R}^3$ is the perpendicular distance; the distance from $\mathbf{p}$ to Π measured along the line through $\mathbf{p}$ perpendicular to Π.

Let Π be given in point-normal form by $\mathbf{n} \cdot (\mathbf{x} - \mathbf{q}) = 0$. Since $\mathbf{n}$ is normal to Π, it provides a parallel vector needed to describe the line through $\mathbf{p}$ perpendicular to Π. Therefore the point-parallel form for this line is $\mathbf{x}(t) = \mathbf{p} + t\mathbf{n}$ (figure 1.29).

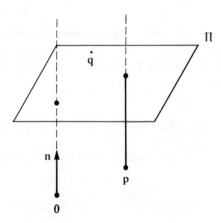

Figure 1.29

To find the point of intersection of this line with Π, we substitute into the equation for Π,

$$\mathbf{n} \cdot ((\mathbf{p} + t\mathbf{n}) - \mathbf{q}) = 0$$
$$\mathbf{n} \cdot (t\mathbf{n} - (\mathbf{q} - \mathbf{p})) = 0$$
$$\mathbf{n} \cdot (t\mathbf{n}) - \mathbf{n} \cdot (\mathbf{q} - \mathbf{p}) = 0$$

$$t(\mathbf{n} \cdot \mathbf{n}) = \mathbf{n} \cdot (\mathbf{q} - \mathbf{p})$$

$$t = \frac{\mathbf{n} \cdot (\mathbf{q} - \mathbf{p})}{\mathbf{n} \cdot \mathbf{n}} = \frac{\mathbf{n} \cdot (\mathbf{q} - \mathbf{p})}{\|\mathbf{n}\|^2}.$$

Thus the point of intersection is

$$\mathbf{x}(t) = \mathbf{p} + \left(\frac{\mathbf{n} \cdot (\mathbf{q} - \mathbf{p})}{\|\mathbf{n}\|^2} \right) \mathbf{n}.$$

Finally, the distance from $\mathbf{p}$ to this point $\mathbf{x}(t)$ is

$$d(\mathbf{p}, \mathbf{x}(t)) = \|\mathbf{x}(t) - \mathbf{p}\| = \left\| \left(\frac{\mathbf{n} \cdot (\mathbf{q} - \mathbf{p})}{\|\mathbf{n}\|^2} \right) \mathbf{n} \right\|$$

$$= \frac{|\mathbf{n} \cdot (\mathbf{q} - \mathbf{p})|}{\|\mathbf{n}\|^2} \|\mathbf{n}\| = \frac{|\mathbf{n} \cdot (\mathbf{q} - \mathbf{p})|}{\|\mathbf{n}\|}. \tag{8}$$

Example 1.24

Find the distance from the point $\mathbf{p} = (2, 1, 3)$ to the plane $2x - 3y - z = 1$.

To use equation (8), we first need to express $2x - 3y - z = 1$ in point-normal form. One such form is $(2, -3, -1) \cdot (\mathbf{x} - (0, 0, -1)) = 0$. Thus we can take $\mathbf{n} = (2, -3, -1)$ and $\mathbf{q} = (0, 0, -1)$. Substituting into equation (8), we have

$$\frac{|\mathbf{n} \cdot (\mathbf{q} - \mathbf{p})|}{\|\mathbf{n}\|} = \frac{|(2, -3, -1) \cdot (-2, -1, -4)|}{\|(2, -3, -1)\|} = \frac{3}{\sqrt{14}}.$$

Two-Point Form for a Line

Another way of determining a line geometrically is by two distinct points. Let $\mathbf{p}$ and $\mathbf{q}$ be distinct points in $\mathbf{R}^2$ or $\mathbf{R}^3$. The **two-point form** for the line determined by $\mathbf{p}$ and $\mathbf{q}$ is

$$\mathbf{x}(t) = (1 - t)\mathbf{p} + t\mathbf{q}, \tag{9}$$

where t ranges over all real numbers.

We now show that the set of all such $\mathbf{x}(t)$ is the line that contains $\mathbf{p}$ and $\mathbf{q}$. Since $\mathbf{x}(0) = \mathbf{p}$ and $\mathbf{x}(1) = \mathbf{q}$, the set does contain the points $\mathbf{p}$ and $\mathbf{q}$. That the set is a line, can be shown by converting it to the form of equation (6). Using properties of scalar multiplication and vector addition we have

$$\mathbf{x}(t) = (1 - t)\mathbf{p} + t\mathbf{q}$$
$$= \mathbf{p} - t\mathbf{p} + t\mathbf{q}$$
$$= \mathbf{p} + t(\mathbf{q} - \mathbf{p}).$$

Thus the set of all such $\mathbf{x}(t)$ is the line that contains the point $\mathbf{p}$ and is parallel to $\mathbf{q} - \mathbf{p}$ (figure 1.30).

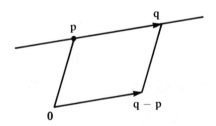

Figure 1.30

Example 1.25

Describe the line through the points $(1, 2, 3)$ and $(-1, 1, 2)$ in both the two-point form and parametrically.

From equation (9), we have the two-point form,

$$\mathbf{x}(t) = (1 - t)(1, 2, 3) + t(-1, 1, 2).$$

To find the parametric equations, let $\mathbf{x} = (x, y, z)$, and perform the indicated operations on the right

$$(x, y, z) = (1 - t, 2 - 2t, 3 - 3t) + (-t, t, 2t)$$
$$= (1 - 2t, 2 - t, 3 - t).$$

Equating coordinates, we have the parametric equations,

$$x = 1 - 2t,$$
$$y = 2 - t,$$
$$z = 3 - t.$$

1.3 Exercises

In exercises 1 through 5 give equations in both point-normal and standard form of the plane described.

1. Through $\mathbf{p} = (2, -1, -4)$ with normal $\mathbf{n} = (3, 1, 2)$

2. Through $\mathbf{p} = (1, 2, 3)$ with normal $\mathbf{n} = (-3, 0, 1)$

3. Through the origin with normal $\mathbf{n} = (2, 1, 3)$

4. Through $\mathbf{p} = (-1, 2, 3)$, $\mathbf{q} = (3, 2, 5)$, and $\mathbf{r} = (-1, 1, 2)$

5. Through $\mathbf{p} = (-1, 2, 3)$, $\mathbf{q} = (0, 2, 0)$, and $\mathbf{r} = (1, 1, 3)$

In exercises 6 through 8 convert the equation in standard form to point-normal form for a plane in $\mathbf{R}^3$.

6. $2x - y + z = 5$ **7.** $2x + 3y = 1$ **8.** $z = 0$

In exercises 9 through 14 give equations in both point-normal and standard form of the $\mathbf{R}^2$ line described.

9. Through $\mathbf{p} = (-1, 2)$ with normal $\mathbf{n} = (2, 1)$

10. Through $\mathbf{p} = (2, -1)$ with normal $\mathbf{n} = (0, 2)$

11. Through $\mathbf{p} = (-1, 2)$ and $\mathbf{q} = (3, -2)$

12. Through $\mathbf{p} = (2, 4)$ and $\mathbf{q} = (2, -6)$

13. Through $\mathbf{p} = (-2, 5)$ and perpendicular to the line $x + 2y = 5$

14. Through $\mathbf{p} = (1, -3)$ and perpendicular to the line $3x - 4y = 1$

In exercises 15 through 23 give a point-parallel or two-point form, and parametric equations for the $\mathbf{R}^2$ or $\mathbf{R}^3$ line described.

15. Through $\mathbf{p} = (2, 1, -3)$ and parallel to $\mathbf{v} = (1, 2, 2)$

16. Through $\mathbf{p} = (3, -1)$ and parallel to $\mathbf{v} = (2, 3)$

17. Through $\mathbf{p} = (2, -3, 1)$ and parallel to the x-axis

18. Through $\mathbf{p} = (1, 2, -1)$ and $\mathbf{q} = (2, -1, 3)$

19. Through $\mathbf{p} = (2, 0, -2)$ and $\mathbf{q} = (1, 4, 2)$

20. Through $\mathbf{p} = (1, 2, 3)$ and perpendicular to the plane $2x - y - 2z = 4$

21. Through $\mathbf{p} = (2, 4, 5)$ and perpendicular to the plane $5x - 5y - 10z = 2$

22. Through $\mathbf{p} = (3, -2)$ and perpendicular to the line $3x - 5y = 1$

23. Through $\mathbf{p} = (1, -1)$ and perpendicular to the line $x - 3y = 4$

24. Let $\mathbf{p} = (p_1, p_2)$, $\mathbf{u} = (u_1, u_2)$, and $\mathbf{x}(t) = \mathbf{x} = (x, y)$. Show that the point-parallel form $\mathbf{x}(t) = \mathbf{p} + t\mathbf{u}$ describes the same line in $\mathbf{R}^2$ as the point-normal form $\mathbf{n} \cdot (\mathbf{x} - \mathbf{p}) = 0$, where $\mathbf{n} = (u_2, -u_1)$.

25. Convert the line $\mathbf{x}(t) = (2, 3) + t(1, -3)$ into point-normal form.

26. Convert the line $2x - y = 5$ into point-parallel form.

27. Find the distance from the point $(3, 2, 3)$ to the plane $x - 2y + z = 5$.

28. Show that in $\mathbf{R}^2$, equation (8) is the formula for the distance from a point to a line.

29. Find the distance from the point $(2, -3)$ to the line $x + 2y = 5$.

30. Find the point of intersection of the line $\mathbf{x}(t) = (3, 0, 1) + t(-2, 1, 0)$ with the plane $3x + 2y - 3z = 3$.

31. Find the point of intersection of the line $\mathbf{x}(t) = (2, 1, 1) + t(-1, 0, 4)$ with the plane $x - 3y - z = 1$.

32. Let Π be the plane described by $\mathbf{n} \cdot (\mathbf{x} - \mathbf{q}) = 0$ and l the line $\mathbf{x}(t) = \mathbf{p} + t\mathbf{u}$ ($\mathbf{n}, \mathbf{u} \neq \mathbf{0}$). Show the following:

(a) If $\mathbf{n} \cdot \mathbf{u} = 0$, then l is contained in Π or *parallel* to it (they have no points in common);

(b) If $\mathbf{n} \cdot \mathbf{u} \neq 0$, then l and Π intersect in the point

$$\mathbf{p} + \left(\frac{\mathbf{n} \cdot (\mathbf{q} - \mathbf{p})}{\mathbf{n} \cdot \mathbf{u}} \right) \mathbf{u}$$

which is at a distance of

$$\left| \frac{\mathbf{n} \cdot (\mathbf{q} - \mathbf{p})}{\mathbf{n} \cdot \mathbf{u}} \right| \|\mathbf{u}\|$$

from $\mathbf{p}$.

33. Use exercise 32 to show that the line $\mathbf{x}(t) = (2, 4, 5) + t(1, 0, 1)$ and the plane $2x - y - 2z = 5$ are parallel.

34. Two planes are *parallel* if they have no points in common. Show that the planes $2x - 3y + z = 1$ and $4x - 6y + 2z = 5$ are parallel. (*Hint:* first show that, if they have one point in common, they are one and the same plane.)

35. Show that the planes $3x - 2y + z = 1$ and $x - y - 2z = 1$ are not parallel *without* finding any point of intersection.

36. Find the distance from the point $\mathbf{p} = (1, -1, 3)$ to the line l given by $\mathbf{x}(t) = (-1, 0, 2) + t(2, 1, 4)$. (*Hint:* find the point of intersection of l with the plane through $\mathbf{p}$ and perpendicular to l.)

37. The lines $\mathbf{x}(t) = (2, 1, 3) + t(0, -2, -1)$ and $\mathbf{y}(s) = (3, 1, 5) + s(1, 0, 2)$ intersect at $(2, 1, 3)$ (let $t = 0$ and $s = -1$). Find the cosine of the angle of intersection. (*Hint:* just use the parallel vectors. The angle θ is restricted $0 \leq \theta \leq \pi/2$ radians, so use exercise 25 of section 1.2.)

38. By considering complementary angles (ones whose sum is $\pi/2$ radians), it can be shown that the angle between two intersecting lines has the same measure as the angle formed by normal vectors to the lines (or its supplement). Find the cosine of the angle between the **R**^{2} lines $2x - y = 3$ and $x + 5y = 9$.

39. The *angle between two intersecting planes* is that of the lines formed by intersecting the planes with a third plane perpendicular to the line of intersection of the planes. As in exercise 38 it suffices to simply use normal vectors to the planes. Find the cosine of the angle between the planes $3x - y + z = 3$ and $2x + 4y - 2z = 1$.

1.4   Euclidean *m*-Space

Everything we have studied in the first three sections can be generalized to Euclidean *m*-space, **R**^{m}. In this section we introduce **R**^{m} and consider many of these generalizations. The terminology and operations presented here will be used throughout the text.

Vectors in $\mathbf{R}^m$

For $m \geq 1$ let $\mathbf{R}^m$ denote the set of all ordered m-tuples,

$$\mathbf{u} = (u_1, u_2, \ldots, u_m),$$

where the u_i are in $\mathbf{R}$, the set of real numbers. An element of $\mathbf{R}^m$ is called a **vector** or **m-vector.** The numbers $u_1, \ldots, u_m$ are called the **components** of $\mathbf{u}$ and, when unspecified, are denoted by the same letter as the vector. In other words, "m-vector $\mathbf{v}$" means the vector $\mathbf{v} = (v_1, v_2, \ldots, v_m)$. However, in case of $m = 2$ or $m = 3$ we will sometimes continue to use $\mathbf{x} = (x, y)$ or $\mathbf{x} = (x, y, z)$.

Two vectors $\mathbf{u}$ and $\mathbf{v}$ are **equal** if their *corresponding components* are equal, if $u_i = v_i$ for each i. The **zero vector** in $\mathbf{R}^m$ is the m-vector $\mathbf{0} = (0, 0, \ldots, 0)$. For a vector $\mathbf{u}$ in $\mathbf{R}^m$, the **negative** of $\mathbf{u}$ is the vector $-\mathbf{u} = (-u_1, -u_2, \ldots, -u_m)$.

Length of a Vector

Definition The **distance** between vectors $\mathbf{u}$ and $\mathbf{v}$ in $\mathbf{R}^m$ is given by the **distance formula,**

$$d(\mathbf{u}, \mathbf{v}) = \sqrt{(v_1 - u_1)^2 + (v_2 - u_2)^2 + \cdots + (v_m - u_m)^2}.$$

The **length (norm, magnitude)** of a vector $\mathbf{u}$ is its distance from $\mathbf{0}$,

$$\|\mathbf{u}\| = d(\mathbf{0}, \mathbf{u}) = \sqrt{u_1^2 + u_2^2 + \cdots + u_m^2}.$$

Note: If $\|\mathbf{u}\| = 0$, then $\mathbf{u} = \mathbf{0}$ (and conversely). A vector $\mathbf{u}$ is called a **unit vector** if $\|\mathbf{u}\| = 1$.

Example 1.26

Find $\|\mathbf{u}\|$ where $\mathbf{u} = (1, 2, 3, -2)$.

$$\|\mathbf{u}\| = \sqrt{1^2 + 2^2 + 3^2 + (-2)^2} = \sqrt{18} = 3\sqrt{2}.$$

We will now extend the definition of operations on vectors given in section 1.1 to vectors in $\mathbf{R}^m$.

Scalar Multiplication

Definition Let c be a scalar and $\mathbf{u}$ a vector in $\mathbf{R}^m$. The product of c and $\mathbf{u}$ is the vector

$$c\mathbf{u} = (cu_1, cu_2, \ldots, cu_m).$$

That is, $c\mathbf{u}$ is the vector obtained by multiplying each component of $\mathbf{u}$ by the scalar c.

If $\mathbf{u}$ and $\mathbf{v}$ are nonzero vectors, and there exists a scalar such that $\mathbf{v} = c\mathbf{u}$, we say $\mathbf{u}$ and $\mathbf{v}$ are **collinear**. If this is the case, we say $\mathbf{u}$ and $\mathbf{v}$ *have the same direction* if c is positive and *have opposite directions* if c is negative.

Theorem 1   Let $\mathbf{u}$ be a vector in **R**^{m} and c a scalar. Then

$$\|c\mathbf{u}\| = |c|\,\|\mathbf{u}\|$$

Proof $\|c\mathbf{u}\| = \|(cu_1, \ldots, cu_m)\| = \sqrt{(cu_1)^2 + \cdots + (cu_m)^2}$

$$= \sqrt{c^2(u_1^2 + \cdots + u_m^2)} = |c|\,\|\mathbf{u}\|. \qquad \blacksquare$$

Notice that this proof is exactly the same one (with more components) as was given for theorem 1 of section 1.1. Furthermore the interpretation of the result is exactly the same: multiplication of a vector by a scalar changes the length by the same factor (in absolute value).

Vector Addition

Definition   Let $\mathbf{u}$ and $\mathbf{v}$ be vectors in **R**^{m}. The **sum** of $\mathbf{u}$ and $\mathbf{v}$ is the vector

$$\mathbf{u} + \mathbf{v} = (u_1 + v_1, u_2 + v_2, \ldots, u_m + v_m).$$

In words, $\mathbf{u} + \mathbf{v}$ is the vector obtained by adding the corresponding components. The **difference** of $\mathbf{u}$ and $\mathbf{v}$ is

$$\mathbf{u} - \mathbf{v} = (u_1 - v_1, u_2 - v_2, \ldots, u_m - v_m).$$

Example 1.27

Letting $c = 3$, $\mathbf{u} = (1, 3, 5, -1)$, and $\mathbf{v} = (1, 0, 2, 1)$, find $3\mathbf{u} - \mathbf{v}$.

$$\begin{aligned} 3\mathbf{u} - \mathbf{v} &= 3(1, 3, 5, -1) - (1, 0, 2, 1) \\ &= (3, 9, 15, -3) - (1, 0, 2, 1) \\ &= (2, 9, 13, -4). \end{aligned}$$

The following properties of vector addition and scalar multiplication follow directly from the definitions of these operations. The proofs are exactly the same as those given for theorem 3 of section 1.1 (with more components) and are left as exercises.

Theorem 2   Let **u**, **v**, and **w** be vectors in $\mathbf{R}^m$, and let c and d be scalars. Then

a. $\mathbf{u} + \mathbf{v} = \mathbf{v} + \mathbf{u}$ (vector addition is commutative),
b. $(\mathbf{u} + \mathbf{v}) + \mathbf{w} = \mathbf{u} + (\mathbf{v} + \mathbf{w})$ (vector addition is associative),
c. $\mathbf{u} + \mathbf{0} = \mathbf{u}$,
d. $\mathbf{u} + (-\mathbf{u}) = \mathbf{0}$,
e. $(cd)\mathbf{u} = c(d\mathbf{u})$,
f. $(c + d)\mathbf{u} = c\mathbf{u} + d\mathbf{u}$
g. $c(\mathbf{u} + \mathbf{v}) = c\mathbf{u} + c\mathbf{v}$ (distributive laws),
h. $1\mathbf{u} = \mathbf{u}$,
i. $(-1)\mathbf{u} = -\mathbf{u}$,
j. $0\mathbf{u} = \mathbf{0}$.

Dot Product

Definition   Let **u** and **v** be vectors in $\mathbf{R}^m$. The **dot product** (or **standard inner product**) of **u** and **v** is

$$\mathbf{u} \cdot \mathbf{v} = u_1 v_1 + u_2 v_2 + \cdots + u_m v_m.$$

In words, $\mathbf{u} \cdot \mathbf{v}$ is the sum of the products of the corresponding components of **u** and **v**.

Because the dot product of two vectors is a scalar, it is often called the **scalar product**. The algebraic properties of dot product are included in the next theorem. The proofs are exactly the same as those given in theorem 1 of section 1.2 (with more components) and are left as exercises.

Theorem 3   Let **u**, **v**, and **w** be vectors in $\mathbf{R}^m$ and let c be a scalar. Then

a. $\mathbf{u} \cdot \mathbf{v} = \mathbf{v} \cdot \mathbf{u}$ (dot product is commutative),
b. $c(\mathbf{u} \cdot \mathbf{v}) = (c\mathbf{u}) \cdot \mathbf{v} = \mathbf{u} \cdot (c\mathbf{v})$ (scalars factor out),
c. $\mathbf{u} \cdot (\mathbf{v} + \mathbf{w}) = \mathbf{u} \cdot \mathbf{v} + \mathbf{u} \cdot \mathbf{w}$ (distributive law),
d. $\mathbf{u} \cdot \mathbf{0} = 0$,
e. $\mathbf{u} \cdot \mathbf{u} = \|\mathbf{u}\|^2$.

Definition   Vectors **u** and **v** in $\mathbf{R}^m$ are **orthogonal** if $\mathbf{u} \cdot \mathbf{v} = 0$.

Everything so far has been so analogous to sections 1.1 and 1.2 that you probably now expect to see the *cross product*. There *is* an analog for the cross product in $\mathbf{R}^m$ but it is more complicated and of less value than the $\mathbf{R}^3$ version. For this reason we shall omit it.

Lines in $\mathbf{R}^m$

The geometry of $\mathbf{R}^m$ for $m \geq 4$ is even more interesting than it is in the case of $m = 2$ and $m = 3$. The problem of course is that we cannot draw pictures of all of it at once. We must focus on subregions for pictures, and often even this device will be of little help.

Let $\mathbf{p}$ and $\mathbf{q}$ be vectors in $\mathbf{R}^m$. Since $\mathbf{p}$ and $\mathbf{q}$ are m-tuples, they seem like points, and it seems reasonable that two points should determine a line. We therefore follow the *two-point form* for a line in $\mathbf{R}^3$, equation (9) of section 1.3, to define the geometric terms that follow.

Definition   Let $\mathbf{p}$ and $\mathbf{q}$ be distinct points in $\mathbf{R}^m$, and let

$$\mathbf{x}(t) = (1 - t)\mathbf{p} + t\mathbf{q}. \tag{1}$$

a. The set of all $\mathbf{x}(t)$ for t in the real numbers is the **line** determined by $\mathbf{p}$ and $\mathbf{q}$.
b. The set of all $\mathbf{x}(t)$ for $0 \leq t \leq 1$ is the **line segment** determined by $\mathbf{p}$ and $\mathbf{q}$. By taking the $\mathbf{x}(t)$ in their natural order (start at $\mathbf{x}(0)$ and finish at $\mathbf{x}(1)$), we have the **directed line segment** from $\mathbf{p}$ to $\mathbf{q}$.

We can now speak of a nonzero vector $\mathbf{u}$ in $\mathbf{R}^m$ as a directed line segment from the origin by letting $\mathbf{p} = \mathbf{0}$ and $\mathbf{q} = \mathbf{u}$. By letting t range over all real numbers, we have the *line determined* by $\mathbf{u}$. Two nonzero vectors $\mathbf{u}$ and $\mathbf{v}$ are collinear if the lines they determine are one and the same, that is, the points $\mathbf{u}$, $\mathbf{v}$, and $\mathbf{0}$ are collinear points.

Since we can view a vector $\mathbf{u}$ as a directed segment from $\mathbf{0}$, we can generalize the *point-parallel form* for a line. The set of all points

$$\mathbf{x}(t) = \mathbf{p} + t\mathbf{v} \tag{2}$$

is the line that contains the point $\mathbf{p}$ and is parallel to $\mathbf{v}$, where of course t is a real number and $\mathbf{v} \neq \mathbf{0}$.

Example 1.28

Give a point-parallel form for the line through $\mathbf{p} = (2, 1, -1, 3)$ and parallel to $\mathbf{v} = (1, 0, -2, 1)$.

$$\mathbf{x}(t) = (2, 1, -1, 3) + t(1, 0, -2, 1).$$

We can also view this function parametrically as before. Let

$$\mathbf{x}(t) = (x, y, z, w).$$

Equating coordinates, we have the *parametric equations*,

$$x = 2 + t,$$

$$y = 1,$$
$$z = -1 - 2t,$$
$$w = 3 + t.$$

In summary, lines are just as easy to discuss in $\mathbf{R}^m$, $m \geq 4$, as they were in $\mathbf{R}^3$.

Linear Equations

Let $\mathbf{p}$ be a point, and let $\mathbf{n}$ be a nonzero vector in $\mathbf{R}^m$, the **point-normal form** for $\mathbf{p}$ and $\mathbf{n}$ is the equation

$$\mathbf{n} \cdot (\mathbf{x} - \mathbf{p}) = 0. \tag{3}$$

Recall that in $\mathbf{R}^2$ this is an equation of the *line* through $\mathbf{p}$ with normal $\mathbf{n}$. In $\mathbf{R}^3$ this is the equation of the *plane* through $\mathbf{p}$ with normal $\mathbf{n}$. In the general case $\mathbf{R}^m$ the set of all solutions to equation (3) is called the **hyperplane** through $\mathbf{p}$ with normal $\mathbf{n}$.

Example 1.29

Find a point-normal form for the hyperplane that contains the point $\mathbf{p} = (2, 1, 0, 3, 1)$ with normal $\mathbf{n} = (2, 1, -1, 3, 0)$.
From equation (3), we have the point-normal form,

$$(2, 1, -1, 3, 0) \cdot (\mathbf{x} - (2, 1, 0, 3, 1)) = 0.$$

Other forms of the equation of a plane are obtained by letting $\mathbf{n} = (n_1, n_2, \ldots, n_m)$, $\mathbf{p} = (p_1, p_2, \ldots, p_m)$, $\mathbf{x} = (x_1, x_2, \ldots, x_m)$, $b = \mathbf{n} \cdot \mathbf{p}$, and expanding equation (3). The alternate point-normal form is

$$n_1(x_1 - p_1) + n_2(x_2 - p_2) + \cdots + n_m(x_m - p_m) = 0. \tag{4}$$

The **standard form** is obtained from equation (3) by distributing $\mathbf{n} \cdot \mathbf{x} - \mathbf{n} \cdot \mathbf{p} = 0$ and transposing to obtain

$$n_1 x_1 + n_2 x_2 + \cdots + n_m x_m = b. \tag{5}$$

An equation that can be put into the form of equation (5) is called a **linear equation** in the variables $x_1, x_2, \ldots, x_m$. This is a generalization of the equation $n_1 x_1 + n_2 x_2 = b$ that describes a line in $\mathbf{R}^2$ and $n_1 x_1 + n_2 x_2 + n_3 x_3 = b$ that describes a plane in $\mathbf{R}^3$ (see section 1.3).

Notice that a linear equation contains no products of variables (such as x_1^2 or $x_1 x_2$) nor functions of variables (such as $\sin x_1$, e^{x_2}, $\sqrt{x_4}$) other than multiplication by a constant scalar.

Given a linear equation in standard form (equation 5) it is an easy matter to put it in point-normal form. Let $\mathbf{n} = (n_1, \ldots, n_m)$ be the vector of coefficients of the variables. For any i from 1 to m with $n_i \neq 0$, let $\mathbf{p} = (0, \ldots, 0, b/n_i, 0, \ldots, 0)$, where the coordinate b/n_i is in the ith position. Then $\mathbf{n} \cdot (\mathbf{x} - \mathbf{p}) = 0$ is equivalent to the original equation as can easily be confirmed by expansion of the dot product.

Example 1.30

Express the linear equation

$$2x_1 + 3x_2 - x_3 - x_4 = 5$$

in point-normal form for a hyperplane in **R**^{4}.

Letting $\mathbf{n} = (2, 3, -1, -1)$, we have $n_3 = -1 \neq 0$, so the point $\mathbf{p} = (0, 0, 5/(-1), 0) = (0, 0, -5, 0)$ is in the hyperplane. The desired equation is

$$(2, 3, -1, -1) \cdot (\mathbf{x} - (0, 0, -5, 0)) = 0.$$

Plane Determined by Two Vectors

We can discuss lines in **R**^{m} (point-parallel form) and hyperplanes in **R**^{m} (point-normal form), but so far we have no means of discussing other types of regions. This can be accomplished in a manner analogous to the point-parallel form by adding more terms, one more for a plane, two more for a three-dimensional region, and so on. As an example we describe a plane in **R**^{m} determined by two noncollinear vectors.

Let $\mathbf{u}$ and $\mathbf{v}$ be noncollinear vectors, that is, one is not a scalar multiple of the other. Let $\mathbf{x}$ be any point in the plane determined by $\mathbf{u}$ and $\mathbf{v}$. The line through $\mathbf{x}$ parallel to the vector $\mathbf{v}$ must intersect the line determined by $\mathbf{u}$. Since any point on that line is a scalar multiple of $\mathbf{u}$, the point of intersection is $s\mathbf{u}$ for some real number s (figure 1.31). Similarly the line

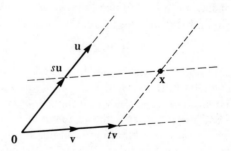

Figure 1.31

through **x** parallel to **u** must intersect the line determined by **v** in some point $t\mathbf{v}$. Then the quadrilateral **0**, $s\mathbf{u}$, **x**, $t\mathbf{v}$ is a parallelogram and by theorem 2 of section 1.1,

$$\mathbf{x} = s\mathbf{u} + t\mathbf{v}.$$

Thus every point in the plane determined by **u** and **v** is representable as the sum of a multiple of **u** and a multiple of **v**. A representation for a vector **x** of this type is called a **linear combination** of **u** and **v**. Conversely, every linear combination $s\mathbf{u} + t\mathbf{v}$ lies in the plane of **u** and **v**. Therefore we may view the plane of **u** and **v** as the set of all points given by the function $\mathbf{x} = \mathbf{x}(s, t)$, where

$$\mathbf{x} = s\mathbf{u} + t\mathbf{v},$$

and s and t range independently over all real numbers.

If we start with three noncoplanar vectors **u**, **v**, and **w**, we have an analogous situation. The set of all linear combinations $\mathbf{x} = \mathbf{x}(r, s, t)$ of **u**, **v**, and **w** given by

$$\mathbf{x} = r\mathbf{u} + s\mathbf{v} + t\mathbf{w},$$

is the 3-space determined by **u**, **v**, and **w**.

The plane and 3-space just described contain the origin. For the general case of a plane determined by three noncollinear points or 3-space determined by four noncoplanar points, translate to the origin, use the procedure described, and then translate back. We demonstrate with an example.

Example 1.31

Describe the plane determined by $\mathbf{p} = (1, 2, 1, 1)$, $\mathbf{q} = (3, -1, 4, 0)$, and $\mathbf{r} = (1, -1, 0, 2)$ in $\mathbf{R}^4$.

We translate **p** to the origin. The vectors $\mathbf{q} - \mathbf{p}$ and $\mathbf{r} - \mathbf{p}$ then determine a parallel plane consisting of all linear combinations $s(\mathbf{r} - \mathbf{p}) + t(\mathbf{q} - \mathbf{p})$. Adding **p** to each such point, we translate back, and the desired plane is given by the set of all points of the form

$$\mathbf{x} = \mathbf{p} + s(\mathbf{r} - \mathbf{p}) + t(\mathbf{q} - \mathbf{p}),$$

where s and t are real numbers. That is,

$$\mathbf{x} = (1, 2, 1, 1) + s(0, -3, -1, 1) + t(2, -3, 3, -1)$$

for all real numbers s and t.

1.4   Exercises

In exercises 1 through 10 let the vectors be given as follows:

$$\mathbf{u} = (2, 1, 3, 0),$$
$$\mathbf{v} = (-1, 1, 2, 1),$$
$$\mathbf{w} = (2, -2, 0, 6).$$

Perform each indicated operation.

1. $\mathbf{u} + \mathbf{v}$ 2. $3\mathbf{u} - \mathbf{v} + 2\mathbf{w}$
3. $2(\mathbf{u} + 2\mathbf{v} + 3\mathbf{w})$ 4. $d(\mathbf{v}, \mathbf{w})$
5. $\mathbf{v} \cdot \mathbf{w}$ 6. $\|\mathbf{w}\|$

7. $\|3\mathbf{w}\|$ 8. $\dfrac{1}{\|\mathbf{w}\|}\mathbf{w}$

9. $\|2(\mathbf{v} - \mathbf{w})\|$ 10. $(3\mathbf{v}) \cdot (5\mathbf{w})$

11. Which pairs of the following vectors are orthogonal?

$\mathbf{u}_1 = (2, 1, 3, -1),\ \mathbf{u}_2 = (3, 1, 0, 2),\ \mathbf{u}_3 = (0, -2, 1, 1),\ \mathbf{u}_4 = (2, 0, 3, -3).$

12. Prove that, for $\mathbf{u}$, $\mathbf{v}$ in $\mathbf{R}^m$, $d(\mathbf{u}, \mathbf{v}) = \|\mathbf{u} - \mathbf{v}\|$.

13. Prove that, for a nonzero vector $\mathbf{u}$ in $\mathbf{R}^m$, $(1/\|\mathbf{u}\|)\mathbf{u}$ is the unit vector in the direction of $\mathbf{u}$.

In exercises 14 and 15 find the unit vector with the same direction as the given one.

14. $(1, 1, 3, 0, 5)$ 15. $(2, 1, -1, 0, 3, 4)$

16. Give two-point and point-parallel forms and parametric equations for the line in $\mathbf{R}^5$ determined by $\mathbf{p} = (2, 1, 0, 3, 1)$ and $\mathbf{q} = (1, -1, 3, 0, 5)$.

17. Give two-point and point-parallel forms and parametric equations for the line in $\mathbf{R}^4$ determined by $\mathbf{p} = (-1, 0, 3, 2)$ and $\mathbf{q} = (-1, 0, 4, 5)$.

18. Give point-normal and standard forms for the hyperplane through $(-2, 1, 4, 0)$ with normal $(1, 2, -1, 3)$.

19. Give point-normal and standard forms for the hyperplane through $(3, 4, 5, 6, 7)$ with normal $(1, -1, 1, -1, 1)$.

20. Find a point-normal form for the linear equation $2x_1 - 3x_2 + x_4 - x_5 = 2$ in $\mathbf{R}^5$.

21. Repeat exercise 20 viewing the equation in $\mathbf{R}^6$.

In exercises 22 through 25 explain why each equation fails to be linear.

22. $x(2y + 3z) = 5$ 23. $x_1 + 2x_2 + x_3 - x_4 x_5 = 3$
24. $x + 2 \sin y + 3z = 0$ 25. $x^2 + y^2 = 0$

26. Find the midpoint of the line segment that joins $(2, 1, 3, 5)$ with $(6, 3, 2, 1)$.

27. Prove the properties of theorem 2.

28. The *angle between two nonzero vectors* $\mathbf{u}$ and $\mathbf{v}$ in $\mathbf{R}^m$ is defined to be the unique number $\theta, 0 \le \theta \le \pi$ radians such that $\cos \theta = \mathbf{u} \cdot \mathbf{v}/(\|\mathbf{u}\|\ \|\mathbf{v}\|)$. Find the cosine of the angle between the vectors $(3, 1, 1, 2, 1)$ and $(0, 2, 1, -2, 0)$.

29. Prove the properties of theorem 3.

30. Describe the plane determined by the points $(3, 1, 0, 2, 1)$, $(2, 1, 4, 2, 0)$, and $(-1, 2, 1, 3, 1)$.

31. Describe the 3-space determined by the points $(3, 1, 0, 2, 1)$, $(2, 1, 4, 2, 0)$, $(-1, 2, 1, 3, 1)$, and $(0, 2, 0, 1, 0)$.

32. Prove the *parallelogram law* in $\mathbf{R}^m$, $\|\mathbf{u} + \mathbf{v}\|^2 + \|\mathbf{u} - \mathbf{v}\|^2 = 2\|\mathbf{u}\|^2 + 2\|\mathbf{v}\|^2$.

33. In metric geometry a point B is **between** A and C if and only if $d(A, B) + d(B, C) = d(A, C)$. For distinct points $\mathbf{p}$ and $\mathbf{q}$ in $\mathbf{R}^m$ prove that $\mathbf{x}(t) = (1 - t)\mathbf{p} + t\mathbf{q}$ is between $\mathbf{p}$ and $\mathbf{q}$ if $0 \le t \le 1$.

34. Use exercise 12 to show that $d(\mathbf{p}, \mathbf{p} + t\mathbf{u}) = |t| \, \|\mathbf{u}\|$.

35. Find a point one-third of the distance from $(2, 1, 3, -4)$ to $(2, -2, 0, 2)$.

36. Let $\mathbf{p}, \mathbf{q}$, and $\mathbf{r}$ be the vertices of a triangle in $\mathbf{R}^m$. Prove that the segment that joins the midpoints of two of the sides is parallel to and half the length of the third side.

37. Let $\mathbf{u}, \mathbf{v}_1$, and $\mathbf{v}_2$ be vectors in $\mathbf{R}^m$ and c_1 and c_2 scalars. If $\mathbf{u}$ is orthogonal to both $\mathbf{v}_1$ and $\mathbf{v}_2$, prove that $\mathbf{u}$ is orthogonal to $c_1\mathbf{v}_1 + c_2\mathbf{v}_2$.

38. Interpret the result of exercise 37 geometrically to show that in $\mathbf{R}^3$ the sum of two vectors is coplanar with the two vectors.

39. Extend the result of exercise 37 to any number of vectors $\mathbf{v}_1, \mathbf{v}_2, \ldots, \mathbf{v}_n$.

Linear Equations and Matrices

2

Linear algebra originated with the study of linear equations, and the topic of linear equations is still of major importance in all disciplines that make use of mathematics. We will begin this chapter by presenting a technique for solving systems of linear equations with the aid of *matrices*. We will then investigate the properties of matrices in some detail and use these properties to develop the theory of linear systems.

2.1 Systems of Linear Equations

Linear equations were introduced in section 1.4, with an emphasis on their geometrical interpretation in Euclidean m-space. In this section we recall some terminology associated with systems of such equations and present a technique for solving them.

A **linear equation** in the variables x_1, x_2, ..., x_n is one that can be put in the form

$$a_1 x_1 + a_2 x_2 + \cdots + a_n x_n = b,$$

where b and the coefficients a_i are constants and not all a_i equal zero. A **system of linear equations** (or **linear system**) is simply a finite set of linear equations. An n-vector $(s_1, s_2, \ldots, s_n)$ is a **solution** to a linear system (in n variables) if it satisfies every equation in the system. The variables in a linear system are also referred to as *unknowns* and of course may be denoted by other symbols such as x, y, z, or w.

Recall from section 1.4 that the set of solutions to a linear equation is a line in $\mathbf{R}^2$, a plane in $\mathbf{R}^3$, and, generally speaking, a hyperplane in $\mathbf{R}^n$. From a geometric point of view a solution to a system of linear equations represents a point that lies in the *intersection* of the regions described by the individual equations.

Example 2.1

The linear system

$$2x = y - 4z$$
$$y = 2z$$

is a system of two equations in the three unknowns x, y, and z. Any 3-vector of the form $(-t, 2t, t)$ is a solution, where t represents an arbitrary real number. Consequently there are infinitely many solutions to this system—one for each value of t. For example, the solutions $(-1, 2, 1)$, $(0, 0, 0)$, and $(1/2, -1, -1/2)$ are obtained by setting t equal to 1, 0, and $-1/2$, respectively. A solution set of this type is called a *one-parameter family* of solutions—here t is the *parameter*.

To check that all vectors of the form $(-t, 2t, t)$ are solutions, simply substitute the components for the corresponding variables in each equation. This substitution results in an identity in both cases, namely, $2(-t) = (2t) - 4(t)$ and $(2t) = 2(t)$. Consequently all such vectors are indeed solutions.

Notice that the solution in example 2.1 may be written as $(x, y, z) = t(-1, 2, 1)$, the point-parallel form of the line through the origin in $\mathbf{R}^3$ determined by the vector $(-1, 2, 1)$. Since each of the two given equations represents a plane in $\mathbf{R}^3$, we see that their intersection is this line (figure 2.1).

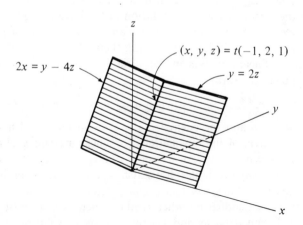

Figure 2.1

Example 2.2

The linear system

$$u + v = 0$$
$$u + v = 1$$

is a system of two equations in two unknowns that does not have a solution. This can be seen by the following argument. Suppose (s_1, s_2) were a solution. Then the first equation states that the two components s_1 and s_2 add up to zero, while the second states that they add up to 1.

Both statements cannot be true at the same time, and consequently no pair of numbers (s_1, s_2) can satisfy both equations simultaneously.

Geometrically speaking, the given equations represent lines in $\mathbf{R}^2$. We have thus shown that these lines do not intersect—they must be parallel (figure 2.2).

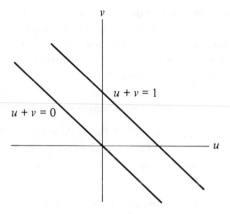

Figure 2.2

Note: If a linear system has no solution, we say that it is **inconsistent.** If it has at least one solution, it is **consistent.** For example, the system of example 2.1 is consistent, while that of example 2.2 is inconsistent.

Example 2.3

The system of equations

$$
\begin{aligned}
x_1 + x_2 \ \ + x_3 + \ \ x_4 &= 4 \\
x_1 - x_2 \ \ \ \ \ \ \ \ \ \ + 2x_4 &= 2 \\
x_1 + x_1 x_2 \ \ \ \ \ \ - \ \ x_4 &= 1
\end{aligned}
$$

is *not* a linear system. Remember from section 1.4 that, if a product of variables occurs in an equation, the equation is not linear. Such a product occurs in the third equation, which contains an $x_1 x_2$ term. Although the first and second equations are linear, a set of equations forms a linear system if and only if *all* equations in the set are linear.

In solving a system of linear equations, it is convenient to first put the system into **standard form,** which means write it in such a way that (1) each equation is in standard form (section 1.4) and (2) like variables are aligned in vertical columns.

Notice that the system of example 2.2 is in standard form but that of example 2.1 is not. To put the latter into standard form, we simply transpose the variables to the left-hand side, obtaining

$$2x - y + 4z = 0$$
$$y - 2z = 0.$$

We now proceed to describe a technique known as an *elimination procedure* for solving linear systems of equations. The basic idea of this technique is to transform the original linear system into simpler and simpler systems that have the same solutions as the given one. The final system should be so simple that its solution set can be easily determined.

Each step of this transformation process is accomplished by applying one of the three following **elementary operations:**

i. Multiply an equation in the system by a nonzero scalar.
ii. Interchange the positions of two equations in the system.
iii. Replace an equation by the sum of itself and a multiple of another equation of the system.

As we shall see at the end of this section, performing any finite sequence of these elementary operations on a linear system of equations results in an **equivalent system,** one with the same set of solutions. Although a description of a *systematic* solution procedure based upon these elementary operations will be deferred until section 2.2, we will use this procedure to solve the examples that follow. In reading through these examples you may begin to see the structure of this systematic procedure. However, in working the exercises for this section, you should feel free to use the elementary operations in any order you wish to arrive at a system with an "obvious" solution. It is often better to be clever than to be systematic!

Example 2.4

Solve

$$
\begin{array}{rl}
-y + z = & 3 \qquad\qquad \text{(A1)} \\
x - y - z = & 0 \qquad\qquad \text{(B1)} \\
-x \quad\quad - z = & -3. \qquad\quad\ \text{(C1)}
\end{array}
$$

Interchange equations (A1) and (B1):

$$
\begin{array}{rl}
x - y - z = & 0 \qquad\qquad \text{(A2)} \\
-y + z = & 3 \qquad\qquad \text{(B2)} \\
-x \quad\quad - z = & -3. \qquad\quad\ \text{(C2)}
\end{array}
$$

Replace (C2) by the sum of (C2) and 1 times (A2):

$$x - y - z = 0 \qquad \text{(A3)}$$
$$- y + z = 3 \qquad \text{(B3)}$$
$$- y - 2z = -3. \qquad \text{(C3)}$$

Multiply (B3) by -1:

$$x - y - z = 0 \qquad \text{(A4)}$$
$$y - z = -3 \qquad \text{(B4)}$$
$$- y - 2z = -3. \qquad \text{(C4)}$$

Replace (A4) by the sum of (A4) and 1 times (B4). Then replace (C4) by the sum of (C4) and 1 times (B4):

$$x \quad - 2z = -3 \qquad \text{(A5)}$$
$$y - z = -3 \qquad \text{(B5)}$$
$$- 3z = -6. \qquad \text{(C5)}$$

Multiply (C5) by $-1/3$:

$$x \quad - 2z = -3 \qquad \text{(A6)}$$
$$y - z = -3 \qquad \text{(B6)}$$
$$z = 2. \qquad \text{(C6)}$$

Replace (A6) by the sum of (A6) and 2 times (C6). Then replace (B6) by the sum of (B6) and 1 times (C6):

$$x \qquad = 1 \qquad \text{(A7)}$$
$$y \qquad = -1 \qquad \text{(B7)}$$
$$z = 2. \qquad \text{(C7)}$$

The only solution of this final system, and hence of the original one, is $(1, -1, 2)$.

Geometrically we have shown that the intersection of the three planes given by equations (A1), (B1), and (C1) is the single point $(1, -1, 2)$ (figure 2.3).

Example 2.5

Solve

$$x_1 \qquad + 3x_3 + x_4 = 0 \qquad \text{(A1)}$$
$$-x_1 + 2x_2 + x_3 + x_4 = 0 \qquad \text{(B1)}$$
$$-x_1 + x_2 - x_3 \qquad = 0. \qquad \text{(C1)}$$

Note: Geometrically the solution set will be the common points of intersection of the following *hyperplanes* in $\mathbf{R}^4$: $x_1 + 3x_3 + x_4 = 0$, $-x_1 + 2x_2 + x_3 + x_4 = 0$, and $-x_1 + x_2 - x_3 = 0$.

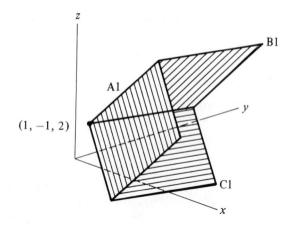

Figure 2.3

To solve the given system, replace (B1) by the sum of (B1) and 1 times (A1). Then replace (C1) by the sum of (C1) and 1 times (A1):

$$x_1 \qquad + 3x_3 + \quad x_4 = 0 \qquad (A2)$$
$$2x_2 + 4x_3 + 2x_4 = 0 \qquad (B2)$$
$$x_2 + 2x_3 + \quad x_4 = 0. \qquad (C2)$$

Multiply (B2) by 1/2:

$$x_1 \qquad + 3x_3 + x_4 = 0 \qquad (A3)$$
$$x_2 + 2x_3 + x_4 = 0 \qquad (B3)$$
$$x_2 + 2x_3 + x_4 = 0. \qquad (C3)$$

Replace (C3) by the sum of (C3) and -1 times (B3):

$$x_1 \qquad + 3x_3 + x_4 = 0 \qquad (A4)$$
$$x_2 + 2x_3 + x_4 = 0 \qquad (B4)$$
$$0 = 0. \qquad (C4)$$

Equation (C4) is an identity. It places no restriction on the values of the unknowns, and we discard it:

$$x_1 \qquad + 3x_3 + x_4 = 0 \qquad (A4)$$
$$x_2 + 2x_3 + x_4 = 0. \qquad (B4)$$

Now solve (A4) for x_1 and (B4) for x_2:

$$x_1 = -3x_3 - x_4$$
$$x_2 = -2x_3 - x_4.$$

Finally, let $x_3 = s$ and $x_4 = t$, obtaining the solution

$$x_1 = -3s - t$$
$$x_2 = -2s - t$$
$$x_3 = s$$
$$x_4 = t,$$

or in vector form $(-3s - t, -2s - t, s, t)$, where the parameters s and t range independently over the real numbers. Once again we get a linear system with infinitely many solutions, but here they form a *two-parameter family*.

Among the solutions are

$(0, 0, 0, 0)$ when $s = 0$, $t = 0$,
$(-1, -1, 0, 1)$ when $s = 0$, $t = 1$,
$(1, 1/2, -1/2, 1/2)$ when $s = -1/2$, $t = 1/2$.

Since any solution may be expressed as $\mathbf{x} = s(-3, -2, 1, 0) + t(-1, -1, 0, 1)$, where $\mathbf{x} = (x_1, x_2, x_3, x_4)$, the set of all such points is the plane in $\mathbf{R}^4$ determined by the vectors $(-3, -2, 1, 0)$ and $(-1, -1, 0, 1)$ (see section 1.4).

We will close this section with a theorem that justifies the elimination procedure used in examples 2.4 and 2.5.

Theorem 1   If any one of the three elementary operations is performed on a given system of linear equations, an equivalent linear system is obtained.

Proof   We will show this to be true for elementary operation iii only. The statement is certainly true for elementary operation ii since it does not alter the system, and we will leave the part of the proof concerning operation i as an exercise.

Let two equations of the given linear system be

$$a_1 x_1 + a_2 x_2 + \cdots + a_n x_n = b \qquad (1)$$

$$c_1 x_1 + c_2 x_2 + \cdots + c_n x_n = d. \qquad (2)$$

Suppose we replace equation (2) by the sum of itself and k times equation (1), and leave all other equations in the system as they stand. We need to verify that $(s_1, s_2, \ldots, s_n)$ is a solution of the given system if and only if it is a solution of the transformed system. Since all the other equations remain unchanged, we need only verify this for equation 2, which has been transformed into

$$(c_1 + ka_1)x_1 + (c_2 + ka_2)x_2 + \cdots + (c_n + ka_n)x_n = d + kb. \qquad (3)$$

First, assume $(s_1, s_2, \ldots, s_n)$ satisfies the original system. Then, in particular it satisfies (1) and (2), so that

$$a_1 s_1 + a_2 s_2 + \cdots + a_n s_n = b,$$
$$c_1 s_1 + c_2 s_2 + \cdots + c_n s_n = d.$$

Multiplying the first of these by k, we have

$$k(a_1 s_1 + a_2 s_2 + \cdots + a_n s_n) = kb,$$

which when added to the second yields

$$k(a_1 s_1 + a_2 s_2 + \cdots + a_n s_n) + (c_1 s_1 + c_2 s_2 + \cdots + c_n s_n) = kb + d.$$

Finally, rearranging terms, we obtain

$$(c_1 + ka_1)s_1 + (c_2 + ka_2)s_2 + \cdots + (c_n + ka_n)s_n = d + kb, \qquad (4)$$

which says that $(s_1, s_2, \ldots, s_n)$ satisfies equation (3).

Now assume $(s_1, s_2, \ldots, s_n)$ satisfies the transformed system. This implies that equation (4) holds. We also know that equation (1) is satisfied, so that $a_1 s_1 + a_2 s_2 + \cdots + a_n s_n = b$. This in turn implies that $k(a_1 s_1 + a_2 s_2 + \cdots + a_n s_n) = kb$. Subtracting this last equation from (4) yields

$$c_1 s_1 + c_2 s_2 + \cdots c_n s_n = d,$$

as desired.   ∎

2.1 Exercises

In exercises 1 through 6 determine whether or not the given system of equations is linear.

1. $x_1 - 3x_2 = x_3 - 4$
$x_4 = 1 - x_1$
$x_1 + x_4 + x_3 - 2 = 0$

2. $2x - \sqrt{y} + 3z = -1$
$x + 2y - z = 2$
$4x - y = -1$

3. $3x - xy = 1$
$x + 2xy - y = 0$

4. $y = 2x - 1$
$y = -x$

5. $2x_1 - \sin x_2 = 3$
$x_2 = x_1 + x_3$
$-x_1 + x_2 - 3x_3 = 0$

6. $x_1 + x_2 + x_3 = 1$
$-2x_1^2 - 2x_3 = -1$
$3x_2 - x_3 = 2$

In exercises 7 and 8 put the given system of equations into standard form.

7. $z = 6 - y$
$z = x + y$
$y + z - 3 = x$

8. $x_2 - x_1 = 3 - x_3 + x_4$
$x_2 = 0$
$x_3 = 1 - x_1 - x_2$

In exercises 9 and 10 show that the given linear system has the solution indicated.

9. $x - 3y + z = 0$
$x + y - 3z = 0$
$x - y - z = 0$

Solution: $(2s, s, s)$

10. $x_1 - 3x_2 - x_3 + x_4 = 1$
$-x_1 + 2x_2 + x_3 + x_4 = 0$

Solution: $(-2 + s + 5t, -1 + 2t, s, t)$

In exercises 11 through 14 solve the given linear system.

11. $x - 4y = 1$
$-2x + 8y = -2$

12. $x_1 + x_2 - x_3 + x_4 = 2$
$-2x_1 + x_3 = -4$
$x_1 - x_2 - 2x_3 + 2x_4 = -1$

13. $x_1 + x_2 + x_3 + x_4 = 1$
$2x_1 + 3x_2 + 3x_3 = 1$
$-x_1 - 2x_2 - 2x_3 + x_4 = 0$
$- x_2 - x_3 + 2x_4 = 1$

14. $2u - v = 0$
$-3u + 2v = 0$
$3u - v = 0$

In exercises 15 through 18 the given equations represent planes in $\mathbf{R}^3$. Describe the region (point, line, or plane), if any, that is the intersection of these planes.

15. $x + y + z = -4$
$x + 2y = 1$
$2y + 3z = -2$

16. $y + z = 1$
$x + y + 2z = -1$
$x + z = 1$

17. $x - y + z = 0$
$y - 2z = 1$
$2x - y = 1$

18. $-x + 2y - z = 1$
$2x - 4y + 2z = -2$
$x - 2y + z = -1$

19. Find the value of c such that the system

$$cx + z = 0$$
$$2y - 4z = 0$$
$$2x - y = 0$$

has a solution other than $(0, 0, 0)$.

20. Find all values of k such that the system

$$x_1 - x_2 + 2x_3 = 0$$
$$x_2 - x_3 = k$$
$$-x_1 + 2x_2 - 3x_3 = 1$$

has no solution.

21. Prove that, if we multiply any equation of an arbitrary linear system by a nonzero scalar, the resultant linear system is equivalent to the original one.

22. Show that the linear system

$$ax + by = e$$
$$cx + dy = f,$$

where a, b, c, d, e, f, are constants, has a unique solution if $ad - bc \neq 0$. Express this solution in terms of a, b, c, d, e, and f.

2.2 Matrices and Row Reduction of Linear Systems

In section 2.1 systems of linear equations were solved by means of an elimination procedure. In this section we will introduce the notion of a *matrix*, and then use *matrices* (the plural of matrix) to write each step of this solution technique in a more concise form.

A **matrix** is a rectangular array of numbers, each of which is called an **entry** of the matrix. We will enclose the rectangular array in brackets. For example,

$$\begin{bmatrix} 2 & 1 \\ -1 & 2 \\ 0 & 5 \end{bmatrix}, \tag{1}$$

$$\begin{bmatrix} 1 & -1 & 0 \\ 0 & 5 & 0 \end{bmatrix}, \tag{2}$$

and

$$\begin{bmatrix} 1 & \sqrt{2} & \pi \end{bmatrix} \tag{3}$$

are all matrices. Each horizontal line of numbers is called a **row** of the matrix; each vertical one is called a **column**. For example, in matrix (1) the third row consists of the entries 0 and 5, while the first column is 2, -1, 0. In other books you may see the rectangular array enclosed in parentheses. For example, matrix (2) might be written

$$\begin{pmatrix} 1 & -1 & 0 \\ 0 & 5 & 0 \end{pmatrix}.$$

Matrices arise naturally in the study of systems of linear equations. If we take a given linear system, put it in standard form, and then delete all unknowns and plus and equals signs, we obtain a rectangular array of numbers. This array is called the **augmented matrix** of the system. It is customary to draw a vertical line separating the column of constants on the right side from the rest of the matrix.

Example 2.6

Find the augmented matrices for the following linear systems:

(a) $2x - 3y + z = 1$
$\quad\quad x \quad\quad + 2z = 0$
$\quad\quad\quad - y - z = -5$

(b) $\quad x_1 - 3x_2 - 1 = x_4 - 4x_3$
$\quad\quad\quad -x_1 + 2x_3 = 0.$

For (a) the augmented matrix is

$$\left[\begin{array}{ccc|c} 2 & -3 & 1 & 1 \\ 1 & 0 & 2 & 0 \\ 0 & -1 & -1 & -5 \end{array}\right].$$

In (b) we first rewrite the system in standard form. This yields

$$\begin{aligned} x_1 - 3x_2 + 4x_3 - x_4 &= 1 \\ -x_1 \qquad\quad + 2x_3 \qquad\;\; &= 0. \end{aligned}$$

Thus the augmented matrix is

$$\left[\begin{array}{cccc|c} 1 & -3 & 4 & -1 & 1 \\ -1 & 0 & 2 & 0 & 0 \end{array}\right].$$

Note: When an unknown is "missing" in the given system (when its coefficient is zero), put a zero in the appropriate place in the augmented matrix corresponding to this system.

In solving the linear systems of examples 2.4 and 2.5 (section 2.1), we made use of the *elementary operations:*

  i. Multiply an equation by a nonzero scalar.
 ii. Interchange the positions of two equations.
iii. Replace an equation by the sum of itself and a multiple of another equation.

In an analogous manner we define for matrices the **elementary row operations:**

  i. Multiply a row by a nonzero scalar.
 ii. Interchange the position of two rows.
iii. Replace a row by the sum of itself and a multiple of another row.

Note: When we say "multiply a row by," we mean more precisely, "multiply each entry of a row by"; and when we say "the sum of two rows," we mean "the sum of the corresponding entries in the two rows."

Example 2.7

Given the matrix

$$\left[\begin{array}{cccc} 1 & -1 & 0 & 4 \\ 2 & 0 & -3 & 1 \\ 5 & -2 & 3 & 0 \end{array}\right],$$

if we multiply the third row by 3 and replace the second row by the sum of itself and -2 times the first, we obtain the matrix

$$\begin{bmatrix} 1 & -1 & 0 & 4 \\ 0 & 2 & -3 & -7 \\ 15 & -6 & 9 & 0 \end{bmatrix}.$$

Performing elementary row operations on the augmented matrix of a given linear system has the same effect as performing the analogous elementary operations on the system itself. In other words, the matrix resulting from the elementary row operations is the same as the augmented matrix of the linear system resulting from the analogous elementary operations.

Example 2.8

Consider the linear system that follows, with augmented matrix at the right:

$$\begin{aligned} x_1 - 2x_2 \quad\quad &= 1 \\ 2x_2 + x_3 &= -4 \\ 3x_1 - 4x_2 + x_3 &= 0 \end{aligned} \qquad \begin{bmatrix} 1 & -2 & 0 & | & 1 \\ 0 & 2 & 1 & | & -4 \\ 3 & -4 & 1 & | & 0 \end{bmatrix}.$$

If we perform the following elementary operations on the system— replace the third equation by the sum of itself and -3 times the first, multiply the second equation by $1/2$, and also perform the analogous elementary row operations on the matrix—replace the third row by the sum of itself and -3 times the first, multiply the second row by $1/2$, then the resulting system and matrix are

$$\begin{aligned} x_1 - 2x_2 \quad\quad &= 1 \\ x_2 + \tfrac{1}{2}x_3 &= -2 \\ 2x_2 + x_3 &= -3 \end{aligned} \qquad \begin{bmatrix} 1 & -2 & 0 & | & 1 \\ 0 & 1 & \tfrac{1}{2} & | & -2 \\ 0 & 2 & 1 & | & -3 \end{bmatrix}.$$

Notice that the resulting matrix is the augmented matrix for the resulting system.

Consequently in implementing the elimination procedure, we can perform the elementary row operations on the augmented matrices instead of on the system itself. This will reduce the amount of writing necessary, as well as the possibilities for error, and lends itself nicely to calculator and computer solution of linear systems. As an illustration of this process we will solve example 2.4 using the same sequence of operations but this time performing them on matrices.

Example 2.9

Use elementary row operations on matrices to solve the linear system (see example 2.4):

$$
\begin{aligned}
- y + z &= 3 \\
x - y - z &= 0 \\
-x \quad\quad - z &= -3.
\end{aligned}
$$

We first write the augmented matrix for the system:

$$
\left[\begin{array}{ccc|c}
0 & -1 & 1 & 3 \\
1 & -1 & -1 & 0 \\
-1 & 0 & -1 & -3
\end{array}\right].
$$

Now we proceed as in example 2.4. Interchange the first and second rows:

$$
\left[\begin{array}{ccc|c}
1 & -1 & -1 & 0 \\
0 & -1 & 1 & 3 \\
-1 & 0 & -1 & -3
\end{array}\right].
$$

Replace the third row by the sum of itself and 1 times the first row:

$$
\left[\begin{array}{ccc|c}
1 & -1 & -1 & 0 \\
0 & -1 & 1 & 3 \\
0 & -1 & -2 & -3
\end{array}\right].
$$

Multiply the second row by -1:

$$
\left[\begin{array}{ccc|c}
1 & -1 & -1 & 0 \\
0 & 1 & -1 & -3 \\
0 & -1 & -2 & -3
\end{array}\right].
$$

Replace the first row by the sum of itself and 1 times the second row. Then replace the third row by the sum of itself and 1 times the second row:

$$
\left[\begin{array}{ccc|c}
1 & 0 & -2 & -3 \\
0 & 1 & -1 & -3 \\
0 & 0 & -3 & -6
\end{array}\right].
$$

Multiply the third row by $-1/3$:

$$
\left[\begin{array}{ccc|c}
1 & 0 & -2 & -3 \\
0 & 1 & -1 & -3 \\
0 & 0 & 1 & 2
\end{array}\right].
$$

Replace the first row by the sum of itself and 2 times the third row. Then replace the second row by the sum of itself and 1 times the third row:

$$\left[\begin{array}{ccc|c} 1 & 0 & 0 & 1 \\ 0 & 1 & 0 & -1 \\ 0 & 0 & 1 & 2 \end{array}\right].$$

Change back to equation form (write the linear system for which this is the augmented matrix):

$$\begin{aligned} x & & & = 1 \\ & y & & = -1 \\ & & z & = 2. \end{aligned}$$

Thus the solution is $(1, -1, 2)$.

Note: When one matrix can be obtained from another by means of a finite sequence of elementary row operations, the two matrices are said to be **row equivalent.** For example, any two of the matrices in example 2.9 are row equivalent.

The final matrix of example 2.9 is an especially simple one. It is said to be in *row-reduced echelon form.* The basic goal of the elimination procedure is to transform the original augmented matrix (by means of elementary row operations) into one that has this special form.

Definition   A matrix is in **row-reduced echelon form** if it satisfies the following:
  i. In each row that does not consist entirely of zeros, the first nonzero entry is a 1 (we call such an element a **leading 1**).
  ii. In each column that contains a leading 1 of some row, all other entries are zero.
  iii. In any two rows with some nonzero entries the leading 1 of the higher row is farther to the left.
  iv. Any row that contains only zeros is lower than any row with some nonzero entries.

Example 2.10

The augmented matrices corresponding to the seventh system of example 2.4 and the fourth of example 2.5, section 2.1, are, respectively

$$\left[\begin{array}{ccc|c} 1 & 0 & 0 & 1 \\ 0 & 1 & 0 & -1 \\ 0 & 0 & 1 & 2 \end{array}\right] \text{ and } \left[\begin{array}{cccc|c} 1 & 0 & 3 & 1 & 0 \\ 0 & 1 & 2 & 1 & 0 \\ 0 & 0 & 0 & 0 & 0 \end{array}\right].$$

Each is in row-reduced echelon form.

Example 2.11

The matrix

$$\begin{bmatrix} 1 & 2 & 0 & -1 & 0 \\ 0 & 0 & 1 & 2 & 1 \\ 0 & 0 & 0 & 0 & 0 \\ 0 & 0 & 0 & 0 & 0 \end{bmatrix}$$

is in row-reduced echelon form.

Example 2.12

The matrices

$$\begin{bmatrix} 1 & 1 & 0 \\ 0 & 1 & 0 \\ 0 & 0 & 1 \end{bmatrix}, \begin{bmatrix} 1 & 0 & 0 \\ 0 & 0 & 1 \\ 0 & 1 & 0 \end{bmatrix}, \begin{bmatrix} 1 & 0 \\ 0 & 2 \end{bmatrix}, \text{ and } \begin{bmatrix} 0 & 0 & 0 \\ 1 & 0 & 0 \end{bmatrix}$$

are *not* in row-reduced echelon form. (They violate conditions ii, iii, i, and iv, respectively, of the definition.)

As we have said, the basic goal of the matrix elimination procedure is to transform by means of a systematic sequence of elementary row operations the original augmented matrix into one that is in row-reduced echelon form. We then change back to the "equation form" of the linear system and complete the solution process as in section 2.1. This procedure is often called **Gauss-Jordan elimination** and will be described in the next two examples.

Example 2.13

Solve the following system by Gauss-Jordan elimination:

$$\begin{aligned} x_3 + 2x_4 &= 3 \\ 2x_1 + 4x_2 - 2x_3 \quad &= 4 \\ 2x_1 + 4x_2 - x_3 + 2x_4 &= 7. \end{aligned}$$

Step 0: Write the augmented matrix for the system:

$$\begin{bmatrix} 0 & 0 & 1 & 2 & | & 3 \\ 2 & 4 & -2 & 0 & | & 4 \\ 2 & 4 & -1 & 2 & | & 7 \end{bmatrix}.$$

Step 1a: *Obtain a leading 1 in the first row, first column. (If there is a 0 in this position, interchange the first row with a row below it so that a nonzero entry appears there.)* Interchange the first and second rows:

$$\begin{bmatrix} 2 & 4 & -2 & 0 & | & 4 \\ 0 & 0 & 1 & 2 & | & 3 \\ 2 & 4 & -1 & 2 & | & 7 \end{bmatrix}.$$

(If the nonzero entry in the first row, first column is not a 1, multiply the first row by the reciprocal of this entry.) Multiply the first row by $1/2$:

$$\begin{bmatrix} 1 & 2 & -1 & 0 & | & 2 \\ 0 & 0 & 1 & 2 & | & 3 \\ 2 & 4 & -1 & 2 & | & 7 \end{bmatrix}.$$

Step 1b: *Obtain zeros in other positions in the first column by adding appropriate multiples of the first row to other rows.* Replace the third row by the sum of itself and -2 times the first:

$$\begin{bmatrix} 1 & 2 & -1 & 0 & | & 2 \\ 0 & 0 & 1 & 2 & | & 3 \\ 0 & 0 & 1 & 2 & | & 3 \end{bmatrix}.$$

Step 2a: *Obtain a leading 1 in the second row, second column. (If there is a 0 in this position, interchange the second row with a row below it so that a nonzero entry appears there. If this is not possible, go to the next column.)* It is not possible since all entries in the second column below the first row are zero.

Step 3a: *Obtain a leading 1 in the second row, third column.* There is one there already.

Step 3b: *Obtain zeros in the other positions in the third column by adding appropriate multiples of the second row to the other rows.* Replace the first row by the sum of itself and 1 times the second row. Then replace the third row by the sum of itself and -1 times the second row:

$$\begin{bmatrix} 1 & 2 & 0 & 2 & | & 5 \\ 0 & 0 & 1 & 2 & | & 3 \\ 0 & 0 & 0 & 0 & | & 0 \end{bmatrix}.$$

Step 3c: *Place any newly created zero rows at the bottom of the matrix.* It is already there. (The matrix is now in row-reduced echelon form.)

Step 4:   *Change back to a system of equations, ignoring any rows that contain only zeros:*

$$x_1 + 2x_2 \qquad + 2x_4 = 5$$
$$x_3 + 2x_4 = 3.$$

Step 5:   Solve each equation for the unknown whose coefficient is a lead-ing 1 in the final augmented matrix:

$$x_1 = 5 - 2x_2 - 2x_4$$
$$x_3 = 3 - 2x_4.$$

The unknowns appearing on the right-hand side are taken to be parameters. Set $x_2 = s$ and $x_4 = t$. We have here a "two-parameter family of solutions" (see example 2.5). All solutions are of the form $(5 - 2s - 2t, s, 3 - 2t, t)$ or, equivalently, $(5, 0, 3, 0) + s(-2, 1, 0, 0) + t(-2, 0, -2, 1)$.

Example 2.14

Solve the following system by Gauss-Jordan elimination:

$$\begin{aligned}
x_1 + 2x_2 \qquad\qquad &= \quad 1 \\
x_1 + 2x_2 + 3x_3 + x_4 &= \quad 0 \\
-x_1 - x_2 + x_3 + x_4 &= -2 \\
x_2 + x_3 + x_4 &= -1 \\
- x_2 + 2x_3 \qquad &= \quad 0.
\end{aligned}$$

Step 0:   Write the augmented matrix for the system:

$$\begin{bmatrix}
1 & 2 & 0 & 0 & | & 1 \\
1 & 2 & 3 & 1 & | & 0 \\
-1 & -1 & 1 & 1 & | & -2 \\
0 & 1 & 1 & 1 & | & -1 \\
0 & -1 & 2 & 0 & | & 0
\end{bmatrix}.$$

Step 1a: Obtain a leading 1 in the first row, first column. There is one there already.

Step 1b: Obtain zeros in the other positions in the first column. Re-place the second row by the sum of itself and -1 times the first row. Then replace the third row by the sum of itself and 1 times the first row:

$$\begin{bmatrix}
1 & 2 & 0 & 0 & | & 1 \\
0 & 0 & 3 & 1 & | & -1 \\
0 & 1 & 1 & 1 & | & -1 \\
0 & 1 & 1 & 1 & | & -1 \\
0 & -1 & 2 & 0 & | & 0
\end{bmatrix}.$$

Step 2a: Obtain a leading 1 in the second row, second column. (If there is a 0 in this position, interchange the second row with a row below it to create a nonzero entry there.) Interchange the second and third rows:

$$\begin{bmatrix} 1 & 2 & 0 & 0 & | & 1 \\ 0 & 1 & 1 & 1 & | & -1 \\ 0 & 0 & 3 & 1 & | & -1 \\ 0 & 1 & 1 & 1 & | & -1 \\ 0 & -1 & 2 & 0 & | & 0 \end{bmatrix}.$$

We now have a 1 in the required position.

Step 2b: *Obtain zeros in the other positions in the second column.* Replace the first row by the sum of itself and -2 times the second row. Next replace the fourth row by the sum of itself and -1 times the second row. Then replace the fifth row by the sum of itself and 1 times the second row:

$$\begin{bmatrix} 1 & 0 & -2 & -2 & | & 3 \\ 0 & 1 & 1 & 1 & | & -1 \\ 0 & 0 & 3 & 1 & | & -1 \\ 0 & 0 & 0 & 0 & | & 0 \\ 0 & 0 & 3 & 1 & | & -1 \end{bmatrix}.$$

Step 2c: *Place any newly created zero rows at the bottom of the matrix.* Interchange the fourth and fifth rows:

$$\begin{bmatrix} 1 & 0 & -2 & -2 & | & 3 \\ 0 & 1 & 1 & 1 & | & -1 \\ 0 & 0 & 3 & 1 & | & -1 \\ 0 & 0 & 3 & 1 & | & -1 \\ 0 & 0 & 0 & 0 & | & 0 \end{bmatrix}.$$

Step 3a: *Obtain a leading 1 in the third row, third column.* (*If there is a nonzero entry there and it is not a 1, multiply the third row by the reciprocal of this entry.*) Multiply the third row by $1/3$:

$$\begin{bmatrix} 1 & 0 & -2 & -2 & | & 3 \\ 0 & 1 & 1 & 1 & | & -1 \\ 0 & 0 & 1 & \frac{1}{3} & | & -\frac{1}{3} \\ 0 & 0 & 3 & 1 & | & -1 \\ 0 & 0 & 0 & 0 & | & 0 \end{bmatrix}.$$

Step 3b: *Obtain zeros in the other positions of the third column.* Replace the first row by the sum of itself and 2 times the third row. Next replace the second row by the sum of itself and -1 times the third row.

Then replace the fourth row by the sum of itself and -3 times the third row:

$$\begin{bmatrix} 1 & 0 & 0 & -\frac{4}{3} & \Big| & \frac{7}{3} \\ 0 & 1 & 0 & \frac{2}{3} & \Big| & -\frac{2}{3} \\ 0 & 0 & 1 & \frac{1}{3} & \Big| & -\frac{1}{3} \\ 0 & 0 & 0 & 0 & \Big| & 0 \\ 0 & 0 & 0 & 0 & \Big| & 0 \end{bmatrix}.$$

Step 3c: Place any newly created zero rows at the bottom of the matrix. It is already there. (The matrix is now in row-reduced echelon form.)

Step 4:   Change back to a system of equations ignoring any zero rows:

$$\begin{aligned} x_1 \qquad\qquad - (4/3)x_4 &= 7/3 \\ x_2 + (2/3)x_4 &= -2/3 \\ x_3 + (1/3)x_4 &= -1/3. \end{aligned}$$

Step 5:   Solve each equation for the unknown whose coefficient is a leading 1:

$$\begin{aligned} x_1 &= 7/3 + (4/3)x_4 \\ x_2 &= -2/3 - (2/3)x_4 \\ x_3 &= -1/3 - (1/3)x_4. \end{aligned}$$

We set $x_4 = s$ and obtain the "one-parameter family of solutions": $(7/3 + (4/3)s, -2/3 - (2/3)s, -1/3 - (1/3)s, s)$ or, equivalently, $(7/3, -2/3, -1/3, 0) + s(4/3, -2/3, -1/3, 1)$.

Before continuing, we will summarize the steps involved in the heart of the Gauss-Jordan elimination procedure: the transformation of the original augmented matrix to row-reduced echelon form.

Procedure to row-reduce a matrix

i. In the leftmost column whose entries are not all zero, obtain a *leading 1* at the top of this column by using elementary row operations i and ii.

ii. Obtain zeros in all other positions in this column by using elementary operation iii.

iii. Place any newly created zero rows at the bottom of the matrix by using elementary operation ii.

iv. Repeat step i for the *submatrix* obtained by deleting all rows containing previously obtained leading 1's, and then do steps ii and iii on the full matrix. Continue until row-reduced echelon form is obtained.

Using the Gauss-Jordan procedure, we can transform any matrix into row-reduced echelon form by means of a sequence of elementary row operations. We could also obtain this form by performing elementary row operations in a different order (see, for example, the *note* following theorem 2). However, we would still arrive at the same row-reduced echelon form. These facts are stated without proof in the following two theorems. The first of these theorems could be proved by giving a precise description of the Gauss-Jordan procedure, which has been informally described in the preceding pages. A proof of the second can be constructed with the aid of the tools introduced in chapter 4.

Theorem 1   Every matrix can be transformed by a finite sequence of elementary row operations into one that is in row-reduced echelon form.

Theorem 2   The row-reduced echelon form of a matrix is unique.

Note: In doing computations by hand, it is often more convenient to alter the sequence of elementary row operations to avoid fractions. This can be accomplished by delaying the placement of a leading 1 in certain rows until the end, together with some additional row multiplications. For example, at the end of step 2 of example 2.14, the augmented matrix has only integral entries:

$$\left[\begin{array}{cccc|c} 1 & 0 & -2 & -2 & 3 \\ 0 & 1 & 1 & 1 & -1 \\ 0 & 0 & 3 & 1 & -1 \\ 0 & 0 & 3 & 1 & -1 \\ 0 & 0 & 0 & 0 & 0 \end{array}\right].$$

To avoid fractions until the last step, we proceed as follows:

Multiply the first row by 3.

Multiply the second row by 3:

$$\left[\begin{array}{cccc|c} 3 & 0 & -6 & -6 & 9 \\ 0 & 3 & 3 & 3 & -3 \\ 0 & 0 & 3 & 1 & -1 \\ 0 & 0 & 3 & 1 & -1 \\ 0 & 0 & 0 & 0 & 0 \end{array}\right].$$

Replace the first row by the sum of itself and 2 times the third row.

Replace the second row by the sum of itself and -1 times the third row.

Replace the fourth row by the sum of itself and -1 times the third row:

$$\begin{bmatrix} 3 & 0 & 0 & -4 & | & 7 \\ 0 & 3 & 0 & 2 & | & -2 \\ 0 & 0 & 3 & 1 & | & -1 \\ 0 & 0 & 0 & 0 & | & 0 \\ 0 & 0 & 0 & 0 & | & 0 \end{bmatrix}.$$

Multiply the first, second, and third rows by $1/3$:

$$\begin{bmatrix} 1 & 0 & 0 & -\frac{4}{3} & | & \frac{7}{3} \\ 0 & 1 & 0 & \frac{2}{3} & | & -\frac{2}{3} \\ 0 & 0 & 1 & \frac{1}{3} & | & -\frac{1}{3} \\ 0 & 0 & 0 & 0 & | & 0 \\ 0 & 0 & 0 & 0 & | & 0 \end{bmatrix}.$$

The matrix is now in row-reduced echelon form. Complete the solution process exactly as in example 2.14.

In example 2.9 the given linear system had a unique solution, while in examples 2.13 and 2.14 there were infinitely many solutions. As we know from section 2.1, a linear system may also have no solution. The following example illustrates the Gauss-Jordan procedure in such a case.

Example 2.15

Solve

$$\begin{aligned} x_1 \quad\quad + x_3 &= \quad 1 \\ x_2 - x_3 &= -1 \\ 2x_1 + x_2 + x_3 &= \quad 2. \end{aligned}$$

The augmented matrix for this system is

$$\begin{bmatrix} 1 & 0 & 1 & | & 1 \\ 0 & 1 & -1 & | & -1 \\ 2 & 1 & 1 & | & 2 \end{bmatrix}.$$

To transform it into row-reduced echelon form, first replace the third row by the sum of itself and -2 times the first row:

$$\begin{bmatrix} 1 & 0 & 1 & | & 1 \\ 0 & 1 & -1 & | & -1 \\ 0 & 1 & -1 & | & 0 \end{bmatrix}.$$

Then replace the third row by the sum of itself and -1 times the second, obtaining

$$\begin{bmatrix} 1 & 0 & 1 & | & 1 \\ 0 & 1 & -1 & | & -1 \\ 0 & 0 & 0 & | & 1 \end{bmatrix}.$$

Although this matrix is not in row-reduced echelon form, we see that the third row represents the equation

$$0x_1 + 0x_2 + 0x_3 = 1,$$

which cannot be satisfied by any choice of x_1, x_2, and x_3. Thus this equation, and hence the system, has no solution.

Geometrically we have shown here that the planes in $\mathbf{R}^3$ whose equations are the given ones have no common points of intersection (see figure 2.4).

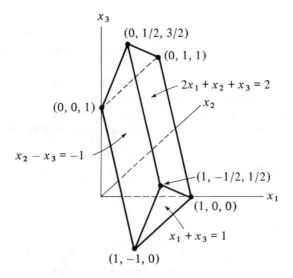

Figure 2.4

Computational note: Gauss-Jordan elimination is not the most efficient matrix elimination procedure for use on computers. This honor belongs to a method known as **Gaussian elimination.** In the latter, elementary row operations are used to transform the original augmented matrix into **echelon form** defined as follows:

i. The leftmost nonzero element in each row is called a **leading entry.**
ii. In each column that contains a leading entry of some row, all elements *below* the leading entry are zero.

iii. In any two rows with leading entries, the leading entry of the higher row is farther to the left.
iv. Any row that contains only zeros is lower than any row with some nonzero entries.

Notice that compared to *row-reduced* echelon form, echelon form does not require the leading entry in a row to be a *one*, nor does it require *all* entries in a column with a leading entry to be zero.
For example, the matrix

$$\begin{bmatrix} 2 & 1 & -1 & 3 & 5 \\ 0 & 0 & -3 & 6 & 1 \\ 0 & 0 & 0 & 2 & 4 \\ 0 & 0 & 0 & 0 & 0 \end{bmatrix}$$

is in echelon form. Once the original augmented matrix is transformed into echelon form, the linear system is solved by a process of *back-substitution* as illustrated in example 2.16.

Example 2.16

Solve the system of example 2.9 by Gaussian elimination:

$$\begin{aligned} -y + z &= 3 \\ x - y - z &= 0 \\ -x \quad\quad - z &= -3. \end{aligned}$$

The augmented matrix is

$$\begin{bmatrix} 0 & -1 & 1 & 3 \\ 1 & -1 & -1 & 0 \\ -1 & 0 & -1 & -3 \end{bmatrix}.$$

Interchange the first and second rows:

$$\begin{bmatrix} 1 & -1 & -1 & 0 \\ 0 & -1 & 1 & 3 \\ -1 & 0 & -1 & -3 \end{bmatrix}.$$

Replace the third row by the sum of itself and the first row:

$$\begin{bmatrix} 1 & -1 & -1 & 0 \\ 0 & -1 & 1 & 3 \\ 0 & -1 & -2 & -3 \end{bmatrix}.$$

Replace the third row by the sum of itself and -1 times the second row:

$$\begin{bmatrix} 1 & -1 & -1 & | & 0 \\ 0 & -1 & 1 & | & 3 \\ 0 & 0 & -3 & | & -6 \end{bmatrix}.$$

The matrix is now in echelon form. Convert back to equational form:

$$\begin{array}{rcr} x - y - z &=& 0 \\ -y + z &=& 3 \\ -3z &=& -6. \end{array}$$

We now solve this system by **back-substitution.** First, solve the last equation for z, obtaining $z = 2$. Next, substitute this value of z into the second equation, and solve it for y, obtaining $y = -1$. Finally, substitute $z = 2$, $y = -1$ into the first equation to get $x = 1$. As before, the solution is $x = 1$, $y = -1$, $z = 2$, or $(1, -1, 2)$.

2.2 Exercises

In exercises 1 and 2 find the augmented matrix for each of the systems.

1. $\begin{array}{rcr} 2x - 3y + z &=& 0 \\ x \quad\quad - 2z &=& 1 \\ -4y + z &=& -1 \end{array}$
  2. $\begin{array}{rcl} x_1 &=& 3 - x_2 + x_3 \\ x_2 &=& x_4 - x_3 \\ x_3 &=& x_1 + x_4 + 1 \end{array}$

In exercises 3 through 8 determine which matrices are in row-reduced echelon form.

3. $\begin{bmatrix} 0 & 1 \\ 1 & 0 \end{bmatrix}$
  4. $\begin{bmatrix} 1 & 2 & 0 \\ 0 & 0 & 1 \end{bmatrix}$
  5. $\begin{bmatrix} 1 & 0 & 2 \\ 0 & 1 & 1 \end{bmatrix}$

6. $\begin{bmatrix} 1 & 2 & 0 & 3 \\ 0 & 1 & 0 & 2 \\ 0 & 0 & 1 & 0 \end{bmatrix}$
  7. $\begin{bmatrix} 1 & 0 & 0 \\ 0 & 0 & 0 \\ 0 & 1 & 0 \end{bmatrix}$
  8. $\begin{bmatrix} 2 & 0 & 0 \\ 0 & 2 & 0 \\ 0 & 0 & 2 \end{bmatrix}$

In exercises 9 through 12 the given matrix is the augmented matrix for a linear system in the variables x_1, x_2, and x_3. Solve the system.

9. $\begin{bmatrix} 1 & 0 & 0 & | & 0 \\ 0 & 1 & 0 & | & 2 \\ 0 & 0 & 1 & | & -1 \end{bmatrix}$
  10. $\begin{bmatrix} 1 & 0 & 1 & | & 0 \\ 0 & 1 & -1 & | & 0 \\ 0 & 0 & 0 & | & 0 \end{bmatrix}$

11. $\begin{bmatrix} 1 & 2 & 3 & | & 4 \\ 0 & 0 & 0 & | & 0 \end{bmatrix}$
  12. $\begin{bmatrix} 1 & 1 & 1 & | & 0 \\ 0 & 0 & 0 & | & 1 \\ 0 & 0 & 0 & | & 0 \end{bmatrix}$

In exercises 13 through 20 solve the linear systems by the Gauss-Jordan matrix elimination method.

13. $x - 2y + z = 5$
          $-2x + 3y + z = 1$
          $x + 3y + 2z = 2$

14. $-x_1 + x_2 - 2x_3 = 1$
        $x_1 + x_2 + 2x_3 = -1$
        $x_1 + 3x_2 + 2x_3 = -1$

15. $2x_1 - 3x_2 + x_3 = 1$
          $-x_1 \qquad + 2x_3 = 0$
          $3x_1 - 3x_2 - x_3 = 1$

16. $x + y = 0$
            $x + y = -z$
            $-x + 1 = y$
            $1 + x + 2z = 0$

17. $w = x + y + z$
        $w = 2x - 3y + z - 1$
        $w = -x + y - 2z + 2$
        $w = 4x - 3y + 4z$

18. $x_1 + 2x_2 \qquad - x_4 = 0$
        $x_1 - 2x_2 + x_3 + x_4 = 0$
        $2x_1 - 3x_2 \qquad - x_4 = 0$

19. $v - 2w + z = 1$
         $2u - v \qquad - z = 0$
         $4u + v - 6w + z = 3$

20. $-x_1 + 3x_2 \qquad + x_4 = 0$
        $2x_1 - 5x_2 + x_3 - x_4 = 1$
        $x_2 + x_3 + x_4 = 1$
        $x_1 + x_2 + x_3 \qquad = 0$

In exercises 21 and 22 solve the given linear system without introducing fractions until the last step (if necessary at all).

21. $2x_1 - x_2 + 3x_3 = 4$
        $-x_1 + 2x_2 - x_3 = 0$
        $2x_1 + x_2 + x_3 = 4$

22. $2x_2 - x_3 = 1$
        $2x_1 - x_2 + x_3 = 1$
        $2x_1 + x_2 \qquad = 2$

23–30. Solve exercises 13 through 20 by the Gaussian elimination procedure.

31. Given a system of linear equations in three variables with the augmented matrix

$$\begin{bmatrix} 1 & 0 & 1 & | & 1 \\ 0 & 1 & 1 & | & 2 \\ 0 & 2 & k & | & k \end{bmatrix},$$

for what value of k is there no solution for the system? Is there any value of k for which there are infinitely many solutions?

32. The possible row-reduced echelon forms of the matrix $\begin{bmatrix} a & b \\ c & d \end{bmatrix}$

are $\begin{bmatrix} 1 & 0 \\ 0 & 1 \end{bmatrix}, \begin{bmatrix} 1 & k \\ 0 & 0 \end{bmatrix}, \begin{bmatrix} 0 & 1 \\ 0 & 0 \end{bmatrix},$ and $\begin{bmatrix} 0 & 0 \\ 0 & 0 \end{bmatrix}.$

List the possible row-reduced echelon forms of the matrix

$$\begin{bmatrix} a & b & c \\ d & e & f \\ g & h & i \end{bmatrix}.$$

Note: A digital computer rounds certain fractions (such as 1/3) to a finite (relatively small) number of decimal places. Consequently "round-off errors" may be introduced at each step of an algorithm such as Gauss-Jordan or Gaussian elimination.

In "small" problems, like those in this exercise set, it is unlikely that these errors will have a significant effect on the accuracy of the result. However, it is possible. The cumulative effect of round-off errors is a function not only of the size of a linear system but also of its coefficients. Exercises 33 and 34 explore this phenomenon.

33. Solve exercises 13 through 20 by using a computer program for Gauss-Jordan or Gaussian elimination. How do the answers compare with the ones you obtained by hand?

34. Solve the linear system whose augmented matrix is given as follows by using a computer program for Gauss-Jordan or Gaussian elimination:

$$\left[\begin{array}{cccccc|c} 1 & 1/2 & 1/3 & 1/4 & 1/5 & 1/6 & 1 \\ 1/2 & 1/3 & 1/4 & 1/5 & 1/6 & 1/7 & 0 \\ 1/3 & 1/4 & 1/5 & 1/6 & 1/7 & 1/8 & 0 \\ 1/4 & 1/5 & 1/6 & 1/7 & 1/8 & 1/9 & 0 \\ 1/5 & 1/6 & 1/7 & 1/8 & 1/9 & 1/10 & 0 \\ 1/6 & 1/7 & 1/8 & 1/9 & 1/10 & 1/11 & 0 \end{array}\right].$$

Compare your result with the "exact" solution (36, -630, 3360, -7560, 7560, -2772).

2.3 Operations on Matrices

In the previous section we used matrices to provide a convenient notation for performing the elimination procedure. Later in this chapter we will use them in another way to provide a second technique for solving certain systems of linear equations. Matrices will also appear frequently throughout the remainder of this text in several different settings. Consequently it will be useful at this point to establish some of the notation and definitions concerning this important mathematical tool.

Basic Terminology

Recall from section 2.2 that a **matrix** is simply a rectangular array of **entries** which are usually real or complex numbers. If a given matrix has m rows and n columns, it is called an **m × n matrix** (read, "m by n matrix"), and the numbers m and n are the **dimensions** of the matrix. The matrices

$$\left[\begin{array}{cc} 1 & 3 \\ -1 & 0 \\ 2 & 4 \end{array}\right], \tag{1}$$

$$\begin{bmatrix} -1 & 0 & 1 \\ 4 & 2 & -3 \end{bmatrix}, \tag{2}$$

and

$$\begin{bmatrix} \frac{1}{2} \\ -6 \\ 4 \end{bmatrix} \tag{3}$$

are, respectively, 3×2, 2×3, and 3×1 matrices. We sometimes refer to an $m \times 1$ matrix (with $m > 1$) as a **column vector** and a $1 \times n$ matrix ($n > 1$) as a **row vector.** For example, the matrix (3) is called a column vector.

Capital (upper-case) letters, such as A, B, and C, will be used to denote matrices, while the entries of a matrix will be denoted by the corresponding lower-case letter with "double-subscripts." For example, to describe a general 2×4 matrix, we could write:

$$A = \begin{bmatrix} a_{11} & a_{12} & a_{13} & a_{14} \\ a_{21} & a_{22} & a_{23} & a_{24} \end{bmatrix}.$$

Notice that the first subscript gives the row number while the second gives the column number. As another example, if

$$B = \begin{bmatrix} 3 & 6 & -1 \\ 4 & 2 & 0 \\ 1 & -3 & -5 \end{bmatrix},$$

then $b_{12} = 6$, $b_{21} = 4$, and $b_{33} = -5$.

Occasionally, when we want to make explicit this upper-case/lower-case convention, we will write something like, "Let $A = [a_{ij}]$ and $B = [b_{ij}]$." All this means is, "Let A be a matrix with entries denoted a_{ij} and let B be a matrix with entries denoted b_{ij}." Moreover, if C is an $m \times n$ matrix, then the subscripts i and j of its entries take on the values $1, 2, \ldots, m$ and $1, 2, \ldots, n$, respectively. However, we will rarely state this explicitly.

Before proceeding to a description of matrix operations, we will give one more definition. Given matrices A and B, the entries a_{ij} and b_{kp} are **corresponding entries** if and only if $i = k$ and $j = p$. In other words, an entry of A *corresponds* to an entry of B if and only if each occupies the same position in the matrices. For example, if

$$A = \begin{bmatrix} 1 & 2 & 3 \\ 4 & 5 & 6 \end{bmatrix} \quad \text{and} \quad B = \begin{bmatrix} 7 & 8 & 9 \\ -1 & -2 & -3 \end{bmatrix},$$

then 1 and 7, 2 and 8, and 4 and -1 are all pairs of corresponding entries. We shall only speak of corresponding entries when the two matrices in question have the same dimensions.

Definition   (equality of matrices)   Two matrices A and B are **equal**, written $A = B$, if they have the same dimensions and their corresponding entries are equal.

More formally, an $m \times n$ matrix A is equal to a $p \times q$ matrix B if and only if $m = p$, $n = q$, and $a_{ij} = b_{ij}$ for all i and j.

Example 2.17

Which of the following matrices are equal:

$$A = \begin{bmatrix} 1 & 2 & -1 \\ 4 & 0 & 1+2 \end{bmatrix}, \quad B = \begin{bmatrix} 1 & 2 \\ 4 & 0 \end{bmatrix},$$

$$C = \begin{bmatrix} 1 & \frac{4}{2} & -1 \\ 4 & 0 & 3 \end{bmatrix}, \quad D = \begin{bmatrix} 1 & 2 & -1 \\ 4 & 0 & 1 \end{bmatrix}?$$

We have $A = C$, but this is the only pair of equal matrices.

Sum, Difference, and Scalar Multiplication

Definition (sum of matrices)   Let A and B be matrices with the same dimensions. The **sum** of A and B, written $A + B$, is the matrix obtained by adding corresponding entries of A and B.

Formally, we have $C = A + B$ if and only if $c_{ij} = a_{ij} + b_{ij}$ for all i and j.

Example 2.18

Find the matrix C that is the sum of

$$A = \begin{bmatrix} 3 & 6 \\ -1 & 5 \\ 0 & 2 \end{bmatrix} \quad \text{and} \quad B = \begin{bmatrix} \frac{1}{2} & -4 \\ -2 & 0 \\ -3 & -2 \end{bmatrix}.$$

$$C = \begin{bmatrix} 3 + \frac{1}{2} & 6 + (-4) \\ -1 + (-2) & 5 + 0 \\ 0 + (-3) & 2 + (-2) \end{bmatrix} = \begin{bmatrix} \frac{7}{2} & 2 \\ -3 & 5 \\ -3 & 0 \end{bmatrix}.$$

Definition (multiplication by a scalar)   Let A be a matrix and c be a scalar. Then the product cA is the matrix obtained by multiplying each entry of A by the scalar c.

Formally, we have $B = cA$ if and only if $b_{ij} = ca_{ij}$ for all i and j.

Example 2.19

Let

$$A = \begin{bmatrix} 1 & -1 & 0 & 3 \\ 2 & 6 & -4 & 8 \end{bmatrix}.$$

Find $3A$ and $(-1)A$.

We have

$$3A = \begin{bmatrix} 3 & -3 & 0 & 9 \\ 6 & 18 & -12 & 24 \end{bmatrix}, \quad \text{and} \quad (-1)A = \begin{bmatrix} -1 & 1 & 0 & -3 \\ -2 & -6 & 4 & -8 \end{bmatrix}.$$

Definition The **negative** of a matrix B, written $-B$, is the matrix $(-1)B$. It is obtained from B by simply reversing the sign of every entry.

Definition (difference of matrices)   Let A and B be matrices with the same dimensions. The **difference** of A and B, written $A - B$, is the matrix C given by $C = A + (-B)$.

Note: The matrix $A - B$ can be obtained by simply subtracting each entry of B from the corresponding entry of A.

Example 2.20

Find $C = 2A - B$, where

$$A = \begin{bmatrix} 1 & 2 \\ -1 & 0 \end{bmatrix}, \quad B = \begin{bmatrix} 0 & 4 \\ -1 & 0 \end{bmatrix}.$$

We have

$$2\begin{bmatrix} 1 & 2 \\ -1 & 0 \end{bmatrix} - \begin{bmatrix} 0 & 4 \\ -1 & 0 \end{bmatrix} = \begin{bmatrix} 2 & 4 \\ -2 & 0 \end{bmatrix} - \begin{bmatrix} 0 & 4 \\ -1 & 0 \end{bmatrix} = \begin{bmatrix} 2 & 0 \\ -1 & 0 \end{bmatrix}.$$

Matrix Multiplication

We will now define the "product" of two matrices. From the way the other matrix operations have been defined, you might guess that we will obtain the product of two matrices by simply multiplying corresponding entries. The standard definition of product given in the definition that follows is considerably more complicated than this but also considerably more useful in applications. Examples of these applications will appear several times throughout the rest of this text, beginning with the next section.

Before getting to the definition, we will introduce some notation. Given an $m \times n$ matrix A, the vector consisting of the entries of the ith row of A will be denoted A_i, while the vector consisting of the entries of the jth column of A will be denoted A^j. For example, if

$$A = \begin{bmatrix} 4 & 3 & -1 \\ 1 & 2 & 5 \\ 0 & 1 & 0 \end{bmatrix},$$

then $A^1 = (4, 1, 0)$, $A_1 = (4, 3, -1)$, and $A_3 = (0, 1, 0)$. Also recall from section 1.4 that the dot product of the vectors $\mathbf{x} = (x_1, x_2, \ldots, x_r)$ and $\mathbf{y} = (y_1, y_2, \ldots, y_r)$, written $\mathbf{x} \cdot \mathbf{y}$, is given by the number $x_1 y_1 + x_2 y_2 + \cdots + x_r y_r$. For example, with A defined as in the matrix just given, we have $A^1 \cdot A_1 = 4 \cdot 4 + 1 \cdot 3 + 0(-1) = 19$ and $A_1 \cdot A^3 = 11$.

Definition (matrix multiplication)   Let A and B be matrices such that the number of columns of A is equal to the number of rows of B. Suppose that the number of rows of A is m and the number of columns of B is q. The **product** AB is the $m \times q$ matrix C defined by

$$c_{ij} = A_i \cdot B^j.$$

In words, to obtain the entry in the ith row and jth column of C, we take the dot product of the ith row of A with the jth column of B.

Note: If A is an $m \times n$ matrix, and B is a $p \times q$ matrix, then the product AB is defined only if $n = p$ and, if so, AB is an $m \times q$ matrix. The entry of AB in the ith row and jth column is obtained by multiplying each entry in the ith row of A by the corresponding entry in the jth column of B, and adding up all such products. In other words,

$$c_{ij} = a_{i1}b_{1j} + a_{i2}b_{2j} + \cdots + a_{in}b_{nj}.$$

Example 2.21

Given the following pairs of matrices, A and B, find the products AB and BA, if defined:

a.
$$A = \begin{bmatrix} 1 & 2 & 3 \\ -3 & -2 & -1 \end{bmatrix}, \quad B = \begin{bmatrix} -4 & 5 \\ 0 & 4 \\ -5 & 0 \end{bmatrix}.$$

Here A is a 2×3 matrix, and B is a 3×2 matrix. Since the number of columns of A is equal to the number of rows of B, the product AB is defined. It has 2 rows (the same number as A) and 2 columns (the same number as B). Let $C = AB$. Then

$$\begin{aligned} c_{11} &= A_1 \cdot B^1 = (1, 2, 3) \cdot (-4, 0, -5) \\ &= 1(-4) + 2(0) + 3(-5) = -19, \\ c_{12} &= A_1 \cdot B^2 = (1, 2, 3) \cdot (5, 4, 0) \\ &= 1(5) + 2(4) + 3(0) = 13, \\ c_{21} &= A_2 \cdot B^1 = (-3, -2, -1) \cdot (-4, 0, -5) \\ &= (-3)(-4) + 0 + (-1)(-5) = 17, \\ c_{22} &= A_2 \cdot B^2 = (-3, -2, -1) \cdot (5, 4, 0) \\ &= (-3)(5) + (-2)(4) + 0 = -23. \end{aligned}$$

Thus
$$C = \begin{bmatrix} -19 & 13 \\ 17 & -23 \end{bmatrix}.$$

In terms of remembering the definition and actually doing the computations to get the ijth entry of AB, it is probably better to think, "go across the ith row of A and down the jth column of B," forming products of corresponding entries and adding the results.

We illustrate this device in finding BA. Now BA is defined since B (now the "first matrix") has 2 columns, which is the same as the number of rows of A (the "second matrix"). Also BA will have 3 rows (as does B) and 3 columns (as does A). Letting $D = BA$, we have

$$D = \begin{bmatrix} -4 & 5 \\ 0 & 4 \\ -5 & 0 \end{bmatrix} \begin{bmatrix} 1 & 2 & 3 \\ -3 & -2 & -1 \end{bmatrix}.$$

We find

d_{11} by "going across first row of B and down first column of A," so
$$d_{11} = (-4)(1) + 5(-3) = -19.$$
d_{12} by "going across first row of B and down second column of A," so
$$d_{12} = (-4)(2) + 5(-2) = -18.$$
d_{21} by "going across second row of B and down first column of A," so
$$d_{21} = 0(1) + 4(-3) = -12.$$

Similarly
$$d_{13} = (-4)(3) + (5)(-1) = -17,$$
$$d_{22} = (0)(2) + (4)(-2) = -8,$$

$$d_{23} = (0)(3) + (4)(-1) = -4,$$
$$d_{31} = (-5)(1) + (0)(-3) = -5,$$
$$d_{32} = (-5)(2) + (0)(-2) = -10,$$
$$d_{33} = (-5)(3) + (0)(-1) = -15.$$

Thus

$$D = BA = \begin{bmatrix} -19 & -18 & -17 \\ -12 & -8 & -4 \\ -5 & -10 & -15 \end{bmatrix}.$$

Notice that $AB \neq BA$ in this example. In fact they don't even have the same dimensions.

b.
$$A = \begin{bmatrix} 1 & 2 & 3 \\ 4 & 5 & 6 \\ 7 & 8 & 9 \end{bmatrix}, \quad B = \begin{bmatrix} x \\ y \\ z \end{bmatrix}.$$

The product AB is defined, since A has 3 columns and B has 3 rows, and it is a 3×1 matrix. We have

$$AB = \begin{bmatrix} 1 & 2 & 3 \\ 4 & 5 & 6 \\ 7 & 8 & 9 \end{bmatrix} \begin{bmatrix} x \\ y \\ z \end{bmatrix} = \begin{bmatrix} x + 2y + 3z \\ 4x + 5y + 6z \\ 7x + 8y + 9z \end{bmatrix}.$$

However, the product BA is not defined, since the number of columns of B (one) is not equal to the number of rows of A (three).

c.
$$A = \begin{bmatrix} 1 & -1 & 0 \\ 2 & 1 & 3 \\ -5 & 0 & 1 \end{bmatrix}, \quad B = \begin{bmatrix} 1 & 0 & 0 \\ 0 & 1 & 0 \\ 0 & 0 & 1 \end{bmatrix}.$$

Here both AB and BA are defined, and both are 3×3 matrices. Performing the computations, we obtain

$$AB = \begin{bmatrix} 1 & -1 & 0 \\ 2 & 1 & 3 \\ -5 & 0 & 1 \end{bmatrix} \quad \text{and} \quad BA = \begin{bmatrix} 1 & -1 & 0 \\ 2 & 1 & 3 \\ -5 & 0 & 1 \end{bmatrix}.$$

Thus not only is $AB = BA$ here but both products are equal to A. In fact with B defined as in the last part of this example, you can check that $CB = C$ and $BC = C$ for *any* 3×3 matrix C! A matrix that has a form like that of B is called an *identity matrix*.

It should be apparent from example 2.21 that multiplication of matrices "acts differently" than multiplication of numbers. The latter has the property of *commutativity* (for all numbers a and b, $ab = ba$), but matrix multiplication does not. However, as we shall soon see, operations on matrices do possess many of the same properties as those on numbers.

Identity and Zero Matrices

A class of matrices that is especially important is that of *square* matrices. An $m \times n$ matrix is **square** if $m = n$, if the number of rows is equal to the number of columns. For a square matrix the number of rows (or columns) is called the **order** of the matrix. (Both matrices in part c of example 2.21 are square of order 3).

The entries a_{ij} of a square matrix A for which $i = j$ are said to comprise the **main diagonal** of A. For example, the main diagonal of

$$B = \begin{bmatrix} 3 & -1 & 4 \\ 0 & 1 & 2 \\ -2 & 5 & -3 \end{bmatrix}$$

consists of 3, 1, and -3. A square matrix in which every element *not* on the main diagonal (every "off diagonal" element) is zero is called a **diagonal matrix.** A special type of diagonal matrix is the **identity matrix,** denoted by I, in which each entry on the main diagonal is 1.

Example 2.22

The matrices

$$\begin{bmatrix} 1 & 0 \\ 0 & 1 \end{bmatrix}, \quad \begin{bmatrix} 1 & 0 & 0 \\ 0 & 1 & 0 \\ 0 & 0 & 1 \end{bmatrix}, \quad \text{and} \quad \begin{bmatrix} 2 & 0 \\ 0 & -3 \end{bmatrix}$$

are all diagonal matrices. The first two are identity matrices of order two and three, respectively. The last is not an identity matrix.

The identity matrix is so named because it has the property that $AI = A$ and $IA = A$ for any matrix A for which the products are defined. Thus it is the matrix analog of the number 1, the "multiplicative identity" for numbers.

Another important type of matrix is the *zero matrix.* A **zero matrix,** denoted by 0, is a matrix of any dimensions consisting of only zero entries.

Example 2.23

The matrices

$$\begin{bmatrix} 0 & 0 & 0 \\ 0 & 0 & 0 \\ 0 & 0 & 0 \end{bmatrix}, \quad \begin{bmatrix} 0 & 0 & 0 \\ 0 & 0 & 0 \end{bmatrix}, \quad \begin{bmatrix} 0 \\ 0 \\ 0 \end{bmatrix}, \quad \text{and} \quad \begin{bmatrix} 0 & 0 & 0 & 0 \end{bmatrix}$$

are all zero matrices. (The first of these is also a diagonal matrix.)

A zero matrix has the property that $A + 0 = A$ and $0 + A = A$ for any matrix A for which the sums are defined. It is the matrix analog of the number zero.

Powers and Transpose of a Matrix

We are now ready to give the final two definitions of this section. They involve the matrix operations of *taking powers* and *transpose*.

Definition (powers of a matrix)   Let A be a square matrix. Then

$$A^0 = I, \text{ the identity matrix of the same order,}$$
$$A^1 = A,$$
$$A^k = A^{k-1}A \text{ for } k > 1.$$

Note: The *inductive* definition $A^k = A^{k-1}A$ is equivalent to multiplying together k factors of A from left to right, $A^k = AA \dots A$. Although matrix multiplication is not commutative in general, this is one time you can multiply in any order and always get the same result. (This fact is a consequence of theorem 1e that follows.) We should remark that the notation for the kth *power* of A duplicates that of the kth *column* of A given earlier. This apparent ambiguity will present no problem since it will be clear from the context, when not explicitly stated, which meaning is intended.

Example 2.24

Let

$$A = \begin{bmatrix} 1 & 0 \\ -1 & 2 \end{bmatrix}.$$

Find the matrices A^0, A^1, A^2, and A^3.

$$A^0 = I = \begin{bmatrix} 1 & 0 \\ 0 & 1 \end{bmatrix},$$

$$A^1 = A = \begin{bmatrix} 1 & 0 \\ -1 & 2 \end{bmatrix},$$

$$A^2 = AA = \begin{bmatrix} 1 & 0 \\ -1 & 2 \end{bmatrix}\begin{bmatrix} 1 & 0 \\ -1 & 2 \end{bmatrix} = \begin{bmatrix} 1 & 0 \\ -3 & 4 \end{bmatrix},$$

$$A^3 = A^2A = \begin{bmatrix} 1 & 0 \\ -3 & 4 \end{bmatrix}\begin{bmatrix} 1 & 0 \\ -1 & 2 \end{bmatrix} = \begin{bmatrix} 1 & 0 \\ -7 & 8 \end{bmatrix}.$$

Note: We found A^3 in example 2.24 by forming the product A^2A^1, the exponents adding just as they do when the base is a number. In general, if A is a square matrix, and m and n are nonnegative integers, we have $A^mA^n = A^{m+n}$. It is also true that under the same conditions, $(A^m)^n = A^{mn}$.

Definition (transpose of a matrix)   The **transpose** of the $m \times n$ matrix A, written A^T, is the $n \times m$ matrix B whose jth column is the jth row of A. Equivalently $B = A^T$ if and only if $b_{ij} = a_{ji}$ for each i and j.

Example 2.25

Find the transpose of

a. $A = \begin{bmatrix} 1 & 2 \\ -1 & 3 \end{bmatrix}$,   b. $B = \begin{bmatrix} 1 & -1 \\ 2 & 0 \\ 3 & 4 \end{bmatrix}$.

We have

a. $A^T = \begin{bmatrix} 1 & -1 \\ 2 & 3 \end{bmatrix}$,   b. $B^T = \begin{bmatrix} 1 & 2 & 3 \\ -1 & 0 & 4 \end{bmatrix}$.

Rules Governing Operations on Matrices

We close this section with a theorem that states some of the rules governing the matrix operations already discussed.

Theorem 1 (properties of matrices and rules of operation)   Let A, B, and C be matrices and a and b scalars. Assume that the dimensions of the matrices are such that each operation is defined. Then

a. $A + B = B + A$ (commutative law for addition),
b. $A + (B + C) = (A + B) + C$ (associative law for addition),
c. $A + 0 = A$,
d. $A + (-A) = 0$,
e. $A(BC) = (AB)C$ (associative law for multiplication),
f. $AI = A, IB = B$,
g. $A(B + C) = AB + AC$
    $(B + C)A = BA + CA$ $\Big\}$ (distributive laws),

h. $a(B + C) = aB + aC$,
i. $(a + b)C = aC + bC$,
j. $(ab)C = a(bC)$,
k. $1A = A$,
l. $A0 = 0, 0B = 0$,
m. $a0 = 0$,
n. $a(AB) = (aA)B = A(aB)$,
o. $(A + B)^T = A^T + B^T$,
p. $(AB)^T = B^T A^T$.

Before we prove some of these assertions, we will give a few illustrative examples.

Example 2.26

Let

$$A = \begin{bmatrix} 1 & 0 & -1 \\ 2 & 1 & -1 \\ 3 & -2 & 0 \end{bmatrix}, \quad B = \begin{bmatrix} -1 & 2 & 3 \\ 0 & 1 & 0 \\ -2 & -1 & 0 \end{bmatrix}, \quad C = \begin{bmatrix} -2 & 1 & 1 \\ 1 & 1 & 0 \\ 3 & 0 & -3 \end{bmatrix}.$$

Find $A(BC)$, $(AB)C$, $A(B + C)$, and $AB + AC$.

$$A(BC) = \begin{bmatrix} 1 & 0 & -1 \\ 2 & 1 & -1 \\ 3 & -2 & 0 \end{bmatrix}\begin{bmatrix} 13 & 1 & -10 \\ 1 & 1 & 0 \\ 3 & -3 & -2 \end{bmatrix} = \begin{bmatrix} 10 & 4 & -8 \\ 24 & 6 & -18 \\ 37 & 1 & -30 \end{bmatrix},$$

$$(AB)C = \begin{bmatrix} 1 & 3 & 3 \\ 0 & 6 & 6 \\ -3 & 4 & 9 \end{bmatrix}\begin{bmatrix} -2 & 1 & 1 \\ 1 & 1 & 0 \\ 3 & 0 & -3 \end{bmatrix} = \begin{bmatrix} 10 & 4 & -8 \\ 24 & 6 & -18 \\ 37 & 1 & -30 \end{bmatrix}.$$

(Notice that $A(BC) = (AB)C$ for these choices of A, B, and C.)

$$A(B + C) = \begin{bmatrix} 1 & 0 & -1 \\ 2 & 1 & -1 \\ 3 & -2 & 0 \end{bmatrix}\begin{bmatrix} -3 & 3 & 4 \\ 1 & 2 & 0 \\ 1 & -1 & -3 \end{bmatrix} = \begin{bmatrix} -4 & 4 & 7 \\ -6 & 9 & 11 \\ -11 & 5 & 12 \end{bmatrix},$$

$$AB + AC = \begin{bmatrix} 1 & 3 & 3 \\ 0 & 6 & 6 \\ -3 & 4 & 9 \end{bmatrix} + \begin{bmatrix} -5 & 1 & 4 \\ -6 & 3 & 5 \\ -8 & 1 & 3 \end{bmatrix} = \begin{bmatrix} -4 & 4 & 7 \\ -6 & 9 & 11 \\ -11 & 5 & 12 \end{bmatrix}.$$

(Notice that $A(B + C) = AB + AC$ for these A, B, and C.)

Of course example 2.26 does not constitute a *proof* of the fact that $A(BC) = (AB)C$ or that $A(B + C) = AB + AC$ for *all* matrices A, B, and C.

Proof of theorem 1 (in part)

Proof of 1b

Let $M = A + (B + C)$ and $M' = (A + B) + C$. Then $m_{ij} = a_{ij} + (b_{ij} + c_{ij}) = (a_{ij} + b_{ij}) + c_{ij} = m'_{ij}$ (since addition of numbers is associative).

Proof of 1g

Let $M = A(B + C)$. Then $m_{ij} = A_i \cdot (B + C)^j = A_i \cdot (B^j + C^j) = A_i \cdot B^j + A_i \cdot C^j$ (by theorem 3c of section 1.4).

Now, let $M' = AB + AC$. Then $m'_{ij} = A_i \cdot B^j + A_i \cdot C^j = m_{ij}$, as desired. In a similar manner we can show that $(B + C)A = BA + CA$.

Proof of 1n

Let $C = a(AB)$, $D = (aA)B$, and $E = A(aB)$. Then $c_{ij} = a(A_i \cdot B^j) = (aA_i) \cdot B^j = d_{ij}$ (by theorem 3b of section 1.4). Moreover $d_{ij} = (aA_i) \cdot B^j = A_i \cdot (aB^j) = e_{ij}$ (same theorem). Thus $c_{ij} = d_{ij} = e_{ij}$, as desired. ∎

2.3 Exercises

In exercises 1 through 12 let

$$A = \begin{bmatrix} -1 & 0 & 1 \\ 2 & -1 & 3 \\ 0 & 1 & -2 \end{bmatrix}, \quad B = \begin{bmatrix} 0 & 4 & -2 \\ 3 & 1 & 2 \\ -1 & 0 & 1 \end{bmatrix},$$

and let I be the identity matrix of order 3. Find each as indicated.

1. $A + B$ 2. $A - B$

3. $2A - 3B$ 4. $A + I$

5. $A - \lambda I$ (λ, a scalar)   6. $A + \lambda B$ (λ, a scalar)

7. A^T 8. A^0

9. AB 10. $(AB)^T$

11. $B^T A^T$ 12. $A^T B^T$

In exercises 13 through 16 find each as indicated, where

$$A = \begin{bmatrix} 2 & -1 & 1/2 \\ 1 & 0 & -2 \end{bmatrix}.$$

13. A^T **14.** $(A^T)^T$

15. The columns A^1 and A^3 and the row A_2

16. $A^1 \cdot A^3$ and $A_1 \cdot A_2$

In exercises 17 through 22 find AB and BA when these products are defined.

17. $A = \begin{bmatrix} 1 & 2 & 0 \\ -2 & 0 & 1 \\ 0 & 0 & 3 \end{bmatrix}$, $B = \begin{bmatrix} 1 & 1 & 1 \\ 3 & -1 & 0 \\ 2 & -2 & 0 \end{bmatrix}$

18. $A = \begin{bmatrix} 1 \\ 0 \\ 0 \end{bmatrix}$, $B = \begin{bmatrix} -3 & 2 & 1 \end{bmatrix}$

19. $A = \begin{bmatrix} -1 & 1 & 0 \\ 2 & 3 & 5 \end{bmatrix}$, $B = \begin{bmatrix} 0 & 4 & -2 \\ 3 & 6 & -1 \end{bmatrix}$

20. $A = \begin{bmatrix} -1 & 1 & 0 \\ 2 & 3 & 5 \end{bmatrix}$, $B = \begin{bmatrix} 0 & 4 & -2 \\ 3 & 6 & -1 \\ 1 & 2 & -3 \end{bmatrix}$

21. $A = \begin{bmatrix} 2 & 1 & 0 \\ -1 & 4 & 3 \end{bmatrix}$, $B = \begin{bmatrix} 1 & 0 & 0 \\ 0 & 1 & 0 \\ 0 & 0 & 1 \end{bmatrix}$

22. $A = \begin{bmatrix} 1 & 2 & 1 \\ -2 & 1 & -2 \end{bmatrix}$, $B = \begin{bmatrix} 0 & 0 \\ 0 & 0 \\ 0 & 0 \\ 0 & 0 \end{bmatrix}$

23. Find

$$\begin{bmatrix} 1 & 3 \\ -1 & 1 \end{bmatrix}\begin{bmatrix} 0 & 1 & -1 \\ 1 & 0 & 1 \end{bmatrix} - 3\begin{bmatrix} 4 & 0 & -2 \\ 1 & 2/3 & 4 \end{bmatrix}.$$

24. Find $3A - 0B + 6I$ if

$$A = \begin{bmatrix} 1/2 & 2 \\ 5 & 1/4 \end{bmatrix}, \quad B = \begin{bmatrix} 1/3 & 1/4 \\ 1/5 & 1/6 \end{bmatrix},$$

and I is the identity matrix of order 2.

In exercises 25 and 26 find the powers A^2 and A^3 for the given matrix A.

25. $A = \begin{bmatrix} 1 & -1 \\ -1 & 2 \end{bmatrix}$ **26.** $A = \begin{bmatrix} 1 & 0 & 2 \\ -2 & -1 & 0 \\ 0 & 0 & 2 \end{bmatrix}$

In exercises 27 through 30 let

$$A = \begin{bmatrix} 3 & 0 & -1 \\ 0 & 0 & 2 \\ 1 & 0 & -2 \end{bmatrix}, \quad B = \begin{bmatrix} 1 & 2 & 1 \\ 0 & 0 & 0 \\ -1 & 1 & 0 \end{bmatrix}, \quad C = \begin{bmatrix} 4 & 0 & -3 \\ -2 & -1 & 0 \\ 1 & 1 & 1 \end{bmatrix},$$

and verify each equation.

27. $(A + B)C = AC + BC$ **28.** $(2A)B = 2(AB)$

29. $A + C = C + A$ **30.** $(A + B) + C = (C + A) + B$

31. Prove the remaining parts of theorem 1.

32. Use theorem 1 to show that $(A + B) + C = (C + A) + B$ for all matrices A, B, and C for which the sums are defined.

33. A matrix A is called *involutory* if $A^2 = I$. Give an example of a 2×2 matrix (not equal to I) that is involutory.

34. A matrix A is *idempotent* if $A^2 = A$. Give an example of a 2×2 matrix (not equal to I or 0) that is idempotent.

35. Prove that $(A + B)^T = A^T + B^T$ for all matrices A and B of the same dimensions.

36. Prove that $(A^T)^T = A$ for all matrices A.

In exercises 37 and 38 let A and B be square matrices of order 10 with

$$A = \begin{bmatrix} 1 & 2 & 3 & \ldots & 10 \\ 2 & 3 & 4 & \ldots & 11 \\ 3 & 4 & 5 & \ldots & 12 \\ \vdots & \vdots & \vdots & & \vdots \\ 10 & 11 & 12 & \ldots & 19 \end{bmatrix}, \quad B = \begin{bmatrix} 1 & -1 & 1 & -1 & \ldots & -1 \\ 0 & 1 & -1 & 1 & \ldots & 1 \\ 0 & 0 & 1 & -1 & \ldots & -1 \\ \vdots & \vdots & \vdots & \vdots & & \vdots \\ 0 & 0 & 0 & 0 & \ldots & 1 \end{bmatrix}.$$

Using a computer program for matrix multiplication, find the following products.

37. AB and BA **38.** $(AB)^T$, $A^T B^T$, and $B^T A^T$

2.4 Matrix Equations and Inverses

In this section we will develop a very concise way of writing a system of linear equations. This concise form will suggest another solution procedure that in turn will lead to the definition and investigation of the notion of the *inverse* of a square matrix.

Matrix Form of a Linear System

A system of m linear equations in the n unknowns $x_1, x_2, \ldots, x_n$ can be written in the general form:

$$a_{11}x_1 + a_{12}x_2 + \cdots + a_{1n}x_n = b_1$$
$$a_{21}x_1 + a_{22}x_2 + \cdots + a_{2n}x_n = b_2$$
$$\vdots$$
$$a_{m1}x_1 + a_{m2}x_2 + \cdots + a_{mn}x_n = b_m.$$

The a_{ij} and the b_i are, respectively, the **coefficients** and **right-side constants** of the system. The "double subscript" notation is particularly convenient for the coefficients. Notice that in the ith equation, the coefficient of the jth unknown is a_{ij}. For example, in the second equation the coefficient of the first unknown is denoted a_{21}.

Now recall from section 2.3 that two matrices are equal if and only if all pairs of corresponding entries are equal. For this reason we can write the linear system just given as an equation between two $m \times 1$ matrices:

$$\begin{bmatrix} a_{11}x_1 + a_{12}x_2 + \cdots + a_{1n}x_n \\ a_{21}x_1 + a_{22}x_2 + \cdots + a_{2n}x_n \\ \vdots \\ a_{m1}x_1 + a_{m2}x_2 + \cdots + a_{mn}x_n \end{bmatrix} = \begin{bmatrix} b_1 \\ b_2 \\ \vdots \\ b_m \end{bmatrix}.$$

But by the definition of matrix multiplication the matrix on the left can be written as the product of an $m \times n$ matrix and $n \times 1$ matrix, so

$$\begin{bmatrix} a_{11} & a_{12} \cdots a_{1n} \\ a_{21} & a_{22} \cdots a_{2n} \\ \vdots & \vdots \quad \vdots \\ a_{m1} & a_{m2} \cdots a_{mn} \end{bmatrix} \begin{bmatrix} x_1 \\ x_2 \\ \vdots \\ x_n \end{bmatrix} = \begin{bmatrix} b_1 \\ b_2 \\ \vdots \\ b_m \end{bmatrix}.$$

Finally, denoting the $m \times n$ matrix with entries a_{ij} by A, the $n \times 1$ matrix with entries x_i by X, and the $m \times 1$ matrix with entries b_i by B, we can write the given system of linear equations in the concise matrix form

$$AX = B.$$

Here A is called the **coefficient matrix** of the system; X is the matrix, or (column) vector, of unknowns; and B is the matrix, or (column) vector, of constants.

Example 2.27

Write in matrix form the linear systems

a. $\quad x \qquad + 2z = 3$
$\quad 2x - 3y - 4z = -1$
$\qquad - y + \;\; z = 0$

b. $x_1 - 3x_2 \qquad + x_4 = 0$
$\qquad - x_2 + x_3 - x_4 = 0.$

The matrix forms of the two systems are

a. $\begin{bmatrix} 1 & 0 & 2 \\ 2 & -3 & -4 \\ 0 & -1 & 1 \end{bmatrix} \begin{bmatrix} x \\ y \\ z \end{bmatrix} = \begin{bmatrix} 3 \\ -1 \\ 0 \end{bmatrix},$

b. $\begin{bmatrix} 1 & -3 & 0 & 1 \\ 0 & -1 & 1 & -1 \end{bmatrix} \begin{bmatrix} x_1 \\ x_2 \\ x_3 \\ x_4 \end{bmatrix} = \begin{bmatrix} 0 \\ 0 \end{bmatrix}.$

Both are written as $AX = B$, where in part a. the matrices are

$$A = \begin{bmatrix} 1 & 0 & 2 \\ 2 & -3 & -4 \\ 0 & -1 & 1 \end{bmatrix}, \quad X = \begin{bmatrix} x \\ y \\ z \end{bmatrix}, \quad B = \begin{bmatrix} 3 \\ -1 \\ 0 \end{bmatrix};$$

and in part b. the matrices are

$$A = \begin{bmatrix} 1 & -3 & 0 & 1 \\ 0 & -1 & 1 & -1 \end{bmatrix}, \quad X = \begin{bmatrix} x_1 \\ x_2 \\ x_3 \\ x_4 \end{bmatrix}, \quad B = \begin{bmatrix} 0 \\ 0 \end{bmatrix}.$$

Note: Since in the matrix equation $AX = B$, both X and B are column vectors, we often write them using vector notation as $\mathbf{x}$ and $\mathbf{b}$. Thus, the matrix form of the general linear system will usually be denoted by $A\mathbf{x} = \mathbf{b}$.

Example 2.28

Write the linear system

$$2x_1 - 3x_2 + 5 = 0$$
$$x_2 = x_1 + 3$$

in the form $A\mathbf{x} = \mathbf{b}$, identifying A, $\mathbf{x}$, and $\mathbf{b}$.

We must first write this linear system in standard form. It becomes

$$2x_1 - 3x_2 = -5$$

$$-x_1 + x_2 = 3$$

or

$$\begin{bmatrix} 2 & -3 \\ -1 & 1 \end{bmatrix} \begin{bmatrix} x_1 \\ x_2 \end{bmatrix} = \begin{bmatrix} -5 \\ 3 \end{bmatrix}.$$

Thus, written in the form $A\mathbf{x} = \mathbf{b}$, the system has coefficient matrix

$$A = \begin{bmatrix} 2 & -3 \\ -1 & 1 \end{bmatrix}$$

and column vectors

$$\mathbf{x} = \begin{bmatrix} x_1 \\ x_2 \end{bmatrix}, \quad \mathbf{b} = \begin{bmatrix} -5 \\ 3 \end{bmatrix}.$$

Note: It is often convenient to write a column vector on a horizontal line, so we establish the convention that $(c_1, c_2, \ldots, c_k)$ will denote the $k \times 1$ column vector

$$\begin{bmatrix} c_1 \\ c_2 \\ \vdots \\ c_k \end{bmatrix}.$$

Notice that parentheses and commas distinguish our column vector written horizontally from the $1 \times k$ row vector $\begin{bmatrix} c_1 & c_2 & \cdots & c_k \end{bmatrix}$. This new convention also conforms with previous usage. Earlier in this chapter we wrote solutions of a linear system with n unknowns as n-tuples $(x_1, x_2, \ldots, x_n)$. We may still write solutions in this manner. For example, the solution of the system

$$\begin{array}{r} 3x_1 - x_2 = 0 \\ 2x_1 + x_2 = 5 \end{array} \quad \text{or} \quad \begin{bmatrix} 3 & -1 \\ 2 & 1 \end{bmatrix} \begin{bmatrix} x_1 \\ x_2 \end{bmatrix} = \begin{bmatrix} 0 \\ 5 \end{bmatrix} \sim$$

may be written as either

$$\begin{bmatrix} 1 \\ 3 \end{bmatrix} \quad \text{or} \quad (1, 3).$$

The Inverse of a Matrix

Writing a linear system in the form $A\mathbf{x} = \mathbf{b}$ suggests a deceptively simple means of solution. Were this a single (scalar) equation, $ax = b$ ($a \neq 0$), with unknown x, we could quickly solve it by multiplying both sides by the *multiplicative inverse* (the *reciprocal*) of a. The solution is $x = a^{-1}b \,(= (1/a)b = b/a)$. Taking our cue from this simpler situation, it is natural to ask if we can solve the matrix equation $A\mathbf{x} = \mathbf{b}$ in the same way. The answer is, "Yes, sometimes," but the process, which is developed next, takes considerably more work.

We must first define the (multiplicative) inverse of a matrix and then determine a procedure for finding it when it exists. Now the multiplicative inverse of a number a is a number b, such that $ab = 1$. (Since multiplication of numbers is commutative, we also have $ba = 1$.) Consequently a natural candidate for the inverse of a matrix A would be a matrix B, such that $AB = I$. However, matrix multiplication is not commutative, so we also require that $BA = I$. Note that, if both products AB and BA are defined and equal to the same matrix I, then A and B must both be square and of the same order as I.

Definition Let A be a square matrix. If there exists a matrix B such that $AB = BA = I$, we say that A is **invertible** (or **nonsingular**) and that B is the **inverse** of A. If A has no inverse, it is said to be **noninvertible** (or **singular**).

Example 2.29

Show that the inverse of the matrix A is B, where

$$A = \begin{bmatrix} 2 & -1 \\ 0 & 1 \end{bmatrix} \quad \text{and} \quad B = \begin{bmatrix} 1/2 & 1/2 \\ 0 & 1 \end{bmatrix}.$$

We have

$$AB = \begin{bmatrix} 2 & -1 \\ 0 & 1 \end{bmatrix}\begin{bmatrix} 1/2 & 1/2 \\ 0 & 1 \end{bmatrix} = \begin{bmatrix} 1 & 0 \\ 0 & 1 \end{bmatrix} = I,$$

and

$$BA = \begin{bmatrix} 1/2 & 1/2 \\ 0 & 1 \end{bmatrix}\begin{bmatrix} 2 & -1 \\ 0 & 1 \end{bmatrix} = \begin{bmatrix} 1 & 0 \\ 0 & 1 \end{bmatrix} = I,$$

as desired.

In the definition and in example 2.29 we have spoken of *the* inverse of a matrix. This terminology implies that a given matrix has at most one inverse. The following theorem justifies the use of this word.

Theorem 1 If a matrix A is invertible, then the inverse is unique.

Proof Suppose that A has inverses B and C. We show that B and C must be equal as follows: $B = BI = B(AC) = (BA)C = IC = C$. ■

Note: We will denote the inverse of an invertible matrix A by the symbol A^{-1}. From the definition it follows that, if $A^{-1} = B$, then B is invertible, and $B^{-1} = A$ as well.

We will now develop a technique for finding the inverse of a square matrix. First we state a theorem that will simplify this task somewhat. Its proof is given in theorem 8 of section 3.2.

Theorem 2   Let A be a square matrix. If a square matrix B exists such that $AB = I$, then $BA = I$ as well, and thus $B = A^{-1}$.

Theorem 2 tells us that the inverse of an $n \times n$ matrix A is the solution X of the matrix equation $AX = I$, where I is the identity matrix of order n. Before establishing a procedure to find this solution, we will consider a particular example that illustrates the basic idea.

Example 2.30

Find the inverse of

$$A = \begin{bmatrix} 1 & 1 & 1 \\ 0 & 2 & 1 \\ 1 & 0 & 1 \end{bmatrix}.$$

We seek a 3×3 matrix X such that $AX = I$, that is,

$$\begin{bmatrix} 1 & 1 & 1 \\ 0 & 2 & 1 \\ 1 & 0 & 1 \end{bmatrix} \begin{bmatrix} x_{11} & x_{12} & x_{13} \\ x_{21} & x_{22} & x_{23} \\ x_{31} & x_{32} & x_{33} \end{bmatrix} = \begin{bmatrix} 1 & 0 & 0 \\ 0 & 1 & 0 \\ 0 & 0 & 1 \end{bmatrix}.$$

Now the first, second, and third columns of the matrix product on the left must be equal to, respectively, the first, second, and third columns of the identity matrix. Symbolically we write this as $(AX)^1 = I^1$, $(AX)^2 = I^2$, and $(AX)^3 = I^3$. But by the definition of matrix multiplication

$$(AX)^1 = \begin{bmatrix} A_1 \cdot X^1 \\ A_2 \cdot X^1 \\ A_3 \cdot X^1 \end{bmatrix} = \begin{bmatrix} (1, 1, 1) \cdot (x_{11}, x_{21}, x_{31}) \\ (0, 2, 1) \cdot (x_{11}, x_{21}, x_{31}) \\ (1, 0, 1) \cdot (x_{11}, x_{21}, x_{31}) \end{bmatrix}$$

$$= \begin{bmatrix} x_{11} + x_{21} + x_{31} \\ 2x_{21} + x_{31} \\ x_{11} \qquad + x_{31} \end{bmatrix}.$$

This is exactly the result we get if we multiply the matrix A by the first column of X that is,

$$AX^1 = \begin{bmatrix} 1 & 1 & 1 \\ 0 & 2 & 1 \\ 1 & 0 & 1 \end{bmatrix} \begin{bmatrix} x_{11} \\ x_{21} \\ x_{31} \end{bmatrix} = \begin{bmatrix} x_{11} + x_{21} + x_{31} \\ 2x_{21} + x_{31} \\ x_{11} \qquad + x_{31} \end{bmatrix}.$$

Thus $(AX)^1 = AX^1$. Similarly $(AX)^2 = AX^2$ and $(AX)^3 = AX^3$. Therefore, to find X, we can solve the three linear systems

$$AX^1 = I^1, \quad AX^2 = I^2, \quad \text{and} \quad AX^3 = I^3.$$

We do this by Gauss-Jordan elimination (section 2.2), forming the augmented matrices

$$\left[\begin{array}{ccc|c} 1 & 1 & 1 & 1 \\ 0 & 2 & 1 & 0 \\ 1 & 0 & 1 & 0 \end{array}\right], \quad \left[\begin{array}{ccc|c} 1 & 1 & 1 & 0 \\ 0 & 2 & 1 & 1 \\ 1 & 0 & 1 & 0 \end{array}\right], \quad \left[\begin{array}{ccc|c} 1 & 1 & 1 & 0 \\ 0 & 2 & 1 & 0 \\ 1 & 0 & 1 & 1 \end{array}\right],$$

and transforming each to its row-reduced echelon form

$$\left[\begin{array}{ccc|c} 1 & 0 & 0 & 2 \\ 0 & 1 & 0 & 1 \\ 0 & 0 & 1 & -2 \end{array}\right], \quad \left[\begin{array}{ccc|c} 1 & 0 & 0 & -1 \\ 0 & 1 & 0 & 0 \\ 0 & 0 & 1 & 1 \end{array}\right], \quad \left[\begin{array}{ccc|c} 1 & 0 & 0 & -1 \\ 0 & 1 & 0 & -1 \\ 0 & 0 & 1 & 2 \end{array}\right].$$

Thus $X^1 = (2, 1, -2)$, $X^2 = (-1, 0, 1)$, and $X^3 = (-1, -1, 2)$, and we have

$$A^{-1} = \begin{bmatrix} 2 & -1 & -1 \\ 1 & 0 & -1 \\ -2 & 1 & 2 \end{bmatrix}.$$

The solution procedure of example 2.30 contains a gross inefficiency. Since the coefficient matrices of the three linear systems $AX^1 = I^1$, $AX^2 = I^2$, and $AX^3 = I^3$ are identical, all three augmented matrices are transformed using the same row operations. We can therefore save a lot of work by doing all the row reductions at the same time. To do this, we form the 3×6 augmented matrix whose first three columns are those of A and whose last three columns are I^1, I^2, and I^3 (that is, the last three columns are those of I):

$$\left[\begin{array}{ccc|ccc} 1 & 1 & 1 & 1 & 0 & 0 \\ 0 & 2 & 1 & 0 & 1 & 0 \\ 1 & 0 & 1 & 0 & 0 & 1 \end{array}\right].$$

We now transform this matrix to the row-reduced echelon form

$$\left[\begin{array}{ccc|ccc} 1 & 0 & 0 & 2 & -1 & -1 \\ 0 & 1 & 0 & 1 & 0 & -1 \\ 0 & 0 & 1 & -2 & 1 & 2 \end{array}\right].$$

Then the fourth, fifth, and sixth columns of the last matrix are X^1, X^2, and X^3, the columns of A^{-1}. Thus (as before)

$$A^{-1} = \begin{bmatrix} 2 & -1 & -1 \\ 1 & 0 & -1 \\ -2 & 1 & 2 \end{bmatrix}.$$

Note: In effect to form the original 3×6 augmented matrix, we adjoined the identity matrix of order three to the coefficient matrix A. We will denote this matrix by $[A|I]$. More generally, if B is an $m \times n$ matrix and C is an $m \times p$ matrix, the $m \times (n + p)$ augmented matrix whose first n columns are those of B and whose last p columns are those of C will be written $[B|C]$. As one important special case the augmented matrix for the system $A\mathbf{x} = \mathbf{b}$ is denoted $[A|\mathbf{b}]$.

We now turn to the general case of attempting to find the inverse of an $n \times n$ matrix A. This can be done by solving the matrix equation $AX = I$ (theorem 2) and (as in example 2.30) the latter is equivalent to solving the n linear systems

$$AX^1 = I^1, \quad AX^2 = I^2, \quad \ldots, \quad AX^n = I^n$$

for the columns X^i of A^{-1}. To solve these systems simultaneously, we transform the augmented matrix $[A|I]$ into row-reduced echelon form. The last n columns of the latter form the inverse of A *provided that each of the systems $AX^i = I^i$ has a solution.* If one of them does not, the matrix equation $AX = I$ has no solution, and hence the inverse of A does not exist.

Example 2.31

Find the inverse of

$$A = \begin{bmatrix} 1 & -2 \\ -2 & 4 \end{bmatrix},$$

or show that no inverse exists.

We form the augmented matrix

$$[A|I] = \begin{bmatrix} 1 & -2 & | & 1 & 0 \\ -2 & 4 & | & 0 & 1 \end{bmatrix},$$

and transform it to the row-reduced echelon form

$$\begin{bmatrix} 1 & 2 & | & 1 & 0 \\ 0 & 0 & | & 2 & 1 \end{bmatrix}.$$

The last row indicates that neither $AX^1 = I^1$, nor $AX^2 = I^2$ has a solution. Therefore A has no inverse.

We now summarize this discussion in the form of a procedure.

Procedure for finding a matrix inverse Let A be a square matrix, and let I be the identity matrix of the same order. Form the augmented matrix $[A|I]$, and transform it into row-reduced echelon form $[C|D]$. Then

i.   if C is the identity matrix, $D = A^{-1}$,
ii. if C is not the identity matrix, A is not invertible.

Example 2.32

Find A^{-1} (if it exists) when

$$A = \begin{bmatrix} 1 & 0 & 1 & 1 \\ 0 & 0 & 1 & 0 \\ 1 & 1 & 1 & 0 \\ 1 & 0 & 0 & 2 \end{bmatrix}.$$

We adjoin the (fourth-order) identity matrix to get $[A|I]$,

$$\left[\begin{array}{cccc|cccc} 1 & 0 & 1 & 1 & 1 & 0 & 0 & 0 \\ 0 & 0 & 1 & 0 & 0 & 1 & 0 & 0 \\ 1 & 1 & 1 & 0 & 0 & 0 & 1 & 0 \\ 1 & 0 & 0 & 2 & 0 & 0 & 0 & 1 \end{array}\right],$$

and transform this matrix to row-reduced echelon form,

$$\left[\begin{array}{cccc|cccc} 1 & 0 & 0 & 0 & 2 & -2 & 0 & -1 \\ 0 & 1 & 0 & 0 & -2 & 1 & 1 & 1 \\ 0 & 0 & 1 & 0 & 0 & 1 & 0 & 0 \\ 0 & 0 & 0 & 1 & -1 & 1 & 0 & 1 \end{array}\right].$$

Consequently A^{-1} exists with

$$A^{-1} = \begin{bmatrix} 2 & -2 & 0 & -1 \\ -2 & 1 & 1 & 1 \\ 0 & 1 & 0 & 0 \\ -1 & 1 & 0 & 1 \end{bmatrix}.$$

Method of Inverses

We now return to the idea of solving the linear system $Ax = b$ by using inverse matrices. Let A be an $n \times n$ nonsingular matrix. We have

$$Ax = b$$

$$A^{-1}A\mathbf{x} = A^{-1}\mathbf{b} \quad \text{(multiplying both sides on the left by } A^{-1})$$

$$I\mathbf{x} = A^{-1}\mathbf{b} \quad \text{(theorem 1e of section 2.3)}$$

$$\mathbf{x} = A^{-1}\mathbf{b} \quad \text{(theorem 1f of section 2.3)}.$$

Hence we can solve a linear system whose coefficient matrix is square and nonsingular by finding the inverse and then multiplying it by the right-side vector. This type of solution procedure is called the **method of inverses.**

Example 2.33

Solve the following system by the method of inverses:

$$
\begin{array}{rcr}
x_1 \quad\quad + x_3 &=& 2 \\
x_1 - x_2 \quad\quad &=& -1 \\
2x_2 + x_3 &=& 1.
\end{array}
$$

This linear system can be written in the form $A\mathbf{x} = \mathbf{b}$ by taking

$$
A = \begin{bmatrix} 1 & 0 & 1 \\ 1 & -1 & 0 \\ 0 & 2 & 1 \end{bmatrix}, \quad \mathbf{x} = \begin{bmatrix} x_1 \\ x_2 \\ x_3 \end{bmatrix}, \quad \text{and} \quad \mathbf{b} = \begin{bmatrix} 2 \\ -1 \\ 1 \end{bmatrix}.
$$

We must first find A^{-1}. We have

$$
[A|I] = \left[\begin{array}{ccc|ccc} 1 & 0 & 1 & 1 & 0 & 0 \\ 1 & -1 & 0 & 0 & 1 & 0 \\ 0 & 2 & 1 & 0 & 0 & 1 \end{array}\right],
$$

which has the row-reduced echelon form

$$
\left[\begin{array}{ccc|ccc} 1 & 0 & 0 & -1 & 2 & 1 \\ 0 & 1 & 0 & -1 & 1 & 1 \\ 0 & 0 & 1 & 2 & -2 & -1 \end{array}\right].
$$

Thus

$$
A^{-1} = \begin{bmatrix} -1 & 2 & 1 \\ -1 & 1 & 1 \\ 2 & -2 & -1 \end{bmatrix}.
$$

Finally, we obtain $\mathbf{x} = A^{-1}\mathbf{b}$ by multiplying

$$
\begin{bmatrix} -1 & 2 & 1 \\ -1 & 1 & 1 \\ 2 & -2 & -1 \end{bmatrix} \begin{bmatrix} 2 \\ -1 \\ 1 \end{bmatrix} = \begin{bmatrix} -3 \\ -2 \\ 5 \end{bmatrix}.
$$

Hence $x_1 = -3$, $x_2 = -2$, and $x_3 = 5$.

Note: The method of inverses cannot be used to solve the linear system $A\mathbf{x} = \mathbf{b}$ if either A is not square or A is singular.

Example 2.34

We wish to solve a number of linear systems of the form $A\mathbf{x} = \mathbf{b}$, all of which have the same fourth-order nonsingular coefficient matrix. If we know that

$$A^{-1} = \begin{bmatrix} 1 & 0 & -1 & 0 \\ 2 & 0 & 0 & 1 \\ 0 & -1 & -2 & 0 \\ 0 & 0 & 1 & 1 \end{bmatrix},$$

give a formula for the solution vector $\mathbf{x}$ in terms of the components of the right-side vector $\mathbf{b}$.

We have $\mathbf{x} = A^{-1}\mathbf{b}$, so

$$\mathbf{x} = \begin{bmatrix} 1 & 0 & -1 & 0 \\ 2 & 0 & 0 & 1 \\ 0 & -1 & -2 & 0 \\ 0 & 0 & 1 & 1 \end{bmatrix} \begin{bmatrix} b_1 \\ b_2 \\ b_3 \\ b_4 \end{bmatrix} = \begin{bmatrix} b_1 - b_3 \\ 2b_1 + b_4 \\ -b_2 - 2b_3 \\ b_3 + b_4 \end{bmatrix}.$$

Thus $\mathbf{x} = (b_1 - b_3,\ 2b_1 + b_4,\ -b_2 - 2b_3,\ b_3 + b_4)$.

Computational note: The method of inverses is not as efficient as Gauss-Jordan elimination for solving linear systems. It requires more operations — additions, subtractions, multiplications, and divisions — to complete the solution process. However, as you can see from these examples, if the inverse is somehow known beforehand, it is an extremely easy task to find the solution by the method of inverses.

2.4 Exercises

In exercises 1 through 4 write the given linear system in the matrix form $A\mathbf{x} = \mathbf{b}$.

1. $\begin{aligned} x - y + 3z &= 1 \\ x \qquad\ - z &= 0 \\ -2x + y \qquad &= -1 \end{aligned}$

2. $\begin{aligned} 2x_1 - 3x_2 + \quad x_4 &= 6 \\ x_2 - x_3 + 3x_4 &= 0 \\ -x_1 + \quad x_4 \quad &= 0 \end{aligned}$

3. $\begin{aligned} x &= y - z \\ y &= x - z \end{aligned}$

4. $\begin{aligned} x_1 &= 2 \\ x_2 &= 3 - x_1 \\ x_1 + x_2 + 1 &= 4 \end{aligned}$

In exercises 5 and 6 verify that matrices A and B are inverses of each other.

5. $A = \begin{bmatrix} 1 & 2 \\ 3 & 5 \end{bmatrix}$, $B = \begin{bmatrix} -5 & 2 \\ 3 & -1 \end{bmatrix}$

6. $A = \begin{bmatrix} 1 & 0 & -1 \\ 3 & -1 & 0 \\ 0 & -1 & 1 \end{bmatrix}$, $B = \frac{1}{2}\begin{bmatrix} -1 & 1 & -1 \\ -3 & 1 & -3 \\ -3 & 1 & -1 \end{bmatrix}$

In exercises 7 through 12 find the inverse, if it exists, of the given matrix. If the matrix has no inverse, state this fact.

7. $\begin{bmatrix} 2 & 0 \\ -3 & 1 \end{bmatrix}$

8. $\begin{bmatrix} -1 & 2 \\ 3 & -4 \end{bmatrix}$

9. $\begin{bmatrix} 1 & 2 & -1 \\ 0 & 1 & 2 \\ 0 & 0 & 1 \end{bmatrix}$

10. $\begin{bmatrix} 1 & 0 & -2 \\ 2 & -1 & 0 \\ 1 & 0 & -2 \end{bmatrix}$

11. $\begin{bmatrix} 1 & 0 & 1 & 0 \\ 0 & 1 & 0 & 1 \\ 0 & 0 & 1 & 1 \\ 1 & 1 & 0 & 0 \end{bmatrix}$

12. $\begin{bmatrix} 1 & -1 & 0 & 0 \\ 0 & 1 & -1 & 0 \\ 0 & 0 & 1 & -1 \\ 1 & 0 & 0 & 1 \end{bmatrix}$

In exercises 13 and 14 find the value of c for which A has no inverse.

13. $A = \begin{bmatrix} 1 & -2 \\ 3 & c \end{bmatrix}$

14. $A = \begin{bmatrix} 1 & -1 & c \\ 0 & 2 & 1 \\ -1 & 0 & 1 \end{bmatrix}$

In exercises 15 and 16 solve the linear system $A\mathbf{x} = \mathbf{b}$ when the inverse, A^{-1}, and $\mathbf{b}$ are as given.

15. $A^{-1} = \begin{bmatrix} -1 & 1/2 & 0 \\ 1/2 & -1 & 1/2 \\ 0 & 1/2 & -1 \end{bmatrix}$, $\mathbf{b} = \begin{bmatrix} -3 \\ 1 \\ 2 \end{bmatrix}$

16. $A^{-1} = \begin{bmatrix} 1 & -1 & 3 & 1 \\ 0 & 1 & 0 & 2 \\ -2 & -1 & -3 & -1 \\ 1 & -1 & 0 & 0 \end{bmatrix}$, $\mathbf{b} = \begin{bmatrix} 1 \\ 2 \\ 3 \\ 4 \end{bmatrix}$.

In exercises 17 through 22 use the method of inverses, if possible, to solve the following systems.

17. $2x - 3y = 3$
$\quad\ 3x - 5y = 1$

18. $A\mathbf{x} = \mathbf{b}$, where $\quad A = \begin{bmatrix} 1 & -3 \\ 2 & 4 \end{bmatrix}$, $\quad \mathbf{x} = \begin{bmatrix} x_1 \\ x_2 \end{bmatrix}$, and $\mathbf{b} = \begin{bmatrix} 3 \\ -1 \end{bmatrix}$

19. $\quad x - 3y + 4z = 4$
$\qquad 2x + 2y \quad\;\; = 0$
$\qquad\qquad y - 2z = 2$

20. $\quad -x_1 + x_3 = \quad 1$
$\qquad x_1 + x_2 = -1$
$\qquad x_2 - x_3 = \quad 0$

21. $A\mathbf{x} = \mathbf{b}$, where $\quad A = \begin{bmatrix} 1 & -1 & 2 \\ 2 & -1 & 1 \\ 0 & -1 & 3 \end{bmatrix}$, $\quad \mathbf{x} = \begin{bmatrix} x_1 \\ x_2 \\ x_3 \end{bmatrix}$, $\quad \mathbf{b} = \begin{bmatrix} 1 \\ 0 \\ 0 \end{bmatrix}$

22. $\quad x - 3y + \;\; z = 2$
$\qquad 2x + \;\; y - 2z = 7$

23. Show that the inverse of the 2×2 matrix

$$\begin{bmatrix} a & b \\ c & d \end{bmatrix} \text{ is the matrix } (1/(ad - bc)) \begin{bmatrix} d & -b \\ -c & a \end{bmatrix},$$

provided that $ad - bc \neq 0$.

In exercises 24 and 25 use the result of exercise 23 to find quickly the inverses.

24. $\begin{bmatrix} 3 & -2 \\ 1 & 4 \end{bmatrix}$
          25. $\begin{bmatrix} 1 & -1 \\ 0 & 4 \end{bmatrix}$

26. Show that, if A and B have inverses and are of the same order, then AB has an inverse and in fact $(AB)^{-1} = B^{-1}A^{-1}$.

27. Show that the inverse of the nth-order diagonal matrix

$$D = \begin{bmatrix} a_1 & 0 & \cdots & 0 \\ 0 & a_2 & \cdots & 0 \\ \vdots & \vdots & \ddots & \vdots \\ 0 & 0 & \cdots & a_n \end{bmatrix} \text{ is } D^{-1} = \begin{bmatrix} 1/a_1 & 0 & \cdots & 0 \\ 0 & 1/a_2 & \cdots & 0 \\ \vdots & \vdots & \ddots & \vdots \\ 0 & 0 & \cdots & 1/a_n \end{bmatrix},$$

provided that the main diagonal entries a_i are all nonzero.

In exercises 28 and 29 use computer programs for finding inverses and multiplying matrices to solve the given system by the method of inverses.

28. $A\mathbf{x} = \mathbf{b}$, where $\quad A = \begin{bmatrix} 1 & 0 & 0 & 0 & 0 \\ 0 & 0 & 0 & 1 & 0 \\ 0 & 1 & 0 & 0 & 0 \\ 0 & 0 & 0 & 0 & 1 \\ 0 & 0 & 1 & 0 & 0 \end{bmatrix}$, $\mathbf{b} = \begin{bmatrix} 1 \\ 0 \\ 0 \\ 0 \\ 0 \end{bmatrix}$

29. $A\mathbf{x} = \mathbf{b}$, where A is the square matrix of order 10 with 1's on and above the main diagonal and 0's below it, and $\mathbf{b}$ is the 10 vector, all of whose entries are 1.

Note: Exercises 28 and 29 can both be easily solved by hand, using Gauss-Jordan elimination rather than the method of inverses.

2.5 Theory of Linear Systems

In this section we will investigate the nature of solutions of systems of linear equations rather than techniques for finding them. In doing so, we will introduce an important type of linear system: the *homogeneous* one.

We will continue to write a linear system in the matrix form, $A\mathbf{x} = \mathbf{b}$, developed in the last section. In such an equation A will be an $m \times n$ matrix, $\mathbf{x}$ an n-vector, and $\mathbf{b}$ an m-vector. If A is square, of course $m = n$.

Theorem 1   Every system of linear equations $A\mathbf{x} = \mathbf{b}$ has no solutions, exactly one solution, or infinitely many solutions.

Proof   We have already seen examples of each of the three cases, so we need only show that there are no other possibilities. Assume that $\mathbf{y}$ and $\mathbf{z}$ are solutions with $\mathbf{y} \neq \mathbf{z}$. Since these are vectors in $\mathbf{R}^n$, the line they determine is given by the two-point form,

$$\mathbf{x}(t) = (1 - t)\mathbf{y} + t\mathbf{z}$$
$$= \mathbf{y} + t(\mathbf{z} - \mathbf{y}).$$

Each point $\mathbf{x}(t)$ on this line is also a solution of $A\mathbf{x} = \mathbf{b}$, as seen by

$$A(\mathbf{x}(t)) = A(\mathbf{y} + t(\mathbf{z} - \mathbf{y}))$$
$$= A\mathbf{y} + A(t(\mathbf{z} - \mathbf{y}))$$
$$= \mathbf{b} + tA(\mathbf{z} - \mathbf{y})$$
$$= \mathbf{b} + t(A\mathbf{z} - A\mathbf{y})$$
$$= \mathbf{b} + t(\mathbf{b} - \mathbf{b}) = \mathbf{b}.$$

Consequently, if there is more than one solution to $A\mathbf{x} = \mathbf{b}$, then there are infinitely many.   ∎

Definition   The linear system $A\mathbf{x} = \mathbf{b}$ is said to be **homogeneous** if $\mathbf{b} = \mathbf{0}$, the vector consisting solely of zeros. If $\mathbf{b} \neq \mathbf{0}$, then the system is said to be **nonhomogeneous**.

For example, the linear systems

$$\begin{bmatrix} 1 & -1 & 2 & 0 \\ 3 & 0 & 1 & 5 \end{bmatrix} \begin{bmatrix} x_1 \\ x_2 \\ x_3 \\ x_4 \end{bmatrix} = \begin{bmatrix} 0 \\ 0 \end{bmatrix}$$

and

$$-x - 3y + z = 0$$

$$2x \qquad - 2z = 0$$
$$3y + \quad z = 0.$$

are *homogeneous* ones, but

$$x_1 - 2x_2 + x_3 = 0$$
$$3x_1 + \quad x_2 - x_3 = 1$$

is *nonhomogeneous*.

Note: The zero vector, **0**, is a solution of every homogeneous linear system $A\mathbf{x} = \mathbf{0}$ since $A\mathbf{0} = \mathbf{0}$. The solution $\mathbf{x} = \mathbf{0}$ is called the **trivial solution** of the homogeneous system. Any other solution is called a **nontrivial** one.

Theorem 2 Every homogeneous linear system $A\mathbf{x} = \mathbf{0}$ has either exactly one solution or infinitely many solutions.

Proof Since every homogeneous linear system has at least one solution (the trivial one), the result follows immediately from theorem 1. ∎

We can probe a little more deeply into this subject by considering the structure of the row-reduced echelon form for the augmented matrix of the given system. The key to this line of investigation is the notion of *rank*.

Definition The **rank** of a matrix A is the number of nonzero rows in the row-reduced echelon form of A.

Note: The rank of a matrix is *well defined* due to the uniqueness of the row-reduced echelon form (theorem 2 of section 2.2). No matter what sequence of elementary row operations is performed to put the given matrix in row-reduced echelon form, there will always be the same number of nonzero rows.

Example 2.35

Find the rank of

$$A = \begin{bmatrix} 1 & -1 & 2 & 1 \\ 0 & 1 & 1 & -2 \\ 1 & -3 & 0 & 5 \end{bmatrix}.$$

The row-reduced echelon form of A is

$$\begin{bmatrix} 1 & 0 & 3 & -1 \\ 0 & 1 & 1 & -2 \\ 0 & 0 & 0 & 0 \end{bmatrix},$$

which has two nonzero rows. Hence the rank of A is 2.

Note: Given the linear system $A\mathbf{x} = \mathbf{b}$, the rank of the coefficient matrix A is always less than or equal to the rank of the augmented matrix for the system $[A|\mathbf{b}]$. This follows from the fact that, when the latter is transformed into row-reduced echelon form, its first n columns are the row-reduced form of the former. It is consequently impossible for the row-reduced echelon form of A to have more nonzero rows than that of $[A|\mathbf{b}]$. In other words, it is impossible for the rank of the coefficient matrix to be greater than that of the augmented matrix.

Theorem 3   The linear system $A\mathbf{x} = \mathbf{b}$ has at least one solution if and only if the rank of the coefficient matrix A is equal to the rank of the augmented matrix $[A|\mathbf{b}]$.

Proof   First, assume that $A\mathbf{x} = \mathbf{b}$ has at least one solution. Suppose that the rank of $[A|\mathbf{b}]$ is *not* equal to that of A. Then, by the note just given, it must be greater than the rank of A. But this means that there is a row in the row-reduced echelon form of $[A|\mathbf{b}]$ that reads

$$[0 \quad 0 \quad \ldots \quad 0 \mid 1].$$

This corresponds to the equation $0 = 1$, a contradiction.

Now assume that the rank of A is equal to that of $[A|\mathbf{b}]$. Let $[C|\mathbf{d}]$ be its row-reduced echelon form. We discard the zero rows in $[C|\mathbf{d}]$ and assume that the first r columns of A are the ones transformed into those that contain the leading ones of C. Then, changing back to a system of equations, we have

$$
\begin{aligned}
x_1 \qquad\qquad\;\; + c_{1,\,r+1}x_{r+1} + \cdots + c_{1n}x_n &= d_1 \\
x_2 \qquad\;\; + c_{2,\,r+1}x_{r+1} + \cdots + c_{2n}x_n &= d_2 \\
\vdots \qquad\qquad \vdots \qquad\qquad\quad \vdots \qquad \vdots \\
x_r + c_{r,\,r+1}x_{r+1} + \cdots + c_{rn}x_n &= d_r.
\end{aligned}
\qquad (1)
$$

Setting $x_{r+1} = x_{r+2} = \cdots = x_n = 0$, we obtain the solution

$$(d_1, d_2, \ldots, d_r, 0, 0, \ldots, 0). \qquad \blacksquare$$

Theorem 4 Let A be an $m \times n$ matrix. The system $A\mathbf{x} = \mathbf{b}$ has a unique solution if and only if the ranks of the coefficient and augmented matrices are both equal to n.

Proof First, assume that $A\mathbf{x} = \mathbf{b}$ has a unique solution. By theorem 3, the ranks of coefficient and augmented matrices are equal. Hence it suffices to show that the rank of A is n. Since every nonzero row in the row-reduced echelon form of A contains a leading 1, there are at most n such rows, so the rank of A is $\leq n$. Suppose it were equal to some $r < n$. Then the equivalent linear system would be as in system (1) and would have infinitely many solutions. (For example, set $x_{r+1} = t, x_{r+2} = \cdots = x_n = 0$ to get the solutions $(d_1 - c_{1,\,r+1}t, \ldots, d_r - c_{r,\,r+1}t, t, 0, 0, \ldots, 0)$.) Thus we must have the rank of A equal to n, as desired.

Now assume that the ranks of the coefficient and augmented matrices are both equal to n. After discarding zero rows of the row-reduced echelon form, we obtain the equivalent system

$$
\begin{aligned}
x_1 \qquad\quad &= d_1 \\
x_2 \quad\; &= d_2 \\
&\;\;\vdots \\
x_n &= d_n,
\end{aligned}
$$

which has the unique solution $(d_1, d_2, \ldots, d_n)$. ∎

Theorems 3 and 4 have several immediate consequences. (The proofs are exercises).

Corollary 1 The linear system $A\mathbf{x} = \mathbf{b}$, where A is an $m \times n$ matrix, has infinitely many solutions if and only if the ranks of the coefficient and augmented matrices are equal and less than n.

Corollary 2 A linear system with more unknowns than equations either has no solution or infinitely many solutions.

Corollary 3 The homogeneous linear system $A\mathbf{x} = \mathbf{0}$, where A is an $m \times n$ matrix, has a unique solution (the trivial one) if and only if the rank of A is equal to n.

Corollary 4 The homogeneous system $A\mathbf{x} = \mathbf{0}$, where A is an $m \times n$ matrix, has a nontrivial solution (and hence infinitely many solutions) if and only if the rank of A is less than n.

Corollary 5 A homogeneous linear system with more unknowns than equations has infinitely many solutions.

Example 2.36

By inspection of the following systems, what can we say concerning the number of solutions they may have?

a. $\begin{aligned} x_1 + 2x_2 - x_3 &= 0 \\ 2x_1 - x_2 + 3x_3 &= 0 \end{aligned}$

b. $\begin{aligned} 2x_1 + x_2 + x_3 &= 1 \\ -x_1 - x_2 + 2x_3 &= 0 \end{aligned}$

c. $\begin{aligned} x_1 + 2x_4 &= 1 \\ x_2 - x_4 &= -1 \\ x_3 + x_4 &= 0. \end{aligned}$

System a is a homogeneous one with more unknowns than equations. By corollary 5 to theorem 4 it has infinitely many solutions. System b is a nonhomogeneous one with more unknowns than equations. By corollary 2 it has either no solution or infinitely many solutions. System c is in row-reduced echelon form. The corresponding augmented matrix is

$$\left[\begin{array}{cccc|c} 1 & 0 & 0 & 2 & 1 \\ 0 & 1 & 0 & -1 & -1 \\ 0 & 0 & 1 & 1 & 0 \end{array}\right].$$

We see that the ranks of the coefficient and augmented matrices are both 3 while the number of variables n is 4. Hence by corollary 1 there are infinitely many solutions.

In example 2.36 it would take little time to solve system c to obtain the one-parameter family of solutions $(1 - 2t, -1 + t, -t, t)$. In general, if the ranks of coefficient and augmented matrices are both r, and the number of variables is $n > r$, the equivalent system in row-reduced echelon form will be as in system (1). The variables $x_{r+1}, \ldots, x_n$ can be taken as parameters and there will be an $(n - r)$-parameter family of solutions. We state this result as the next theorem.

Theorem 5 Given the system $A\mathbf{x} = \mathbf{b}$, where A is an $m \times n$ matrix with rank equal to $r < n$. If the augmented matrix $[A|\mathbf{b}]$ also has rank $= r$, then the system has an $(n - r)$-parameter family of solutions.

Computational note: One should be wary of using a computer to determine the rank of a matrix—this may lead to an erroneous answer. Because a computer has to round-off nonterminating decimals (such as the fraction $1/3$), small errors may be introduced in transforming a given matrix into row-reduced echelon form. These small errors could possibly result in nonzero entries occurring in the latter where in fact there should be zeros, or vice-versa. It is therefore possible for the computed rank to be either smaller or larger than it actually is.

We will now restrict ourselves to consideration of linear systems with the same number of equations as unknowns. Since in such cases the coefficient matrix is square, we call these **square linear systems.**

Theorem 6   If A is an $n \times n$ matrix, the following conditions are equivalent:

a. A is invertible,
b. $A\mathbf{x} = \mathbf{b}$ has a unique solution for any $\mathbf{b}$,
c. $A\mathbf{x} = \mathbf{0}$ has only the trivial solution,
d. the rank of A is n,
e. the row-reduced echelon form of A is I, the identity matrix.

Proof   We will show the equivalence of these statements by showing that a implies b, b implies c, c implies d, d implies e, and e implies a. By doing so, any two statements will imply each other by the transitivity of implication.

Proof of a implies b.   We first note that the vector $A^{-1}\mathbf{b}$ is a solution since $A(A^{-1}\mathbf{b}) = (AA^{-1})\mathbf{b} = I\mathbf{b} = \mathbf{b}$, as desired. Now suppose $\mathbf{y}$ is also a solution. Then $A\mathbf{y} = \mathbf{b}$, so $A^{-1}(A\mathbf{y}) = A^{-1}\mathbf{b}$, or $(A^{-1}A)\mathbf{y} = A^{-1}\mathbf{b}$. But this yields $\mathbf{y} = A^{-1}\mathbf{b}$ as well. Thus the solution is unique.

Proof of b implies c.   Since $A\mathbf{x} = \mathbf{0}$ is a special case of $A\mathbf{x} = \mathbf{b}$, the former has a unique solution. Since $\mathbf{0}$ is a solution, it is the only solution.

Proof of c implies d.   This is corollary 3 of theorem 4.

Proof of d implies e.   Since A has n rows, and its rank is n, its row-reduced echelon form has n nonzero rows, each with a leading one and no zero rows. But since the matrix has n columns as well, the only nonzero element in column j is a 1 in the jth row. Thus the row-reduced echelon form is the identity matrix.

Proof of   e implies a. This follows immediately from the procedure for matrix inverses presented in section 2.4.   ∎

2.5 Exercises

In exercises 1 and 2 determine which systems are homogeneous.

1. $x_1 = 2x_2 + x_3$
$x_2 = x_3 - x_4$
$x_3 = x_1 - 4x_4$

2. $x + y + 3z = 0$
$x - y + 2 = 0$
$z - 3 = 0$

In exercises 3 through 6 find the rank of the given matrix.

3. $\begin{bmatrix} 1 & 2 & -1 \\ 3 & -6 & 2 \end{bmatrix}$

4. $\begin{bmatrix} 1 & 0 & -1 \\ 3 & 1 & 1 \\ -1 & -1 & -3 \end{bmatrix}$

5. $\begin{bmatrix} 1 & 0 & 1 \\ -2 & 1 & 1 \\ 1 & 1 & 2 \end{bmatrix}$

6. $\begin{bmatrix} 2 & 4 & 0 & -2 \\ 0 & 1 & -1 & 0 \\ 1 & 2 & -1 & -1 \end{bmatrix}$

In exercises 7 and 8 consider the linear system $A\mathbf{x} = \mathbf{b}$, where

$$A = \begin{bmatrix} a & b \\ c & d \end{bmatrix}, \quad \mathbf{b} = \begin{bmatrix} e \\ f \end{bmatrix}, \quad \text{and} \quad A \neq 0.$$

7. Determine conditions on the constants $a, b, c, d, e,$ and f so that
(a) the rank of A is 2,
(b) the rank of A is 1, but the rank of $[A|\mathbf{b}]$ is 2,
(c) the rank of A and the rank of $[A|\mathbf{b}]$ are each 1.

8. Using the results of exercise 7, give conditions on $a, b, c, d, e,$ and f so that the linear system has
(a) no solution,
(b) one solution,
(c) infinitely many solutions.

In exercises 9 through 12 use the results of section 2.5 to say as much as possible concerning the number of solutions of the given linear system without actually solving it.

9. $2x + 3y - z \qquad = 0$
$\qquad y + 2z - 3w = 0$
$2x \qquad - 4z + w = 0$

10. $2x + 3y - z \qquad = 1$
$\qquad y + 2z - 3w = 0$
$2x \qquad - 4z + w = 0$

11. $2x_1 + x_2 - x_3 = 0$
$3x_1 - 2x_2 + x_3 = 0$
$x_1 - x_2 + x_3 = 0$

12. $2x - 3y = 2$
$-x + 4y = 3$
$4x + 3y = 7$

13. Let $A, X,$ and B be $n \times n$ matrices with A nonsingular. If $AX = B$, show that $(AXA^{-1})A = B$.

14. Let A and B be invertible matrices with $AB = I$. Show that $(A^T)^{-1} = B^T$.

15–18. Use a computer program to transform a matrix to row-reduced echelon form to find the rank of the matrices in exercises 3 through 6. Do your results agree with the ones obtained by hand?

19. Find the rank of the matrix

$$A = \begin{bmatrix} 1 & 1/2 & 1/3 & 1/4 & 1/5 \\ 1/2 & 1/3 & 1/4 & 1/5 & 1/6 \\ 1/3 & 1/4 & 1/5 & 1/6 & 1/7 \\ 1/4 & 1/5 & 1/6 & 1/7 & 1/8 \\ 1/5 & 1/6 & 1/7 & 1/8 & 1/9 \end{bmatrix},$$

using a computer program to transform a matrix to row-reduced echelon form. (The rank of A is 5, but your computed answer may differ. See the computational note after theorem 5.)

Determinants

3

Associated with each square matrix is a number known as the *determinant* of the matrix. This number carries a great deal of information about the matrix, most important of which is whether or not the matrix is invertible. When it is invertible, use of determinants will allow us to express the inverse in terms of the original matrix, which serves as an alternative to the inversion procedure described in section 2.4. Similarly the solution to a square system of linear equations can sometimes be expressed using determinants in a procedure known as Cramer's rule. Other uses of determinants will be encountered in later chapters.

3.1 Definition of Determinant

In section 1.2 we defined the determinant of an order 2 matrix for the sole purpose of providing a simple means of remembering the definition of cross product. In this section we will extend the definition to square matrices of any order $n \geq 1$. We first give a convention that will make this description easier.

Notation: For an $m \times n$ matrix A, the *submatrix* A_{ij} is obtained by deleting the ith row and jth column of A.

Example 3.1

Find A_{23}, where

$$A = \begin{bmatrix} 1 & 1 & 0 & 2 \\ 2 & 1 & 1 & 1 \\ 3 & 0 & 0 & -1 \\ 1 & 1 & 2 & 1 \end{bmatrix}.$$

We mentally line out the second row and the third column,

$$\begin{bmatrix} 1 & 1 & 0 & 2 \\ -2--1--1--1- \\ 3 & 0 & 0 & -1 \\ 1 & 1 & 2 & 1 \end{bmatrix},$$

105

obtaining

$$A_{23} = \begin{bmatrix} 1 & 1 & 2 \\ 3 & 0 & -1 \\ 1 & 1 & 1 \end{bmatrix}.$$

Definition   Let $A = [a_{ij}]$ be an $n \times n$ matrix. The **determinant** of A is given by

i. $\det A = a_{11}$ if $n = 1$,

ii. $\det A = a_{11}a_{22} - a_{12}a_{21}$ if $n = 2$,

iii. $\det A = a_{11} \det A_{11} - a_{12} \det A_{12} + \cdots + (-1)^{1+n}a_{1n} \det A_{1n}$, if $n > 2$.

Case iii says that for each entry a_{1j} in the first row, multiply a_{1j} by $\det A_{1j}$, the determinant of the submatrix formed by deleting the row and the column that contain a_{1j}. Starting with a positive sign, alternate the signs of these terms—plus, minus, plus, minus, and so on. Finally, add all such terms.

Note: The definition of determinant is an *inductive definition*. An explicit rule is given for $n = 1$ and $n = 2$. For $n = 3$ each A_{1j} is of order 2, so $\det A_{1j}$ has been defined. For $n = 4$ each A_{1j} is of order 3, so $\det A_{1j}$ can be computed by the $n = 3$ case. This process may be continued for all positive integers n.

Example 3.2

Compute $\det A$, where A is given by

$$A = \begin{bmatrix} 2 & 1 & -3 \\ -3 & -2 & 0 \\ 2 & 1 & 2 \end{bmatrix}.$$

Since A is order 3, the definition of determinant yields

$$\det A = a_{11} \det A_{11} - a_{12} \det A_{12} + a_{13} \det A_{13}$$

$$= (2)\det \begin{bmatrix} -2 & 0 \\ 1 & 2 \end{bmatrix} - (1)\det \begin{bmatrix} -3 & 0 \\ 2 & 2 \end{bmatrix} + (-3)\det \begin{bmatrix} -3 & -2 \\ 2 & 1 \end{bmatrix}$$

$$= (2)[(-2)(2) - (1)(0)] - (1)[(-3)(2) - (2)(0)]$$
$$\quad + (-3)[(-3)(1) - (2)(-2)]$$

$$= 2(-4) - (1)(-6) + (-3)(1)$$

$$= -5.$$

Example 3.3

Compute det A, where

$$A = \begin{bmatrix} 2 & 1 & -3 & 1 \\ -3 & -2 & 0 & 2 \\ 2 & 1 & 0 & -1 \\ 1 & 0 & 1 & 2 \end{bmatrix}.$$

We have

$$\det A = (2)\det A_{11} - (1)\det A_{12} + (-3)\det A_{13} - (1)\det A_{14}.$$

This looks innocent enough but note that each A_{1j} is of order 3 and computation of det A_{1j} requires repeating the process of example 3.2. This must be done four times to finish the job. These values are as follows: det $A_{11} = 0$, det $A_{12} = 1$, det $A_{13} = 2$, and det $A_{14} = 1$. The final result is det $A = -8$.

Fortunately there are easier ways to compute det A. The next section will be devoted to the basic properties of the determinant, some of which will simplify the process considerably.

If A is a square matrix of order n, the submatrix A_{ij} is a square matrix of order $n - 1$. The determinant of this matrix, det A_{ij}, is called the ijth **minor** of A. The ijth **cofactor** of A is defined to be

$$(-1)^{i+j}\det A_{ij}.$$

Example 3.4

Find the 2, 3-minor and the 2, 3-cofactor of

$$A = \begin{bmatrix} 2 & -2 & 1 & 1 \\ 1 & 3 & 3 & 2 \\ 1 & 0 & 9 & 1 \\ 3 & 4 & 2 & 0 \end{bmatrix}.$$

The 2, 3-minor is

$$\det A_{23} = \det \begin{bmatrix} 2 & -2 & 1 \\ 1 & 0 & 1 \\ 3 & 4 & 0 \end{bmatrix} = -10.$$

The 2, 3-cofactor is

$$(-1)^{2+3} \det A_{23} = -\det \begin{bmatrix} 2 & -2 & 1 \\ 1 & 0 & 1 \\ 3 & 4 & 0 \end{bmatrix} = 10.$$

Notice that equation iii in the definition of the determinant may be written as

$$\det A = a_{11}[(-1)^{1+1} \det A_{11}] + a_{12}[(-1)^{1+2} \det A_{12}]$$
$$+ \cdots + a_{1n}[(-1)^{1+n} \det A_{1n}]_1.$$

In other words, det A is the sum of the products of each entry in the first row with the cofactor for that position. For this reason equation iii is called **cofactor expansion** along the first row (or simply **expansion** along the first row).

The next theorem states that cofactor expansion along *any* row (or down any column) of a square matrix will also yield the determinant. The theorem will be assumed without proof.

Theorem 1 Let A be a matrix of order n with $n \geq 2$. Let i be a fixed row number. Then

$$\det A = (-1)^{i+1}a_{i1} \det A_{i1} + (-1)^{i+2}a_{i2} \det A_{i2}$$
$$+ \cdots + (-1)^{i+n}a_{in} \det A_{in}.$$

Let j be a fixed column number. Then

$$\det A = (-1)^{1+j}a_{1j} \det A_{1j} + (-1)^{2+j}a_{2j} \det A_{2j}$$
$$+ \cdots + (-1)^{n+j}a_{nj} \det A_{nj}.$$

In computations, it is convenient to choose a row or column with as many zeros as possible because it is not necessary to compute the cofactors of zero entries since the product will always be zero.

Example 3.5

Compute det A by expansion along the row or column with the maximum number of zeros, where

$$A = \begin{bmatrix} 2 & 1 & -3 & 1 \\ -3 & -2 & 0 & 2 \\ 2 & 1 & 0 & -1 \\ 1 & 0 & 1 & 2 \end{bmatrix}.$$

This is just example 3.3 again but expanded down the third column:

$$\det A = (-1)^{1+3}(-3) \det A_{13} + (-1)^{2+3}(0) \det A_{23}$$
$$+ (-1)^{3+3}(0) \det A_{33} + (-1)^{4+3}(1) \det A_{43}$$
$$= -3 \det A_{13} - \det A_{43}.$$

The computation still requires evaluation of the determinants of two

3×3 matrices but that is half as much work as was required before. Of course the numerical value of det A is still -8.

Note: It is easy to obtain the sign factors $(-1)^{i+j}$ in the terms of the expansion of a determinant since the signs alternate when moving horizontally or vertically from one position to another as indicated in the following arrays for $n = 4$ and $n = 5$:

$$
\begin{bmatrix}
+ & - & + & - \\
- & + & - & + \\
+ & - & + & - \\
- & + & - & +
\end{bmatrix},
\begin{bmatrix}
+ & - & + & - & + \\
- & + & - & + & - \\
+ & - & + & - & + \\
- & + & - & + & - \\
+ & - & + & - & +
\end{bmatrix}.
$$

For example, if you wish to expand down the *j*th column, start in the upper left corner and proceed plus, minus, plus, minus, and so on, along the first row to the top of the *j*th column. That establishes the first position sign, and they alternate thereafter. If you are expanding along the *i*th row, alternate signs down the first column to the *i*th row and proceed from there.

Example 3.6

Compute det A, where

$$
A = \begin{bmatrix}
2 & 9 & 3 & 2 \\
4 & 0 & 0 & 6 \\
3 & -1 & 1 & 2 \\
5 & 0 & 0 & 1
\end{bmatrix}.
$$

For the simplest computation we should expand along the second row, the fourth row, the second column, or the third column. We'll use the fourth row. Going down the first column—plus, minus, plus, minus—we see the first position is negative and therefore

$$
\det A = -(5) \det A_{41} + (1) \det A_{44}
$$

$$
= -5 \det \begin{bmatrix} 9 & 3 & 2 \\ 0 & 0 & 6 \\ -1 & 1 & 2 \end{bmatrix} + \det \begin{bmatrix} 2 & 9 & 3 \\ 4 & 0 & 0 \\ 3 & -1 & 1 \end{bmatrix}.
$$

The second row is the best for each of these third-order determinants,

$$
\det \begin{bmatrix} 9 & 3 & 2 \\ 0 & 0 & 6 \\ -1 & 1 & 2 \end{bmatrix} = -(6) \det \begin{bmatrix} 9 & 3 \\ -1 & 1 \end{bmatrix} = -6(12) = -72,
$$

$$\det \begin{bmatrix} 2 & 9 & 3 \\ 4 & 0 & 0 \\ 3 & -1 & 1 \end{bmatrix} = -(4)\det \begin{bmatrix} 9 & 3 \\ -1 & 1 \end{bmatrix} = -4(12) = -48.$$

Therefore

$$\det A = -5(-72) + (-48) = 312.$$

The freedom to compute determinants along any row or column makes the computation of determinants of certain special matrices extremely simple.

Theorem 2   If a square matrix has a zero row or column, then its determinant is zero.

Proof   Expand along the zero row or column.   ∎

Theorem 3   If a square matrix has two equal rows (or columns), its determinant is zero.

Proof   Let n be the order of A. If $n = 2$, A has the form

$$A = \begin{bmatrix} a_{11} & a_{12} \\ a_{11} & a_{12} \end{bmatrix},$$

so that $\det A = a_{11}a_{12} - a_{11}a_{12} = 0$.

For $n > 2$ let A be a matrix of order n with two equal rows. Since n is at least 3, we can choose a row, say the ith, that is not one of the two equal rows. Expand along this row to obtain

$$\det A = (-1)^{i+1}a_{i1}\det A_{i1} + (-1)^{i+2}a_{i2}\det A_{i2}$$
$$+ \cdots + (-1)^{i+n}a_{in}\det A_{in}.$$

Now each of the matrices A_{ij} has two equal rows (deleting a column does not disturb the equality of the rows). If $n = 3$, each of the A_{ij} is of order $n - 1 = 2$ and by the first case, $\det A_{ij} = 0$ for each j. But then $\det A = 0$ as well. If $n > 3$, repeat the process just described as often as needed until you get down to the 2×2 case in each term. Thus $\det A = 0$.

A similar argument holds if two columns are equal.   ∎

Example 3.7

Compute $\det B$ for

$$B = \begin{bmatrix} 2 & 1 & 3 & 2 \\ 1 & 4 & 2 & 1 \\ 3 & -2 & 1 & 3 \\ 2 & 1 & 0 & 2 \end{bmatrix}.$$

The first and last columns are equal, so $\det B = 0$.

Recall from section 2.3 that the *main diagonal* of a square matrix A of order n consists of the entries a_{ii} for $i = 1$ to n. A square matrix is called **upper triangular** if all the entries below the main diagonal are zero and **lower triangular** if all the entries above the main diagonal are zero. Of course a **diagonal matrix** (section 2.3) is both upper and lower triangular.

As examples, let

$$A = \begin{bmatrix} 1 & 2 \\ 0 & 3 \end{bmatrix} \quad \text{and} \quad B = \begin{bmatrix} 2 & 0 & 0 \\ 1 & 2 & 0 \\ -1 & 1 & 3 \end{bmatrix}.$$

Here A is upper triangular and B is lower triangular.

Note: When displaying a general triangular matrix of order greater than 2, we will sometimes not print those zeros that determine its triangularity; other zeros will be printed to fill out the array.

Theorem 4 If a matrix A of order n is upper triangular, lower triangular, or diagonal, then $\det A = a_{11}a_{22} \ldots a_{nn}$, the product of the entries on the main diagonal.

Proof Suppose A is upper triangular. Then A has the form

$$\begin{bmatrix} a_{11} & a_{12} \cdots a_{1n} \\ & a_{22} \cdots a_{2n} \\ & & \ddots & \vdots \\ & & & a_{nn} \end{bmatrix}.$$

Expansion down the first column gives at most one nonzero term

$$\det A = (-1)^{1+1} a_{11} \det A_{11} = a_{11} \det \begin{bmatrix} a_{22} & \cdots & a_{2n} \\ & \ddots & \vdots \\ & & a_{nn} \end{bmatrix}$$

The resulting matrix A_{11} is still upper triangular, so the same argument gives

$$\det A = a_{11}a_{22} \det \begin{bmatrix} a_{33} & \cdots & a_{3n} \\ & \ddots & \vdots \\ & & a_{nn} \end{bmatrix}.$$

Continue this process until the submatrix is 2×2,

$$\det A = a_{11}a_{22}\cdots a_{n-2,\,n-2} \det \begin{bmatrix} a_{n-1,\,n-1} & a_{n-1,\,n} \\ 0 & a_{nn} \end{bmatrix}$$

$$= a_{11}a_{22}\cdots a_{nn}.$$

Since a diagonal matrix is upper triangular, the same result holds in case the matrix is diagonal. In case A is lower triangular, the same argument applies except that expansion is done along first rows. ∎

Example 3.8

Compute $\det A$, where

$$A = \begin{bmatrix} 2 & 4 & 2 & 1 \\ 0 & 1 & 0 & 3 \\ 0 & 0 & 3 & 1 \\ 0 & 0 & 0 & -2 \end{bmatrix}.$$

The answer is just $\det A = (2)(1)(3)(-2) = -12$.

Theorem 5   If I is an identity matrix of any order, then $\det I = 1$.

Proof   Since I is diagonal, $\det I$ is the product of the main diagonal entries. Since each of these is simply 1, the product is also 1. ∎

3.1 Exercises

1. Find the 3,1 and 3,2 minors and cofactors of

$$A = \begin{bmatrix} 2 & -1 & 3 \\ 1 & 2 & 1 \\ 0 & -3 & -4 \end{bmatrix}.$$

2. Find the 2,2 and 2,3 minors and cofactors of the matrix A in exercise 1.

In exercises 3 through 17 evaluate the determinant of each matrix.

3. $[3]$ 4. $[-5]$ 5. $\begin{bmatrix} 2 & 1 \\ 3 & -1 \end{bmatrix}$

6. $\begin{bmatrix} 4 & -6 \\ -2 & 3 \end{bmatrix}$

7. $\begin{bmatrix} 2 & 1 & 1 \\ 3 & 0 & -1 \\ 4 & 5 & 2 \end{bmatrix}$

8. $\begin{bmatrix} 1 & -2 & 3 \\ 2 & -2 & 4 \\ 0 & 3 & 2 \end{bmatrix}$

9. $\begin{bmatrix} 2 & 4 & 2 \\ 1 & 5 & 1 \\ 3 & -7 & 3 \end{bmatrix}$

10. $\begin{bmatrix} 1 & 3 & -1 & 2 \\ 2 & 1 & 0 & -4 \\ 3 & 1 & 0 & 4 \\ -1 & 2 & 0 & 3 \end{bmatrix}$

11. $\begin{bmatrix} 2 & 3 & 4 & 5 \\ 0 & 3 & 4 & 5 \\ 0 & 0 & 4 & 5 \\ 0 & 0 & 0 & 5 \end{bmatrix}$

12. $\begin{bmatrix} 2 & 0 & 2 & 0 \\ 1 & 4 & 0 & 0 \\ 2 & 3 & 1 & 0 \\ 2 & 5 & -1 & 0 \end{bmatrix}$

13. $\begin{bmatrix} a & ab \\ b & a^2 + b^2 \end{bmatrix}$

14. $\begin{bmatrix} ak & c \\ bk & d \end{bmatrix}$

15. $\begin{bmatrix} ak & ck \\ bk & dk \end{bmatrix}$

16. $\begin{bmatrix} \lambda - 2 & 3 \\ 4 & \lambda - 1 \end{bmatrix}$

17. $\begin{bmatrix} t - 3 & -1 \\ 3 & t \end{bmatrix}$

18. Verify one special case of theorem 1 by proving that for matrices of order 3, computation of the determinant by expansion down the second column yields the same result as expansion along the first row.

19. Determine a formula analogous to theorem 4 for the determinant of a matrix A of order n such that $a_{ij} = 0$ if $i + j > n + 1$. Such a matrix is upper triangular in appearance but not upper triangular with respect to the main diagonal.

20. Let a_{ij} be constants for all integers i and j from 1 to n, let x be a variable, let

$$A(x) = \begin{bmatrix} x + a_{11} & a_{12} & \cdots & a_{1n} \\ a_{21} & x + a_{22} & \cdots & a_{2n} \\ \vdots & \vdots & & \vdots \\ a_{n1} & a_{n2} & \cdots & x + a_{nn} \end{bmatrix},$$

and let $f(x) = \det A(x)$ for each x. Prove that $f(x)$ is an nth degree polynomial.

21. Find all zeros (values of x such that $f(x) = 0$) of the polynomial

$$f(x) = \det \begin{bmatrix} x - 2 & 1 \\ 3 & x \end{bmatrix}.$$

22. Find all zeros of the polynomial

$$f(t) = \det \begin{bmatrix} t - 2 & 4 & -3 \\ 0 & t - 3 & 2 \\ 0 & 0 & t \end{bmatrix}.$$

23. For matrices of *order 3 only*, there is another convenient way to compute determinants that is sometimes called the "basket-weave method." First construct a 3×5 array by writing down the entries of the given matrix and adjoining (on the right) its first and second columns (see figure 3.1). Then form the products of the three numbers on each of the six diagonals indicated,

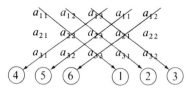

Figure 3.1

and set det A equal to the sum of the circled products 1, 2, and 3 minus the sum of circled products 4, 5, and 6. Show that this procedure does indeed yield the determinant of A.

24. In chapter 1 the *cross product* of two 3-vectors $\mathbf{v} = (a_1, a_2, a_3)$, $\mathbf{w} = (b_1, b_2, b_3)$ was defined to be

$$\mathbf{v} \times \mathbf{w} = (a_2 b_3 - a_3 b_2, \, a_3 b_1 - a_1 b_3, \, a_1 b_2 - a_2 b_1)$$

or using $\mathbf{i}, \mathbf{j}, \mathbf{k}$ notation

$$\mathbf{v} \times \mathbf{w} = (a_2 b_3 - a_3 b_2)\mathbf{i} + (a_3 b_1 - a_1 b_3)\mathbf{j} + (a_1 b_2 - a_2 b_1)\mathbf{k}.$$

Show that this vector can be conveniently remembered as the "determinant" of the matrix

$$\begin{bmatrix} \mathbf{i} & \mathbf{j} & \mathbf{k} \\ a_1 & a_2 & a_3 \\ b_1 & b_2 & b_3 \end{bmatrix}.$$

(The word *determinant* is in quotes because $\mathbf{i}, \mathbf{j}, \mathbf{k}$ are vectors, so the result is a vector, not a scalar.)

25. Prove that $(A^T)_{ij} = (A_{ji})^T$, for any i, j, and square matrix A.

26. Let $\mathbf{u}, \mathbf{v}$, and $\mathbf{w}$ be vectors in $\mathbf{R}^3$, and let A be the matrix with these vectors as rows:

$$A = \begin{bmatrix} u_1 & u_2 & u_3 \\ v_1 & v_2 & v_3 \\ w_1 & w_2 & w_3 \end{bmatrix}.$$

Show that $|\det A|$ is the *volume of the parallelepiped* determined by the vectors $\mathbf{u}, \mathbf{v}$, and $\mathbf{w}$ (see exercise 29 of section 1.2).

3.2   Properties of Determinants

In the last section we looked at some special types of square matrices, ones with determinants that require almost no computation to establish. In this section we will examine properties of determinants that will allow us to perform operations that convert arbitrary square matrices to ones with more easily computed determinants. We will also give an important relationship between the value of the determinant of a matrix and the inverse of that matrix.

Theorem 1   Let A and A' be square matrices that are the same except that in one row (or in one column) of A' each entry of A is multiplied by a scalar c. Then det $A' = c$ det A. (That is, c can be "factored out" of the row or column.)

Proof   Suppose the difference occurs in the ith row. Then

$$A = \begin{bmatrix} a_{11} & \cdots & a_{1n} \\ \vdots & & \vdots \\ a_{i1} & \cdots & a_{in} \\ \vdots & & \vdots \\ a_{n1} & \cdots & a_{nn} \end{bmatrix} \quad \text{and} \quad A' = \begin{bmatrix} a_{11} & \cdots & a_{1n} \\ \vdots & & \vdots \\ ca_{i1} & \cdots & ca_{in} \\ \vdots & & \vdots \\ a_{n1} & \cdots & a_{nn} \end{bmatrix}.$$

Compute det A' by expansion along the ith row. Then

$$\begin{aligned} \det A' &= (-1)^{i+1}ca_{i1} \det A'_{i1} + (-1)^{i+2}ca_{i2} \det A'_{i2} \\ &\quad + \cdots + (-1)^{i+n}ca_{in} \det A'_{in} \\ &= c[(-1)^{i+1}a_{i1} \det A_{i1} + (-1)^{i+2}a_{i2} \det A_{i2} \\ &\quad + \cdots + (-1)^{i+n}a_{in} \det A_{in}], \end{aligned}$$

since c factors out of each term and since $A_{ij} = A'_{ij}$ for each j (the ith row has been deleted). The last expression is just c det A. A similar argument holds for columns. ∎

Example 3.9

Compute det A, given that

$$A = \begin{bmatrix} 2 & 0 & 4 \\ 1 & -6 & 2 \\ 3 & 9 & 12 \end{bmatrix}.$$

In the computation of det A, 2 can be factored out of the first row, 3 out of the third row, 3 out of the second column, and 2 out of the third column:

$$\det \begin{bmatrix} 2 & 0 & 4 \\ 1 & -6 & 2 \\ 3 & 9 & 12 \end{bmatrix} = 2 \det \begin{bmatrix} 1 & 0 & 2 \\ 1 & -6 & 2 \\ 3 & 9 & 12 \end{bmatrix} = 2(3) \det \begin{bmatrix} 1 & 0 & 2 \\ 1 & -6 & 2 \\ 1 & 3 & 4 \end{bmatrix}$$

$$= 2(3^2) \det \begin{bmatrix} 1 & 0 & 2 \\ 1 & -2 & 2 \\ 1 & 1 & 4 \end{bmatrix} = 2^2(3^2) \det \begin{bmatrix} 1 & 0 & 1 \\ 1 & -2 & 1 \\ 1 & 1 & 2 \end{bmatrix}.$$

Expanding, say, along the first row, we still need to compute two 2×2 determinants as in the original problem. Now, however, the numbers are easier to work with than before. The answer is $36(-2) = -72$.

The same idea can be used in reverse to clear the fractions in a matrix before computing its determinant. The advantage of course is that it is easier to work with integers than fractions, and there is less chance of making an error. The rule here is to examine each row (or column) for the smallest integer that will clear the fractions (for the *least common denominator*), multiply the row by that number, and compensate by multiplying the determinant by the reciprocal of that number.

Example 3.10

Compute det A, where

$$A = \begin{bmatrix} 1/3 & 0 & 3/4 \\ 2/5 & -1 & 3/2 \\ 1/8 & -3/4 & 5/4 \end{bmatrix}.$$

Since 12 will clear the first row, 10 the second, and 8 the last, we multiply the determinant by $1/12$ and the first row by 12, and proceed similarly with the other rows.

$$\det A = (1/12)(1/10)(1/8) \det \begin{bmatrix} 4 & 0 & 9 \\ 4 & -10 & 15 \\ 1 & -6 & 10 \end{bmatrix}.$$

Now -2 can be factored out of the second column,

$$\det A = (1/12)(1/10)(1/8)(-2) \det \begin{bmatrix} 4 & 0 & 9 \\ 4 & 5 & 15 \\ 1 & 3 & 10 \end{bmatrix}.$$

The final value is det $A = (-1/480)(83) = -83/480$.

Example 3.11

Compute det A for the matrix

$$A = \begin{bmatrix} 2 & 1 & 3 & -4 \\ -1 & 0 & 1 & 2 \\ -3 & 2 & -1 & 6 \\ 4 & 1 & 4 & -8 \end{bmatrix}.$$

Your first thought might be to expand along the second row or column because of the zero. This is a reasonable approach, but, if you happen to notice that the last column is -2 times the first, using theorem 1 to factor out -2, we have

$$\det A = -2 \det \begin{bmatrix} 2 & 1 & 3 & 2 \\ -1 & 0 & 1 & -1 \\ -3 & 2 & -1 & -3 \\ 4 & 1 & 4 & 4 \end{bmatrix}.$$

Now the first and last columns are equal, and by theorem 3 of section 3.1 $\det A = 0$.

The method of example 3.11 is important enough to be identified as a theorem.

Theorem 2 If a square matrix A has a row that is a scalar multiple of another row (or a column that is a scalar multiple of another column), then $\det A = 0$.

Proof Use theorem 1 to factor out the scalar multiple, and apply theorem 3 of the last section. ∎

Theorem 3 If two rows (or columns) of a square matrix are interchanged, then the determinant changes by a factor of -1, that is, its sign is reversed.

Proof Let A and A' be matrices of order n such that A' is obtained from A by interchanging two of its rows.

If $n = 2$, there are only two rows to interchange

$$A = \begin{bmatrix} a_{11} & a_{12} \\ a_{21} & a_{22} \end{bmatrix} \quad \text{and} \quad A' = \begin{bmatrix} a_{21} & a_{22} \\ a_{11} & a_{12} \end{bmatrix}.$$

Then $\det A' = a_{21}a_{12} - a_{11}a_{22} = -(a_{11}a_{22} - a_{21}a_{22}) = -\det A$.

For $n > 2$ expand $\det A'$ along any row not involved in the interchange. Suppose the ith row is one such, then

$$\det A' = (-1)^{i+1}a_{i1} \det A'_{i1} + (-1)^{i+2}a_{i2} \det A'_{i2}$$
$$+ \cdots + (-1)^{i+n}a_{in} \det A'_{in}.$$

Each submatrix A'_{ij} is the same as the corresponding submatrix A_{ij} of A, except that two rows of A_{ij} are interchanged to obtain A'_{ij}. (Taking out a row not involved in the interchange leaves two rows still interchanged.)

If $n = 3$, each of these A'_{ij} is order 2, and by the $n = 2$ case each term in the expansion of det A' is the opposite of the corresponding term in the expansion of det A along the same row. But then det $A' = -\det A$. If $n > 3$, repeat the process just given as often as necessary to reduce to the order 2 case. A similar argument holds if two columns are interchanged. ■

So far we have seen how determinants of square matrices behave with respect to two of the three elementary row operations:

i. If a row is multiplied by a scalar, the determinant is multiplied by the same scalar.
ii. If two rows are interchanged, the sign of the determinant is reversed.

The third elementary row operation, addition of a scalar multiple of one row to another, has the simplest effect of all. It is also the most useful for simplifying computation of determinants.

Theorem 4   If a scalar multiple of one row (or column) is added to another row (or column) of a square matrix, the determinant is unchanged.

Proof   Let A and A' be matrices of order n, where A and A' agree except in the ith row, and $A'_i = cA_k + A_i$, in other words, the ith row of A' is the sum of the ith row of A and c times the kth row of A, where $k \neq i$. We compute det A' by expansion along the ith row:

$$\det A' = (-1)^{i+1}a'_{i1}\det A'_{i1} + \cdots + (-1)^{i+n}a'_{in}\det A'_{in}.$$

By hypothesis $a'_{ij} = ca_{kj} + a_{ij}$ for each j, and since the ith row has been deleted, $A'_{ij} = A_{ij}$ for each j. Then

$$\begin{aligned}
\det A' &= (-1)^{i+1}(ca_{k1} + a_{i1})\det A_{i1} \\
&\quad + \cdots + (-1)^{i+n}(ca_{kn} + a_{in})\det A_{in} \\
&= (-1)^{i+1}ca_{k1}\det A_{i1} + (-1)^{i+1}a_{i1}\det A_{i1} \\
&\quad + \cdots + (-1)^{i+n}ca_{kn}\det A_{in} + (-1)^{i+n}a_{in}\det A_{in} \\
&= (-1)^{i+1}ca_{k1}\det A_{i1} + \cdots + (-1)^{i+n}ca_{kn}\det A_{in} \\
&\quad + (-1)^{i+1}a_{i1}\det A_{i1} + \cdots + (-1)^{i+1}a_{in}\det A_{in} \\
&= c[(-1)^{i+1}a_{k1}\det A_{i1} + \cdots + (-1)^{i+n}a_{kn}\det A_{in}] \\
&\quad + [(-1)^{i+1}a_{i1}\det A_{i1} + \cdots + (-1)^{i+1}a_{in}\det A_{in}] \\
&= c\det A'' + \det A,
\end{aligned}$$

where A'' is obtained from A by replacing the ith row by its own kth row. But then A'' has two equal rows (the ith and the kth), and by theorem 3 of section 3.1, det $A'' = 0$. This leaves just det $A' = \det A$.

A similar proof holds for columns. ■

As we noted in the last section, det A is easy to compute if a row or column has lots of zeros. The usefulness of theorem 4 is that, even if A does not have many zeros, we can "create them"—we can find a new matrix with the same determinant that does have many zeros.

Example 3.12

One last time we compute the determinant of the matrix given in example 3.3

$$A = \begin{bmatrix} 2 & 1 & -3 & 1 \\ -3 & -2 & 0 & 2 \\ 2 & 1 & 0 & -1 \\ 1 & 0 & 1 & 2 \end{bmatrix}.$$

We have already noticed (example 3.5) that it is most convenient to expand down the third column. Before doing so, however, the remaining work can be cut in half if we first add 3 times the last row to the first. Then we have

$$A' = \begin{bmatrix} 5 & 1 & 0 & 7 \\ -3 & -2 & 0 & 2 \\ 2 & 1 & 0 & -1 \\ 1 & 0 & 1 & 2 \end{bmatrix},$$

and by theorem 4

$$\det A = \det A' = (-1)(1) \det A'_{43} = -\det \begin{bmatrix} 5 & 1 & 7 \\ -3 & -2 & 2 \\ 2 & 1 & -1 \end{bmatrix}.$$

To finish the computation as simply as possible, we use theorem 4 again to obtain a new matrix with the same determinant as A'_{43} but with two zeros in some row or column. Any row or column would work, but a quick inspection shows that the second column or the third row requires minimal computation. We choose the third row. Adding -2 times the second column to the first, and just adding the second column to the third, we obtain

$$\det \begin{bmatrix} 5 & 1 & 7 \\ -3 & -2 & 2 \\ 2 & 1 & -1 \end{bmatrix} = \det \begin{bmatrix} 3 & 1 & 8 \\ 1 & -2 & 0 \\ 0 & 1 & 0 \end{bmatrix} = -(1) \det \begin{bmatrix} 3 & 8 \\ 1 & 0 \end{bmatrix} = 8.$$

Then det $A = -8$.

Computational note: It is somewhat surprising to realize that even with the aid of high-speed computers similar simplifications must be made to compute det A when A is of even modest size. For example, computation time of an order 25 matrix using only the definition of determinant would be astronomically long even using the fastest computers available (or on the drawing board). For this reason, programs to compute the determinant of a matrix by computer are based on the following idea. Use theorems 3 and 4 to bring the matrix into upper triangular form, keeping track of the number of row interchanges used. By theorem 4 of section 3.1 the determinant is then just ε times the product of the main diagonal entries, where $\varepsilon = 1$ if there are an even number of row interchanges, and $\varepsilon = -1$ otherwise.

Example 3.13

Find det A by using the process described in the *computational note* above, where

$$A = \begin{bmatrix} 1 & 2 & 0 & -2 \\ 0 & 0 & 2 & -1 \\ 0 & -1 & 1 & 0 \\ 1 & 3 & 4 & 1 \end{bmatrix}.$$

By interchanging the second and third rows, and then adding multiples of each row to the bottom row, we obtain the sequence of equations

$$\det A = -\det \begin{bmatrix} 1 & 2 & 0 & -2 \\ 0 & -1 & 1 & 0 \\ 0 & 0 & 2 & -1 \\ 1 & 3 & 4 & 1 \end{bmatrix} = -\det \begin{bmatrix} 1 & 2 & 0 & -2 \\ 0 & -1 & 1 & 0 \\ 0 & 0 & 2 & -1 \\ 0 & 1 & 4 & 3 \end{bmatrix}$$

$$= -\det \begin{bmatrix} 1 & 2 & 0 & -2 \\ 0 & -1 & 1 & 0 \\ 0 & 0 & 2 & -1 \\ 0 & 0 & 5 & 3 \end{bmatrix} = -\det \begin{bmatrix} 1 & 2 & 0 & -2 \\ 0 & -1 & 1 & 0 \\ 0 & 0 & 2 & -1 \\ 0 & 0 & 0 & 11/2 \end{bmatrix}.$$

Finally, det $A = -(1)(-1)(2)(11/2) = 11$.

Perhaps the most important single property of the determinant of a matrix is that it tells us whether or not the matrix in question is invertible.

Theorem 5   A square matrix A is invertible if and only if det $A \neq 0$.

Proof   Let B be the row-reduced echelon form for A. Since B is obtained from A by elementary row operations, repeated application of theorems 1, 3, and

4 implies that det $B = c$ det A for some nonzero scalar c. This implies that det $A \neq 0$ if and only if det $B \neq 0$. But B is a row-reduced square matrix and as such is either the identity matrix or has at least one row of zeros. In other words, det $B = 1$ if B is the identity matrix and det $B = 0$ otherwise. Thus we have det $A \neq 0$ if and only if B is the identity matrix. By theorem 6 of section 2.5 this is equivalent to A being invertible. ∎

Example 3.14

Determine whether or not the matrix A is invertible, where A is given by

$$A = \begin{bmatrix} 2 & 1 & -3 & 1 \\ -3 & -2 & 0 & 2 \\ 2 & 1 & 0 & -1 \\ 1 & 0 & 1 & 2 \end{bmatrix}.$$

Since A is the matrix of example 3.12, det $A = -8 \neq 0$, and thus A is invertible.

Note: The fact that det $A \neq 0$ implies that A is invertible does not tell us what the matrix A^{-1} looks like. Determinants can be used for this purpose as we will see in section 3.3.

The next theorem is important to the development. Its proof, however, would require the introduction of notions not needed for working with determinants but essential in a formal treatment. Since our aim is understanding the basics, possibly at the expense of formality, the proof is omitted.

Theorem 6 If A and B are square matrices of the same order,

$$\text{det } AB = (\text{det } A)(\text{det } B).$$

Recall that the *transpose* of a matrix is the matrix obtained by interchanging its rows and columns; in other words, for each i the ith *row* of a matrix A is the ith *column* of the matrix A^T and the jth *column* of A is the jth *row* of A^T. The relationship between their determinants is as nice as possible.

Theorem 7 If A is a square matrix, det $A^T = $ det A.

Proof   Let n be the order of A. If $n = 2$,

$$A = \begin{bmatrix} a_{11} & a_{12} \\ a_{21} & a_{22} \end{bmatrix} \quad \text{and} \quad A^T = \begin{bmatrix} a_{11} & a_{21} \\ a_{12} & a_{22} \end{bmatrix}.$$

Hence det $A = a_{11}a_{22} - a_{21}a_{12} =$ det A^T.

For larger values of n the proof is based on the following fact: the expansion of det A^T repeatedly along first *rows* yields exactly the same terms and in the same order as the expansion of det A repeatedly down first *columns*.   ■

Theorems 6 and 7 together can be used to prove theorem 2 of section 2.4. We restate the result here and prove it.

Theorem 8   Let A be a square matrix. If a square matrix B exists such that $AB = I$, then $BA = I$ as well, and thus $B = A^{-1}$.

Proof   We first show that there is some square matrix C, such that $CA = I$. Consider the system

$$A^T X = I.$$

Since (det A)(det B) = det $(AB) =$ det $I = 1$, we see that det $A \neq 0$. However, det $A^T =$ det A, so det $A^T \neq 0$ as well. Then the row-reduced echelon form for A^T is I, and there is a solution, say D, to this matrix equation. That is $A^T D = I$. But then $(A^T D)^T = I^T$, and from properties of transpose

$$D^T A = I.$$

Thus there is a matrix C, namely $C = D^T$, such that $CA = I$.

Finally, we show that $C = B$ (which implies that $B = A^{-1}$):

$$C = CI = C(AB) = (CA)B = IB = B.$$ ■

3.2   **Exercises**

Evaluate the determinant of each matrix in exercises 1 through 9. Use the theorems of this section to simplify the work.

1. $\begin{bmatrix} 1 & 2 & 3 \\ 1 & 3 & 7 \\ 1 & 4 & 13 \end{bmatrix}$ 2. $\begin{bmatrix} 1 & 0 & 0 & 3 \\ 0 & 3 & 9 & 21 \\ 0 & 0 & 2 & 1 \\ 2 & 0 & 0 & 7 \end{bmatrix}$ 3. $\begin{bmatrix} 1 & 0 & 0 & 0 \\ 0 & 0 & 3 & 0 \\ 0 & 2 & 0 & 0 \\ 0 & 0 & 0 & 4 \end{bmatrix}$

4. $\begin{bmatrix} 2 & -1 & 3 & -4 \\ 1 & 0 & 2 & 1 \\ -4 & 2 & -6 & 8 \\ 3 & 1 & 2 & 0 \end{bmatrix}$ **5.** $\begin{bmatrix} 1/3 & 3/5 & 2/5 \\ 3/8 & 1/2 & 1/4 \\ 1/3 & 2/3 & 1/2 \end{bmatrix}$

6. $\begin{bmatrix} 0.02 & -0.10 & 0.08 \\ 0.03 & 0.06 & -0.09 \\ 0 & 0.07 & 0.20 \end{bmatrix}$ **7.** $\begin{bmatrix} x & 2x & -3x \\ x & x-1 & -3 \\ 0 & 0 & 2x-1 \end{bmatrix}$

8. $\begin{bmatrix} t & -1 & 3 \\ 1 & t-3 & 1 \\ 2 & 3 & t-1 \end{bmatrix}$ **9.** $\begin{bmatrix} \lambda-1 & 0 & 0 & 0 \\ 2 & 0 & \lambda+1 & 0 \\ 1 & \lambda-2 & 0 & 0 \\ 2 & 3 & 9 & \lambda+2 \end{bmatrix}$

10. Determine which of the matrices in exercises 1 through 6 are invertible. Do *not* compute the inverse.

11. For which values of the variables in exercises 7 through 9 are the matrices invertible?

12. Prove that, for a square matrix A of order n and scalar c, det $(cA) = c^n$ det A.

13. Without expanding the determinant, show that for any values of s, t, and u

$$\det \begin{bmatrix} s & s+1 & s+2 \\ t & t+1 & t+2 \\ u & u+1 & u+2 \end{bmatrix} = 0.$$

14. Let A' be obtained from the square matrix A by interchanging pairs of rows (columns) m times. Express det A' in terms of det A and m.

15. Use exercise 14 to find a formula for det A' in terms of det A, where A' is obtained from A by reversing the order in which the rows appear. That is, $A'_i = A_{n+1-i}$ for each i.

16. Compute and leave in simplest factored form

$$\det \begin{bmatrix} 1 & s & s^2 \\ 1 & t & t^2 \\ 1 & u & u^2 \end{bmatrix}.$$

17. We know that in general $AB \neq BA$. However, prove that det AB = det BA for all square nth-order matrices A and B.

18. Prove that det $A^m = ($det $A)^m$, where A^m denotes the mth power of A.

19. If $A^T A = I$, prove that det $A = \pm 1$.

20. If S is an invertible matrix, and $B = S^{-1}AS$, prove that det A = det B.

Exercises 21 through 23 give an extremely simple method of finding a linear equation of the hyperplane that contains n independent points in $\mathbf{R}^n$ (see section 1.4).

21. Let (a_1, a_2) and (b_1, b_2) be distinct points in $\mathbf{R}^2$. Prove that the equation

$$\det \begin{bmatrix} x & y & 1 \\ a_1 & a_2 & 1 \\ b_1 & b_2 & 1 \end{bmatrix} = 0$$

is a linear equation that describes the line that contains (a_1, a_2) and (b_1, b_2). (*Hint:* what happens when you make the substitution $x = a_1$ and $y = a_2$? Similarly for $x = b_1$ and $y = b_2$? Finally, expand along the top row.)

22. Let (a_1, a_2, a_3), (b_1, b_2, b_3), and (c_1, c_2, c_3) be noncollinear points in $\mathbf{R}^3$. Prove that the equation

$$\det \begin{bmatrix} x & y & z & 1 \\ a_1 & a_2 & a_3 & 1 \\ b_1 & b_2 & b_3 & 1 \\ c_1 & c_2 & c_3 & 1 \end{bmatrix} = 0$$

is a linear equation that describes the plane containing (a_1, a_2, a_3), (b_1, b_2, b_3), and (c_1, c_2, c_3).

23. Generalize the result of exercises 21 and 22 to the case of n independent points in $\mathbf{R}^n$.

24. Compute the determinant of the following matrix by three different methods: (a) the definition, (b) triangularization, (c) any other. Which seems to be the best method?

$$A = \begin{bmatrix} 2 & 1 & -1 & 3 \\ 1 & 2 & 0 & 3 \\ -1 & 2 & -2 & 1 \\ 0 & 3 & 1 & 4 \end{bmatrix}.$$

25. Find the determinant of the matrix A using a computer.

$$A = \begin{bmatrix} 0.21 & -2.31 & 2.01 & 3.00 \\ 0.03 & -1.20 & 1.02 & 0.00 \\ 1.21 & 1.00 & -0.64 & -0.54 \\ 1.02 & 0.04 & 0.57 & 0.19 \end{bmatrix}.$$

3.3 Cramer's Rule

In this section we give new techniques for solving two old problems: solving a square linear system and inverting a matrix. In sections 2.2 and 2.4 these problems were dealt with by using row reduction. Here we solve them with the aid of determinants.

Cramer's Rule

Consider a system of linear equations $A\mathbf{x} = \mathbf{b}$, where A is a *square* matrix of order n. We will denote by $A(i)$ the matrix obtained from A by replacing its ith column A^i by $\mathbf{b}$. For example, if the given linear system is

$$\begin{aligned} x_1 - x_2 + x_3 &= 6 \\ 2x_1 \quad\quad + 3x_3 &= 1 \\ 2x_2 - x_3 &= 0, \end{aligned}$$

then

$$A = \begin{bmatrix} 1 & -1 & 1 \\ 2 & 0 & 3 \\ 0 & 2 & -1 \end{bmatrix},$$

$$A(1) = \begin{bmatrix} 6 & -1 & 1 \\ 1 & 0 & 3 \\ 0 & 2 & -1 \end{bmatrix}, \quad A(2) = \begin{bmatrix} 1 & 6 & 1 \\ 2 & 1 & 3 \\ 0 & 0 & -1 \end{bmatrix}, \quad \text{and } A(3) = \begin{bmatrix} 1 & -1 & 6 \\ 2 & 0 & 1 \\ 0 & 2 & 0 \end{bmatrix}.$$

The following procedure for solving a square linear system is called **Cramer's rule** and will be justified later in the section.

Theorem 1   Let $A\mathbf{x} = \mathbf{b}$ be a square linear system of n equations with $\det A \neq 0$. Then the solution to this system is given by

$$x_1 = \frac{\det A(1)}{\det A}, \quad x_2 = \frac{\det A(2)}{\det A}, \ldots, x_n = \frac{\det A(n)}{\det A}.$$

Example 3.15

Solve the linear system of equations

$$\begin{aligned} 2x_1 + 3x_2 - x_3 &= 1 \\ x_1 + 4x_2 + 2x_3 &= 2 \\ 3x_1 - x_2 - x_3 &= 3. \end{aligned}$$

Letting A be the coefficient matrix,

$$A = \begin{bmatrix} 2 & 3 & -1 \\ 1 & 4 & 2 \\ 3 & -1 & -1 \end{bmatrix},$$

and, noting that $\det A = 30 \neq 0$, Cramer's rule asserts that

$$x_1 = \frac{\det A(1)}{\det A} = \frac{\det \begin{bmatrix} 1 & 3 & -1 \\ 2 & 4 & 2 \\ 3 & -1 & -1 \end{bmatrix}}{\det A} = \frac{36}{30} = \frac{6}{5},$$

$$x_2 = \frac{\det A(2)}{\det A} = \frac{\det \begin{bmatrix} 2 & 1 & -1 \\ 1 & 2 & 2 \\ 3 & 3 & -1 \end{bmatrix}}{\det A} = -\frac{6}{30} = -\frac{1}{5},$$

$$x_3 = \frac{\det A(3)}{\det A} = \frac{\det \begin{bmatrix} 2 & 3 & 1 \\ 1 & 4 & 2 \\ 3 & -1 & 3 \end{bmatrix}}{\det A} = \frac{24}{30} = \frac{4}{5}.$$

Example 3.16

Solve the linear system of equations

$$\begin{aligned} x_1 + 2x_2 + x_3 &= 3 \\ 3x_1 + x_2 - x_3 &= -1 \\ 5x_1 \qquad\quad - 3x_3 &= -5. \end{aligned}$$

Letting A be the coefficient matrix, we see that $\det A = 0$. Cramer's rule does not apply. The row-reduction methods of section 2.2 should be used.

Adjoint Form of Inverse

From theorem 5 of section 3.2 we know that, if a square matrix A is invertible, then $\det A \neq 0$. If $\det A \neq 0$, we can not only invert A, but we can obtain an explicit formula for the inverse with the aid of determinants. Toward this goal we introduce the following matrix.

Definition   For a square matrix A, the (**classical**) **adjoint** of A is the matrix

$$\text{Adj } A = [c_{ij}], \quad \text{where } c_{ij} = (-1)^{i+j} \det A_{ji}.$$

In other words, the ijth entry of Adj A is the jith cofactor of A.

Example 3.17

Find Adj A, where

$$A = \begin{bmatrix} 1 & -1 & 1 \\ 2 & 0 & 2 \\ 1 & 2 & 3 \end{bmatrix}.$$

We compute three of the entries of $C = \text{Adj } A$ explicitly:

$$c_{11} = (-1)^{1+1} \det A_{11} = (-1)^{1+1} \det \begin{bmatrix} 0 & 2 \\ 2 & 3 \end{bmatrix} = (-1)^2(-4) = -4,$$

$$c_{12} = (-1)^{1+2} \det A_{21} = (-1)^{1+2} \det \begin{bmatrix} -1 & 1 \\ 2 & 3 \end{bmatrix} = (-1)^3(-5) = 5,$$

$$c_{23} = (-1)^{2+3} \det A_{32} = (-1)^{2+3} \det \begin{bmatrix} 1 & 1 \\ 2 & 2 \end{bmatrix} = (-1)^5(0) = 0.$$

The other entries are computed in a similar manner to give

$$\text{Adj } A = \begin{bmatrix} -4 & 5 & -2 \\ -4 & 2 & 0 \\ 4 & -3 & 2 \end{bmatrix}.$$

Theorem 2 For a square matrix A, Adj A can be formed by replacing each entry of A^T by the corresponding cofactor of A^T.

Proof This fact is an immediate consequence of exercise 25 of section 3.1 and theorem 7 of section 3.2. That is, $\det (A^T)_{ij} = \det (A_{ji})^T = \det A_{ji}$. ∎

As an example of the preceding theorem we look again at example 3.17. First form the transpose

$$A^T = \begin{bmatrix} 1 & 2 & 1 \\ -1 & 0 & 2 \\ 1 & 2 & 3 \end{bmatrix}.$$

Then replace each entry by its cofactor.

$$\text{Adj } A = \begin{bmatrix} -4 & 5 & -2 \\ -4 & 2 & 0 \\ 4 & -3 & 2 \end{bmatrix}.$$

The major importance of Adj A stems from the very simple product that results when it is multiplied by the original matrix A. As a numerical example we again use the matrix of example 3.17. We have

$$A(\text{Adj } A) = \begin{bmatrix} 1 & -1 & 1 \\ 2 & 0 & 2 \\ 1 & 2 & 3 \end{bmatrix} \begin{bmatrix} -4 & 5 & -2 \\ -4 & 2 & 0 \\ 4 & -3 & 2 \end{bmatrix} = \begin{bmatrix} 4 & 0 & 0 \\ 0 & 4 & 0 \\ 0 & 0 & 4 \end{bmatrix}$$

$$= 4 \begin{bmatrix} 1 & 0 & 0 \\ 0 & 1 & 0 \\ 0 & 0 & 1 \end{bmatrix} = 4I$$

$$(\text{Adj } A)A = \begin{bmatrix} -4 & 5 & -2 \\ -4 & 2 & 0 \\ 4 & -3 & 2 \end{bmatrix} \begin{bmatrix} 1 & -1 & 1 \\ 2 & 0 & 2 \\ 1 & 2 & 3 \end{bmatrix} = \begin{bmatrix} 4 & 0 & 0 \\ 0 & 4 & 0 \\ 0 & 0 & 4 \end{bmatrix}$$

$$= 4 \begin{bmatrix} 1 & 0 & 0 \\ 0 & 1 & 0 \\ 0 & 0 & 1 \end{bmatrix} = 4I.$$

In addition note that det $A = 4$. But then

$$A((1/4)\text{Adj } A) = ((1/4)\text{Adj } A)A = I.$$

Therefore, $(1/4)$Adj A is the inverse of A. This fact suggests the following theorem that yields the *adjoint form* of the inverse of a matrix. The proof will be delayed until later in the section.

Theorem 3   Let A be a square matrix, with det $A \neq 0$. Then A is invertible, and

$$A^{-1} = \frac{1}{\det A} \text{Adj } A.$$

We have already seen one example of this theorem. As another we show how to compute the inverse of a 2×2 matrix by inspection.

Example 3.18

Find the inverse of A, where

$$A = \begin{bmatrix} a_{11} & a_{12} \\ a_{21} & a_{22} \end{bmatrix}.$$

If det $A = a_{11}a_{22} - a_{21}a_{12} = 0$, A has no inverse. If det $A \neq 0$, we proceed to compute $C = $ Adj A,

$$c_{11} = (-1)^{1+1} \det A_{11} = (1) \det [a_{22}] = a_{22}$$
$$c_{12} = (-1)^{1+2} \det A_{21} = (-1) \det [a_{12}] = -a_{12}$$
$$c_{21} = (-1)^{2+1} \det A_{12} = (-1) \det [a_{21}] = -a_{21}$$
$$c_{22} = (-1)^{2+2} \det A_{22} = (1) \det [a_{11}] = a_{11}.$$

Clearly these entries can all be found with no computation. Now by theorem 3

$$A^{-1} = \frac{1}{\det A} \text{Adj } A = \frac{1}{\det A} \begin{bmatrix} a_{22} & -a_{12} \\ -a_{21} & a_{11} \end{bmatrix}.$$

Computational note: Solving a square system of equations by Cramer's rule and likewise computing the inverse of a matrix by using the classical adjoint of the matrix are numerically inefficient methods. Although the situation is not nearly as bad as for computing determinants directly from the definition, the number of multiplications required to solve an $n \times n$ system by Cramer's rule is approximately n times the number needed to do this by row operations. For this reason Cramer's rule and the adjoint form of the inverse are more useful as theoretical tools than for computational purposes.

Theory of Cramer's Rule

We turn now to the proofs of the theorems we have been discussing. In a square linear system $A\mathbf{x} = \mathbf{b}$ the notation $A(i)$ used in Cramer's rule is sometimes not explicit enough. When there is any doubt, we will write the matrix explicitly as a matrix of columns:

$$A(i) = [A^1 \ldots A^{i-1} \, \mathbf{b} \, A^{i+1} \ldots A^n].$$

Expressed as columns in this manner, theorem 4 of section 3.2 can be written as follows. Let A be an $n \times n$ matrix, and let d be a scalar. Then for any $k \neq i$

$$\det [A^1 \ldots A^{i-1} \, A^i + dA^k \, A^{i+1} \ldots A^n] = \det A.$$

Combined with theorem 1 of section 3.2, we can extend this as follows: let A be an $n \times n$ matrix and c and d be scalars. Then for any $k \neq i$

$$\det [A^1 \ldots A^{i-1} \, cA^i + dA^k \, A^{i+1} \ldots A^n] = c \det A.$$

Repeated application of this fact gives the proof of the following result.

Lemma 1   Let A be a square matrix of order n, and let $\mathbf{b}$ be a column vector with $\mathbf{b} = c_1 A^1 + \cdots + c_n A^n$, a linear combination of the columns of A. Then for each i

$$\det A(i) = c_i \det A.$$

As an illustration of lemma 1 let

$$A = \begin{bmatrix} 2 & 3 & -1 \\ 1 & 2 & 4 \\ 3 & -1 & 0 \end{bmatrix},$$

and

$$\mathbf{b} = \begin{bmatrix} 1 \\ -5 \\ 11 \end{bmatrix} = 3\begin{bmatrix} 2 \\ 1 \\ 3 \end{bmatrix} - 2\begin{bmatrix} 3 \\ 2 \\ -1 \end{bmatrix} - \begin{bmatrix} -1 \\ 4 \\ 0 \end{bmatrix}.$$

Then by lemma 1

$$\det \begin{bmatrix} 2 & 1 & -1 \\ 1 & -5 & 4 \\ 3 & 11 & 0 \end{bmatrix} = -2 \det \begin{bmatrix} 2 & 3 & -1 \\ 1 & 2 & 4 \\ 3 & -1 & 0 \end{bmatrix}.$$

Consider the square system of linear equations $A\mathbf{x} = \mathbf{b}$:

$$
\begin{aligned}
a_{11}x_1 + a_{12}x_2 + \cdots + a_{1n}x_n &= b_1 \\
a_{21}x_1 + a_{22}x_2 + \cdots + a_{2n}x_n &= b_2 \\
&\vdots \\
a_{n1}x_1 + a_{n2}x_2 + \cdots + a_{nn}x_n &= b_n .
\end{aligned}
\tag{1}
$$

Another way of viewing this system is to note that each element in the first column is multiplied by x_1, each in the second by x_2, and so on. The system can then be written as

$$A^1 x_1 + A^2 x_2 + \cdots + A^n x_n = \mathbf{b}. \tag{2}$$

In words, $\mathbf{x} = (x_1, x_2, \ldots, x_n)$ is a solution to the system if and only if it is a vector such that x_1 times the first column of A plus x_2 times the second, and so on, yields the column $\mathbf{b}$. We now restate and prove Cramer's rule.

Theorem 1 Let $A\mathbf{x} = \mathbf{b}$ be a square system of linear equations. Each component x_i of $\mathbf{x}$ satisfies the equation $x_i \det A = \det A(i)$. If $\det A \neq 0$, the solution is unique and for each i,

$$x_i = \frac{\det A(i)}{\det A}.$$

Proof From equation (2), any solution $\mathbf{x}$ gives the column vector $\mathbf{b}$ as a linear combination of the columns of A,

$$\mathbf{b} = x_1 A^1 + \cdots + x_n A^n.$$

From lemma 1, $\det A(i) = x_i \det A$. If $\det A \neq 0$, divide both sides by $\det A$ to obtain the final equation. That the solution is *unique* in case $\det A \neq 0$ follows from the expression for each x_i, which does not involve $\mathbf{x}$ in any way. To see this, suppose that $\mathbf{y}$ is also a solution. Then $y_i = \det A(i)/\det A = x_i$, for each i, and hence $\mathbf{y} = \mathbf{x}$. ∎

The proof of theorem 3 requires no special preparation, only a careful look at the products (Adj A)A and A(Adj A). We restate theorem 3 in a slightly more general manner and prove it.

Theorem 3 For a square matrix A,

$$A(\text{Adj } A) = (\text{Adj } A)A = (\det A)I.$$

Proof We compute $(\text{Adj } A)A$ and leave $A(\text{Adj } A)$ as an exercise. Let $A = [a_{ij}]$ be of order n, and let $C = (\text{Adj } A)A$. Then by the definition of matrix multiplication,

$$c_{ij} = (-1)^{i+1}(\det A_{1i})a_{1j} + \cdots + (-1)^{i+n}(\det A_{ni})a_{nj}$$
$$= (-1)^{i+1}a_{1j} \det A_{1i} + \cdots + (-1)^{i+n}a_{nj} \det A_{ni}.$$

If $i = j$ (for an element on the main diagonal),

$$c_{ii} = (-1)^{i+1}a_{1i} \det A_{1i} + \cdots + (-1)^{i+n}a_{ni} \det A_{ni}$$
$$= \det A,$$

since this is just the expansion of $\det A$ down the ith column. If $i \neq j$ (for an off diagonal element),

$$c_{ij} = (-1)^{i+1}a_{1j} \det A_{1i} + \cdots + (-1)^{i+n}a_{nj} \det A_{ni}$$
$$= \det A',$$

where A' is obtained from A by replacing the ith column by the jth column. (The expansion of $\det A'$ is down the ith column.) However, A' has two equal columns (the ith and jth), and therefore $\det A' = 0$. That is, $c_{ij} = 0$ for any off diagonal element. ∎

Note: The fact that $A(\text{Adj } A) = (\text{Adj } A)A$ can be used to offer another proof of theorem 8 of section 3.2. That is, if A and B are square matrices with $AB = I$, then $BA = I$ as well, and thus $B = A^{-1}$.

3.3 Exercises

In exercises 1 through 6 determine whether or not Cramer's rule can be used to solve the given system. If it can be used, solve the system by Cramer's rule. If it cannot be used, explain why not.

1. $x - 2y = 3$
$ 2x + y = 1$

2. $x - y + z = 1$
$ 2x + 3y + z = 2$
$ x + 2y = 3$

3. $3x - y + z = 2$
$ 2x + y - z = 1$
$ x + 5y - 3z = 3$

4. $x + 2y - z = 3$
$ 2x - z = 1$
$ x + 6y - z = 5$

5. $3w +\ x +\ y +\ z = 2$
$x + 2y + 3z = 1$
$3w\qquad\ \ - 2z = 2$
$2x + 4y + 7z = 1$

6. $2w + x\qquad\ - z = 1$
$x + y\qquad\ \ = 2$
$y + z = 2$

In exercises 7 through 9 compute the adjoint of each matrix, and find the inverse of the matrix if it exists.

7. $\begin{bmatrix} 1 & 2 \\ -3 & 4 \end{bmatrix}$

8. $\begin{bmatrix} 2 & 1 & 0 \\ 1 & 2 & 2 \\ 3 & -1 & 4 \end{bmatrix}$

9. $\begin{bmatrix} 1 & 2 & -1 \\ 2 & 1 & 4 \\ 1 & 5 & -7 \end{bmatrix}$

10. Prove directly that $A(\text{Adj } A) = (\det A)I$ for any square matrix A.

11. A matrix A is *symmetric* if $A = A^T$. Prove that, if A is symmetric and invertible, then A^{-1} is also symmetric and invertible.

12. Prove that the inverse of an invertible upper triangular matrix of order 3 is invertible and upper triangular. (The statement is in fact true for invertible upper triangular matrices in general.)

13. Prove that $\det A \neq 0$ if and only if $\det (\text{Adj } A) \neq 0$. (*Hint:* use theorem 3.)

14. Prove that $\det (\text{Adj } A) = (\det A)^{n-1}$, $n > 1$. (*Note:* you need exercise 13 to prove the $\det A = 0$ case.)

15. Solve the system of equations by Cramer's rule using a computer:

$$0.02w - 0.23x + 0.40y -\qquad z = 3.02$$
$$2.10w +\qquad x - 1.30y\qquad\ \ = 4.11$$
$$1.16w - 0.15x\qquad\ \ - 2.35z = 0$$
$$3.91x + 0.63y -\qquad z = 2.34.$$

16. Find the adjoint of the matrix using a computer:

$$\begin{bmatrix} 2.11 & 2.11 & -3.04 & 1.11 \\ -0.02 & 1.23 & 2.22 & 1.02 \\ 0.14 & -0.06 & 1.21 & -1.08 \\ 1.32 & 0.20 & 0 & 3.90 \end{bmatrix}.$$

(*Hint:* if A is invertible, $\text{Adj } A = (\det A)A^{-1}$. Otherwise use the definition of Adj A and a program for determinants.)

Independence and Basis in $\mathbf{R}^m$

4

We have sometimes referred to $\mathbf{R}^m$, the set of all m-vectors, as "m-dimensional space." In this chapter we will make the notion of *dimension* more precise. To do so, we shall first introduce the concepts of *linear independence* and *basis* in $\mathbf{R}^m$. The ideas and tools developed here will be used throughout the rest of this text.

4.1 Linear Dependence and Independence

As we have seen in section 1.4, two 3-vectors, $\mathbf{v}_1$ and $\mathbf{v}_2$, determine a plane in $\mathbf{R}^3$, as long as one of them is not a scalar multiple of the other. Moreover a third 3-vector, $\mathbf{w}$, will lie in the plane of $\mathbf{v}_1$ and $\mathbf{v}_2$ if in some sense it is "dependent" upon these vectors. More precisely this will occur if (and only if) we can find scalars c_1 and c_2 so that $\mathbf{w} = c_1\mathbf{v}_1 + c_2\mathbf{v}_2$. Figure 4.1 illustrates this statement. In this section we will investigate such dependence relations in the more general setting of $\mathbf{R}^m$.

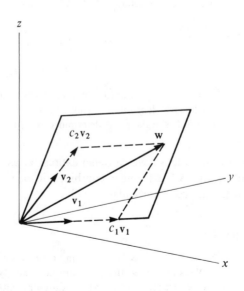

Figure 4.1

Definition A **linear combination** of the vectors $\mathbf{v}_1, \mathbf{v}_2, \ldots, \mathbf{v}_n$ in $\mathbf{R}^m$ is an expression of the form

$$c_1\mathbf{v}_1 + c_2\mathbf{v}_2 + \cdots + c_n\mathbf{v}_n,$$

where the c_i are scalars.

Example 4.1

Let $\mathbf{v} = (1, 2, 3)$, $\mathbf{v}_1 = (1, 0, 0)$, $\mathbf{v}_2 = (1, 1, 0)$, and $\mathbf{v}_3 = (1, 1, 1)$. Since

$$(1, 2, 3) = (-1)(1, 0, 0) + (-1)(1, 1, 0) + 3(1, 1, 1),$$

we can express $\mathbf{v}$ as a linear combination of the given vectors, $\mathbf{v} = c_1\mathbf{v}_1 + c_2\mathbf{v}_2 + c_3\mathbf{v}_3$, by taking $c_1 = -1$, $c_2 = -1$, and $c_3 = 3$.

Definition A set of two or more vectors is **linearly dependent** in $\mathbf{R}^m$ if one of them can be expressed as a linear combination of the others. A set consisting of a single vector is **linearly dependent** if that vector is equal to $\mathbf{0}$. If a set of vectors is *not* linearly dependent, it is said to be **linearly independent.**

Example 4.2

Show that the vectors

$$\mathbf{v}_1 = (-2, 0, 1), \quad \mathbf{v}_2 = (1, -1, 2), \quad \text{and} \quad \mathbf{v}_3 = (4, -2, 3)$$

are linearly dependent.

Since
$$(4, -2, 3) = (-1)(-2, 0, 1) + (2)(1, -1, 2),$$
we have
$$\mathbf{v}_3 = (-1)\mathbf{v}_1 + (2)\mathbf{v}_2,$$

which says $\mathbf{v}_3$ is a linear combination of $\mathbf{v}_1$ and $\mathbf{v}_2$, so the given vectors are linearly dependent. (From a geometric point of view we have shown that $\mathbf{v}_3$ lies in the plane determined by $\mathbf{v}_1$ and $\mathbf{v}_2$ as shown in figure 4.2.)

Note: In example 4.2 we can also express $\mathbf{v}_1$ as a linear combination of $\mathbf{v}_2$ and $\mathbf{v}_3$, and $\mathbf{v}_2$ as a linear combination of $\mathbf{v}_1$ and $\mathbf{v}_3$. You might try to find the appropriate scalars involved.

Example 4.3

Show that the vectors $(1, 0)$ and $(0, 1)$ are linearly independent.

We must show that they are not linearly dependent. If they were, then either $(1, 0)$ would be a linear combination of $(0, 1)$ or $(0, 1)$ would be a linear combination of $(1, 0)$. That is either

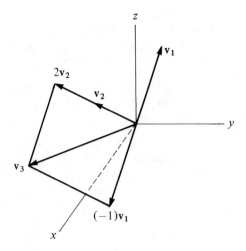

Figure 4.2

i. $(1, 0) = c_1(0, 1)$ or

ii. $(0, 1) = c_2(1, 0)$.

In the first case the vector equation is equivalent to the linear system

$$1 = c_1(0)$$
$$0 = c_1(1),$$

which has no solution.

In the second case we get the linear system

$$0 = c_2(1)$$
$$1 = c_2(0),$$

which also fails to have a solution.

Thus neither c_1 nor c_2 exist that satisfy i or ii, and consequently the given vectors are *not* linearly dependent—they are *linearly independent*.

Note: There is a much shorter geometric solution to this example. The two vectors will be linearly dependent if and only if they lie on the same line (since one must be a scalar multiple of the other). That this is impossible can either be seen by plotting the vectors, or more formally by taking their dot product. Since $(1, 0) \cdot (0, 1) = 0$, the given vectors are orthogonal—they meet at right angles.

The following theorem gives another characterization of linear dependence/independence. This characterization leads to a more systematic way of testing a set of *m*-vectors for linear dependence.

The vector equation $c_1\mathbf{v}_1 + c_2\mathbf{v}_2 + \cdots + c_n\mathbf{v}_n = \mathbf{0}$, where the $\mathbf{v}_i$ are given, always has the solution $c_1 = c_2 = \cdots = c_n = 0$. Theorem 1 states that this is the *only* solution if the vectors are linearly independent.

Theorem 1   Let $\mathscr{S} = \{\mathbf{v}_1, \mathbf{v}_2, \ldots, \mathbf{v}_n\}$. Then $\mathscr{S}$ is a linearly dependent set if and only if the vector equation $c_1\mathbf{v}_1 + c_2\mathbf{v}_2 + \cdots + c_n\mathbf{v}_n = \mathbf{0}$ has a solution with at least one $c_i \neq 0$.

Proof   When $n = 1$, the vector equation is $c_1\mathbf{v}_1 = \mathbf{0}$. This has a solution $c_1 \neq 0$ if and only if $\mathbf{v}_1 = \mathbf{0}$, as desired. Now consider the case when $n > 1$.

First, assume the set $\mathscr{S}$ is linearly dependent. Then we can express one of the vectors, say, $\mathbf{v}_1$, as a linear combination of the others. In other words, there exist $k_2, \ldots, k_n$ such that $\mathbf{v}_1 = k_2\mathbf{v}_2 + \cdots + k_n\mathbf{v}_n$. This implies that $1\mathbf{v}_1 - k_2\mathbf{v}_2 - \cdots - k_n\mathbf{v}_n = \mathbf{0}$. By letting $c_1 = 1$ and $c_i = -k_i$ for $i > 1$, we have a solution of the vector equation $c_1\mathbf{v}_1 + c_2\mathbf{v}_2 + \cdots + c_n\mathbf{v}_n = \mathbf{0}$ with at least one $c_i \neq 0$, namely, $c_1 = 1$.

Now assume the vector equation $c_1\mathbf{v}_1 + c_2\mathbf{v}_2 + \cdots + c_n\mathbf{v}_n = \mathbf{0}$ has a solution with at least one $c_i \neq 0$. Say, it is c_1 that is not zero. Then dividing both sides by c_1 yields

$$\mathbf{v}_1 + (c_2/c_1)\mathbf{v}_2 + \cdots + (c_n/c_1)\mathbf{v}_n = \mathbf{0},$$

and transposing all terms except $\mathbf{v}_1$ to the right-hand side, we obtain

$$\mathbf{v}_1 = -(c_2/c_1)\mathbf{v}_2 - \cdots - (c_n/c_1)\mathbf{v}_n.$$

Thus we have expressed one of the vectors as a linear combination of the others. (The same argument holds if some other c_i is not zero.) Hence the set is linearly dependent.   ∎

Note: The solution $c_1 = c_2 = \cdots = c_n = 0$ of the vector equation $c_1\mathbf{v}_1 + c_2\mathbf{v}_2 + \cdots + c_n\mathbf{v}_n = \mathbf{0}$ is called the **trivial solution.** Using this terminology, we can state the following corollary to theorem 1.

Corollary   The set $\{\mathbf{v}_1, \mathbf{v}_2, \ldots, \mathbf{v}_n\}$ is linearly independent if and only if the vector equation $c_1\mathbf{v}_1 + c_2\mathbf{v}_2 + \cdots + c_n\mathbf{v}_n = \mathbf{0}$ has only the trivial solution.

Example 4.4

Determine whether or not the set of vectors $\{\mathbf{v}_1, \mathbf{v}_2, \mathbf{v}_3\}$ is linearly dependent, where

$$\mathbf{v}_1 = (1, 0, -1), \quad \mathbf{v}_2 = (0, 1, 2), \quad \mathbf{v}_3 = (1, 1, 3).$$

We wish to determine whether or not the vector equation $c_1\mathbf{v}_1 + c_2\mathbf{v}_2 + c_3\mathbf{v}_3 = \mathbf{0}$ has a nontrivial solution. This equation is equivalent to

$$c_1(1, 0, -1) + c_2(0, 1, 2) + c_3(1, 1, 3) = (0, 0, 0),$$

$$(c_1, 0, -c_1) + (0, c_2, 2c_2) + (c_3, c_3, 3c_3) = (0, 0, 0),$$

or

$$(c_1 + c_3, c_2 + c_3, -c_1 + 2c_2 + 3c_3) = (0, 0, 0).$$

Now, equating corresponding components, we get the system of equations:

$$
\begin{aligned}
c_1 \quad\quad + \ c_3 &= 0 \\
c_2 + \ c_3 &= 0 \\
-c_1 + 2c_2 + 3c_3 &= 0.
\end{aligned}
$$

In matrix form this system is

$$
\begin{bmatrix} 1 & 0 & 1 \\ 0 & 1 & 1 \\ -1 & 2 & 3 \end{bmatrix}
\begin{bmatrix} c_1 \\ c_2 \\ c_3 \end{bmatrix}
=
\begin{bmatrix} 0 \\ 0 \\ 0 \end{bmatrix}.
$$

So we may restate the original problem as, "Does the linear system, $A\mathbf{c} = \mathbf{0}$, where

$$
A = \begin{bmatrix} 1 & 0 & 1 \\ 0 & 1 & 1 \\ -1 & 2 & 3 \end{bmatrix}, \quad
\mathbf{c} = \begin{bmatrix} c_1 \\ c_2 \\ c_3 \end{bmatrix}, \quad \text{and} \quad
\mathbf{0} = \begin{bmatrix} 0 \\ 0 \\ 0 \end{bmatrix},
$$

have a nontrivial solution?" If so, the set $\{v_1, v_2, v_3\}$ is linearly dependent.

Transforming the augmented matrix for this system to row-reduced echelon form, we obtain

$$
\left[\begin{array}{ccc|c} 1 & 0 & 0 & 0 \\ 0 & 1 & 0 & 0 \\ 0 & 0 & 1 & 0 \end{array} \right].
$$

Hence the system has the unique solution $c_1 = 0$, $c_2 = 0$, $c_3 = 0$, and the set $\{v_1, v_2, v_3\}$ is linearly independent by the corollary to theorem 1.

Notice that in example 4.4 the first, second, and third columns of the matrix A are, respectively, the given vectors v_1, v_2, and v_3. This is no coincidence and suggests the next procedure, which will be justified by theorem 4 given toward the end of this section. We first introduce the following terminology.

Definition An **elementary vector** in $\mathbf{R}^m$ is one that has one component equal to 1 and all other components equal to 0. If the 1 occurs as the ith component, the elementary vector is denoted e_i. A column of a matrix whose entries form an elementary vector is called an **elementary column.**

For example if

$$A = \begin{bmatrix} 1 & 2 & 0 & 0 & 1 \\ 0 & 0 & 1 & 0 & 2 \\ 0 & 0 & 0 & 1 & 3 \end{bmatrix},$$

then the first, third, and fourth columns of A are *elementary columns*. Specifically $A^1 = \mathbf{e}_1$, $A^3 = \mathbf{e}_2$, and $A^4 = \mathbf{e}_3$.

Test for linear dependence/independence   Let $\mathscr{S} = \{\mathbf{v}_1, \mathbf{v}_2, \dots, \mathbf{v}_n\}$ be a set of m-vectors. Construct the $m \times n$ matrix A whose ith column A^i consists of the components of $\mathbf{v}_i$. The set $\mathscr{S}$ is linearly independent if and only if the row-reduced echelon form of A contains only distinct elementary columns.

Example 4.5

Determine whether or not the set

$$\{(-1, 0, 2, 1), (3, 1, -2, 0), (0, 1, 4, 3)\}$$

is linearly dependent.

The 4×3 matrix A with these vectors as columns is

$$A = \begin{bmatrix} -1 & 3 & 0 \\ 0 & 1 & 1 \\ 2 & -2 & 4 \\ 1 & 0 & 3 \end{bmatrix}.$$

The row-reduced echelon form of A is

$$\begin{bmatrix} 1 & 0 & 3 \\ 0 & 1 & 1 \\ 0 & 0 & 0 \\ 0 & 0 & 0 \end{bmatrix}.$$

Since the third column is not an elementary column, this set is linearly dependent.

Note: Since the set of vectors of example 4.5 is linearly dependent, one of them can be expressed as a linear combination of the others. In fact we have

$$(0, 1, 4, 3) = (3)(-1, 0, 2, 1) + (1)(3, 1, -2, 0).$$

Notice that the coefficients (3 and 1) of the linear combination appear in the third column of the row-reduced echelon matrix. When a set of m-vectors is linearly dependent, these *coefficients of dependence* can always be found in this way. To be more explicit, we state the following procedure (justified by theorem 5 at the end of this section).

Determining coefficients of dependence   Let $\mathscr{S} = \{v_1, v_2, \ldots, v_n\}$ be a set of vectors in $\mathbf{R}^m$. Construct the $m \times n$ matrix A with $A^i = v_i$. Transform A to its row-reduced echelon form, B. Suppose B^j either is not an elementary column or is a duplication of a preceding elementary column. Then we have

$$v_j = b_{1j}w_1 + b_{2j}w_2 + \cdots + b_{kj}w_k,$$

where $w_1, w_2, \ldots, w_k$ are the vectors in $\mathscr{S}$ that have been transformed into the distinct elementary columns of B that precede the jth column.

Example 4.6

Let $v_1 = (1, -1, 2)$, $v_2 = (-2, 3, 0)$, $v_3 = (0, 1, 4)$, $v_4 = (1, 0, 1)$, and $v_5 = (2, 1, 1)$. Show that the set $\{v_1, v_2, v_3, v_4, v_5\}$ is linearly dependent, and express at least one of the v_i as a linear combination of the others.
   The matrix A with $A^i = v_i$ is

$$A = \begin{bmatrix} 1 & -2 & 0 & 1 & 2 \\ -1 & 3 & 1 & 0 & 1 \\ 2 & 0 & 4 & 1 & 1 \end{bmatrix}.$$

The row-reduced echelon form of A is the matrix

$$B = \begin{bmatrix} 1 & 0 & 2 & 0 & -1 \\ 0 & 1 & 1 & 0 & 0 \\ 0 & 0 & 0 & 1 & 3 \end{bmatrix}.$$

We see that v_1, v_2, and v_4 have been transformed into the elementary vectors e_1, e_2, and e_3, respectively. (Thus v_1, v_2, and v_4 become w_1, w_2, and w_3 in the terminology of the procedure.) We conclude that

$$v_3 = 2v_1 + 1v_2 \qquad \text{(from the third column of } B\text{)}$$

and

$$v_5 = (-1)v_1 + (0)v_2 + 3v_4 \quad \text{(from the fifth column of } B\text{)}.$$

This can be easily checked by computing

$$2(1, -1, 2) + (-2, 3, 0) = (0, 1, 4)$$

and

$$(-1)(1, -1, 2) + 3(1, 0, 1) = (2, 1, 1).$$

Had example 4.6 been phrased simply, "Is the set $\{v_1, v_2, v_3, v_4, v_5\}$ linearly independent?" the following theorem would have enabled us to immediately answer, "No." We leave the proof as exercise 23.

Theorem 2   Let $\mathscr{S}$ be a set of n vectors in $\mathbf{R}^m$ with $n > m$. Then $\mathscr{S}$ is linearly dependent.

Example 4.7

Show that the vectors $v_1 = (1, -1)$, $v_2 = (0, 1)$, and $v_3 = (2, 1)$ form a linearly dependent set.
Here the given set consists of three 2-vectors. So by theorem 2 (with $m = 2$, $n = 3$), it is linearly dependent.

If we are given a subset of $\mathbf{R}^m$ containing exactly m vectors, the following theorem supplies us with a relatively simple alternative to our previous test for linear independence. It is a consequence of theorem 4, (following example 4.8) since for a square matrix A, det $A \neq 0$ if and only if A is invertible, which is true if and only if its row-reduced form is the identity matrix.

Theorem 3   Let $\mathscr{S} = \{v_1, v_2, \ldots, v_m\}$ be a set of vectors in $\mathbf{R}^m$, and let A be the $m \times m$ matrix with $A^i = v_i$. Then $\mathscr{S}$ is linearly independent if and only if det $A \neq 0$.

Example 4.8

Determine whether or not the set $\mathscr{S} = \{v_1, v_2, v_3\}$ is linearly dependent, where $v_1 = (1, 2, 3)$, $v_2 = (-1, 0, 1)$, and $v_3 = (0, 0, 1)$.
We form the matrix A with $A^i = v_i$:

$$A = \begin{bmatrix} 1 & -1 & 0 \\ 2 & 0 & 0 \\ 3 & 1 & 1 \end{bmatrix}.$$

Now det $A = 2 \neq 0$, so by theorem 3, $\mathscr{S}$ is linearly independent.

We close this section with two theorems, the results of which have already been used in our computational work.

Theorem 4   Let $\mathscr{S} = \{v_1, v_2, \ldots, v_n\}$ be a set of m-vectors. Let A be the $m \times n$ matrix with $A^i = v_i$. Let B be the row-reduced echelon form of A. Then the set $\mathscr{S}$ is linearly independent if and only if the columns of B are the elementary vectors $e_1, e_2, \ldots, e_n$.

Proof   The set $\mathcal{S}$ will be linearly independent if and only if the vector equation $c_1\mathbf{v}_1 + c_2\mathbf{v}_2 + \cdots + c_n\mathbf{v}_n = \mathbf{0}$ has a unique solution, the trivial one. By the definition of matrix multiplication and equality of vectors this will be true if and only if $A\mathbf{c} = \mathbf{0}$ (where $\mathbf{c} = (c_1, c_2, \ldots, c_n)$) has a unique solution. But by theorem 4, corollary 3 of section 2.5, the latter is true if and only if the rank of A is equal to n, that is, if and only if the number of nonzero rows of B is equal to n. Since in row-reduced echelon form each nonzero row has a leading 1, and each column with a leading 1 has all other entries equal to 0, the rank of A is n if and only if the columns of B are the m-vectors $\mathbf{e}_1, \mathbf{e}_2, \ldots, \mathbf{e}_n$, as desired.   ∎

Theorem 5   Let $\mathcal{S} = \{\mathbf{v}_1, \mathbf{v}_2, \ldots, \mathbf{v}_n\}$ be a set of m-vectors. Let A be the $m \times n$ matrix with columns $A^i = \mathbf{v}_i$. Let B be the row-reduced echelon form of A, and suppose that one of its columns, B^j, is not an elementary vector or is a duplication of a preceding elementary vector. Then $\mathcal{S}$ is linearly dependent, and we have $\mathbf{v}_j = b_{1j}\mathbf{w}_1 + b_{2j}\mathbf{w}_2 + \cdots + b_{kj}\mathbf{w}_k$, where $\mathbf{w}_1, \mathbf{w}_2, \ldots, \mathbf{w}_k$ are vectors in S that have been transformed into $\mathbf{e}_1, \mathbf{e}_2, \ldots, \mathbf{e}_k$, the distinct elementary columns of B that precede the jth column.

Proof   By theorem 4 the set $\mathcal{S}$ is lineafly dependent. Now the vector equation $c_1\mathbf{v}_1 + c_2\mathbf{v}_2 + \cdots + c_n\mathbf{v}_n = \mathbf{0}$ is equivalent to $A\mathbf{c} = \mathbf{0}$, which in turn is equivalent to $B\mathbf{c} = \mathbf{0}$. For the sake of notational convenience suppose $\mathbf{w}_i = \mathbf{v}_i (i = 1, 2, \ldots, k)$. Then the first k equations of $B\mathbf{c} = \mathbf{0}$ become

$$c_1 \quad\quad\quad + \cdots + b_{1j}c_j + \cdots = 0$$
$$c_2 \quad\quad + \cdots + b_{2j}c_j + \cdots = 0$$
$$\vdots$$
$$c_k + \cdots + b_{kj}c_j + \cdots = 0.$$

Letting $c_j = 1$ and $c_i = 0$ for all variables other than $c_1, \ldots, c_k$, we have a solution $c_1 = -b_{1j}, c_2 = -b_{2j}, \ldots, c_k = -b_{kj}, c_j = 1$ and $c_i = 0$ for $i > k$ and $i \neq j$. But the c_i were the coefficients needed to express

$$c_1\mathbf{v}_1 + c_2\mathbf{v}_2 + \cdots + c_n\mathbf{v}_n = \mathbf{0}.$$

On replacing the c_i by the solution values, we have

$$-b_{1j}\mathbf{v}_1 - b_{2j}\mathbf{v}_2 - \cdots - b_{kj}\mathbf{v}_k + 0\mathbf{v}_{k+1} + \cdots + 0\mathbf{v}_{j-1} + 1\mathbf{v}_j + 0\mathbf{v}_{j+1}$$
$$+ \cdots + 0\mathbf{v}_n = \mathbf{0}$$

or, solving for $\mathbf{v}_j$,

$$\mathbf{v}_j = b_{1j}\mathbf{v}_1 + b_{2j}\mathbf{v}_2 + \cdots + b_{kj}\mathbf{v}_k,$$

as desired.   ∎

4.1 Exercises

In exercises 1 through 4 let $\mathbf{u} = (1, 0, -1)$ and $\mathbf{v} = (-2, 1, 1)$.

1. Write $\mathbf{w}_1 = (-1, 2, -1)$ as a linear combination of $\mathbf{u}$ and $\mathbf{v}$.

2. Show that $\mathbf{w}_2 = (-1, 1, 1)$ *cannot* be written as a linear combination of $\mathbf{u}$ and $\mathbf{v}$.

3. For what value of c is the vector $(1, 1, c)$ a linear combination of $\mathbf{u}$ and $\mathbf{v}$?

4. If the vector (x_1, x_2, x_3) is a linear combination of $\mathbf{u}$ and $\mathbf{v}$, find an equation relating x_1, x_2, and x_3.

In exercises 5 through 14 determine whether the given set of vectors is linearly dependent or independent. If it is linearly dependent, express one of the vectors in the set as a linear combination of the others.

5. $\{(1, 2, 3), (-1, 0, 1), (0, 1, 2)\}$

6. $\{(-1, 1, 2), (3, 3, 1), (1, 2, 2)\}$

7. $\{(0, 1, 2, 3), (0, -1, 2, -1), (0, 1, 0, 1)\}$

8. $\{(-1, -2, 2, 1), (0, 0, 0, 0), (1, 2, 3, 4)\}$

9. $\{(1, 2), (-1, 2)\}$

10. $\{(1, -1, 0), (2, 4, 0)\}$

11. $\{(1, 0), (0, 1), (1, 1)\}$

12. $\{(2, 0, 1), (1, -2, 0), (4, -4, 1), (1, 1, 1)\}$

13. $\{(1, -1, 1, -1), (0, 1, 0, 1), (1, 1, 0, 0), (2, 1, 1, 1)\}$

14. $\{(1, 1, 1, 1), (0, 1, 2, 3), (0, -1, -2, -3), (1, 0, -1, -2)\}$

In exercises 15 through 22 use the definition or theorems 2 or 3 to determine whether or not the given set is linearly dependent.

15. $\{(1, 2), (0, 2), (1, 0), (-1, 1)\}$

16. $\{(1, -1, 0), (1, 1, 1), (0, -2, 1), (1, 4, -2)\}$

17. $\{(1, 2), (3, -1)\}$

18. $\{(1, 0, 1), (-1, 1, 0), (0, 1, 1)\}$

19. $\{(1, -1, 3, 0), (1, 0, 1, 0), (0, 0, 0, 0), (4, 2, 3, -6)\}$

20. $\{(2, 0, 0, 0), (2, 1, 0, 0), (-1, 3, -2, 0), (1, -2, 4, -3)\}$

21. $\{(1, 1, 1), (2, 2, 2)\}$

22. $\{(1, 0, -1, 0), (3, 6, 1, 2), (-2, 0, 2, 0)\}$

23. Prove theorem 2: If $\mathscr{S}$ is a set of n vectors in $\mathbf{R}^m$ with $n > m$, then $\mathscr{S}$ is linearly dependent. (*Hint:* apply theorem 4, corollary 5 of section 2.5.)

24. Let $\mathbf{v}_1, \mathbf{v}_2, \ldots, \mathbf{v}_n$ be arbitrary vectors in $\mathbf{R}^m$. Show that $\{\mathbf{0}, \mathbf{v}_1, \mathbf{v}_2, \ldots, \mathbf{v}_n\}$ is linearly dependent.

25. Show that any set of distinct elementary vectors in $\mathbf{R}^m$ is linearly independent.

26. Let $\{\mathbf{v}_1, \mathbf{v}_2, \mathbf{v}_3, \mathbf{v}_4\}$ be a linearly independent set in $\mathbf{R}^m$. Show that $\{\mathbf{v}_1, \mathbf{v}_2, \mathbf{v}_3\}$ is also linearly independent.

27. Let **u** and **v** be nonzero orthogonal vectors in $\mathbf{R}^m$. Show that they are linearly independent. (*Hint:* use the fact that $\mathbf{u} \cdot (c\mathbf{u} + d\mathbf{v}) = c\mathbf{u} \cdot \mathbf{u}$ and $\mathbf{v} \cdot (c\mathbf{u} + d\mathbf{v}) = d\mathbf{v} \cdot \mathbf{v}$.)

28. Show that a set $\{\mathbf{u}_1, \mathbf{u}_2, \ldots, \mathbf{u}_n\}$ of mutually orthogonal nonzero vectors in $\mathbf{R}^m$ is linearly independent. (See exercise 27.)

In exercises 29 through 31, using a computer program to transform a matrix to row-reduced echelon form, determine whether or not the following sets of vectors are linearly dependent. If a set is linearly dependent, express one of its vectors as a linear combination of the others.

29. $\{(1, 2, 3, 4, 5), (3, 4, 5, 2, 1), (2, 3, 1, 5, 4), (4, 5, 2, 1, 3), (5, 1, 4, 3, 2)\}$

30. $\{(1, 1, 1, 1, 1, 1), (2, 2, 2, 2, 2, 3), (3, 3, 3, 3, 3, 5), (4, 4, 4, 4, 4, 8)\}$

31. $\mathscr{S} = \{\mathbf{v}_1, \mathbf{v}_2, \mathbf{v}_3, \mathbf{v}_4\}$, where $\mathbf{v}_1 = (0.216, 0.531, 0.870, 1.213)$, $\mathbf{v}_2 = (0.760, 1.432, 1.614, 0.666)$, $\mathbf{v}_3 = (0.228, 0.370, -1.026, -1.760)$, $\mathbf{v}_4 = (-0.050, 0, -0.450, 0)$.

4.2 Subspaces of $\mathbf{R}^m$

As we have seen in chapter 1, lines, planes, and certain other regions in $\mathbf{R}^m$ can be described by considering linear combinations of vectors. In this section we see that these regions are *subspaces* of $\mathbf{R}^m$ formed by the *span* of a set of vectors.

Subspaces of $\mathbf{R}^m$

Definition   A **subspace** of $\mathbf{R}^m$ is a nonempty set, **S**, of vectors satisfying the following two conditions:

   i. if $\mathbf{v}_1$ and $\mathbf{v}_2$ are both in **S**, then $\mathbf{v}_1 + \mathbf{v}_2$ is in **S**,
   ii. if **v** is in **S**, then $c\mathbf{v}$ is in **S** for any scalar c.

Note: If $\mathbf{v}_1$ and $\mathbf{v}_2$ are vectors in a subspace, **S**, and c_1 and c_2 are scalars, then $c_1\mathbf{v}_1 + c_2\mathbf{v}_2$ is also in **S**. This results from the fact that both $c_1\mathbf{v}_1$ and $c_2\mathbf{v}_2$ are in **S** due to condition ii of the definition, and hence their sum is in **S** from condition i. More generally, if $\mathbf{v}_1, \mathbf{v}_2, \ldots, \mathbf{v}_n$ are vectors in a subspace **S**, then every linear combination of the $\mathbf{v}_i$ is in **S** (exercise 21).

   Taking $c = 0$ in condition ii, we see that every subspace must contain the vector **0**. In fact it is easily checked (exercise 4) that $\mathbf{S} = \{\mathbf{0}\}$, the set containing only the zero vector, is itself a subspace of $\mathbf{R}^m$. Moreover, since $\mathbf{R}^m$ is a subset of itself, $\mathbf{R}^m$ is a subspace as well. Thus for each m, $\mathbf{R}^m$ has at least two subspaces, $\{\mathbf{0}\}$ and $\mathbf{R}^m$. Other examples of subspaces are given in examples 4.9 through 4.11.

Example 4.9

Show that the set $\mathbf{S}$ consisting of all 4-vectors whose first and last components are equal to zero is a subspace of $\mathbf{R}^4$.

All vectors in $\mathbf{S}$ are of the form $(0, a, b, 0)$, where a and b are scalars. To verify that $\mathbf{S}$ is a subspace, we check conditions i and ii of the definition.

i. We have $(0, a_1, b_1, 0) + (0, a_2, b_2, 0) = (0, a_1 + a_2, b_1 + b_2, 0)$. Since this is a 4-vector whose first and last components are zero, the sum of two vectors in $\mathbf{S}$ is in $\mathbf{S}$.

ii. We have $c(0, a, b, 0) = (0, ca, cb, 0)$. Thus any scalar multiple of a vector in $\mathbf{S}$ is in $\mathbf{S}$.

Consequently $\mathbf{S}$ is a subspace of $\mathbf{R}^4$.

Example 4.10

Let A be a $m \times n$ matrix. Show that the set, $\mathbf{S}$, of all solutions of the homogeneous linear system $A\mathbf{x} = \mathbf{0}$ is a subspace of $\mathbf{R}^n$.

To verify that $\mathbf{S}$ is a subspace, we check conditions i and ii of the definition.

i. Let $\mathbf{x}_1$ and $\mathbf{x}_2$ be vectors in $\mathbf{R}^n$ that are solutions of $A\mathbf{x} = \mathbf{0}$, that is, $A\mathbf{x}_1 = \mathbf{0}$ and $A\mathbf{x}_2 = \mathbf{0}$. Now $A(\mathbf{x}_1 + \mathbf{x}_2) = A\mathbf{x}_1 + A\mathbf{x}_2 = \mathbf{0} + \mathbf{0} = \mathbf{0}$ (by theorem 1g of section 2.3). So $\mathbf{x}_1 + \mathbf{x}_2$ is a solution of $A\mathbf{x} = \mathbf{0}$.

ii. Let $\mathbf{x}$ be a solution of $A\mathbf{x} = \mathbf{0}$, and let c be a scalar. Then $A(c\mathbf{x}) = c(A\mathbf{x}) = c\mathbf{0} = \mathbf{0}$ (by theorem 1n of section 2.3). Hence $c\mathbf{x}$ is a solution of $A\mathbf{x} = \mathbf{0}$.

Thus $\mathbf{S}$ is a subspace of $\mathbf{R}^n$.

Note: The subspace of example 4.10 is called the **solution space** of the given homogeneous linear system. It is *not* true (exercise 12) that the set of all solutions of a *nonhomogeneous* linear system forms a subspace.

Example 4.11

Show that the set $\mathbf{S}$ of all linear combinations of the vectors $\mathbf{u}_1 = (1, 2, 3)$ and $\mathbf{u}_2 = (1, 0, -1)$ is a subspace of $\mathbf{R}^3$.

A vector is in $\mathbf{S}$ if and only if it is of the form $a\mathbf{u}_1 + b\mathbf{u}_2$ for some scalars a and b. We check conditions i and ii.

i. Let $\mathbf{v}_1$ and $\mathbf{v}_2$ be vectors in $\mathbf{S}$. Then there are scalars a_1, b_1 and

a_2, b_2 so that $\mathbf{v}_1 = a_1\mathbf{u}_1 + b_1\mathbf{u}_2$ and $\mathbf{v}_2 = a_2\mathbf{u}_1 + b_2\mathbf{u}_2$. Now $\mathbf{v}_1 + \mathbf{v}_2 = (a_1 + a_2)\mathbf{u}_1 + (b_1 + b_2)\mathbf{u}_2$, a linear combination of $\mathbf{u}_1$ and $\mathbf{u}_2$, so $\mathbf{v}_1 + \mathbf{v}_2$ is in **S**.

ii. Let **v** be a vector in **S**, $\mathbf{v} = a\mathbf{u}_1 + b\mathbf{u}_2$, and let c be a scalar. Then $c\mathbf{v} = (ca)\mathbf{u}_1 + (cb)\mathbf{u}_2$, a linear combination of $\mathbf{u}_1$ and $\mathbf{u}_2$, so $c\mathbf{v}$ is in **S**.

Thus **S** is a subspace of $\mathbf{R}^3$.

Note: As we have seen in section 1.4, the set of all linear combinations of two noncollinear vectors in $\mathbf{R}^m$ describes a plane. The plane generated in example 4.11 is pictured in figure 4.3. Notice that it contains the origin, as must all subspaces of $\mathbf{R}^m$.

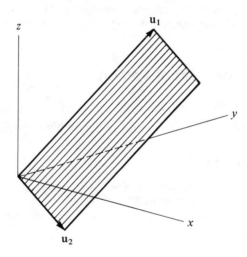

Figure 4.3

In the solution to example 4.11 we made no use of the fact that $\mathbf{u}_1$ and $\mathbf{u}_2$ were specific vectors in $\mathbf{R}^3$. Consequently a similar argument can be used to prove the following theorem.

Theorem 1   Let $\mathcal{T} = \{\mathbf{v}_1, \mathbf{v}_2, \ldots, \mathbf{v}_n\}$ be a set of vectors in $\mathbf{R}^m$. The set of all linear combinations, $c_1\mathbf{v}_1 + c_2\mathbf{v}_2 + \cdots + c_n\mathbf{v}_n$, of the vectors in $\mathcal{T}$ is a subspace of $\mathbf{R}^m$.

Proof   Exercise 22.   ∎

Note: The subspace described in theorem 1 is called the **subspace generated by** (or **spanned by**) $\mathcal{T}$. For example, the subspace of **R**^{3} described in example 4.11 is the *subspace generated by* $(1, 2, 3)$ *and* $(1, 0, -1)$.

We now give an example of a subset of **R**^{m} that is *not* a subspace.

Example 4.12

Show that the set **S** of all vectors of the form $(a, 1)$ is not a subspace of **R**^{2}.

We will show that condition i of the definition does not hold for **S**. To do so, we need only find two vectors in **S** whose sum is not in **S**. Let $v_1 = (2, 1)$ and $v_2 = (3, 1)$. Then v_1 and v_2 are each in **S** (since their second components are equal to 1), but $v_1 + v_2 = (5, 2)$ is not in **S**. (You can also show that condition ii fails to hold.)

Span of a Set of Vectors

Definition   Let **S** be a subspace of **R**^{m}. A subset $\mathcal{T}$ of **S** **spans S** if every vector in **S** can be written as a linear combination of elements of $\mathcal{T}$.

Example 4.13

Show that the set of m distinct elementary vectors in **R**^{m}, $\mathcal{T} = \{e_1, e_2, \ldots, e_m\}$, spans **R**^{m}.

Let $w = (w_1, w_2, \ldots, w_m)$ be an arbitrary vector in **R**^{m}. Then

$$w = w_1 e_1 + w_2 e_2 + \cdots + w_m e_m$$

is a linear combination of the e_i, as desired.

Example 4.14

Let **S** be the subspace of **R**^{4}, consisting of all vectors whose first and last components are zero (see example 4.9). Show that the vectors $v_1 = (0, 1, -1, 0)$ and $v_2 = (0, 2, -1, 0)$ span **S**.

Let $w = (0, a, b, 0)$ be an arbitrary vector in **S**. We must show that there are constants c_1 and c_2 such that $w = c_1 v_1 + c_2 v_2$, that is,

$$(0, a, b, 0) = c_1(0, 1, -1, 0) + c_2(0, 2, -1, 0).$$

Performing the scalar multiplications and additions on the right side and equating corresponding components yields the two equations:

$$c_1 + 2c_2 = a,$$
$$-c_1 - c_2 = b.$$

(Equating the first, as well as the last, components just gives $0 = 0$.) Since this system has the solution $c_1 = -(a + 2b)$, $c_2 = a + b$, $\{\mathbf{v}_1, \mathbf{v}_2\}$ spans S. For example, to express the vector $(0, 3, 7, 0)$ as a linear combination of $\mathbf{v}_1$ and $\mathbf{v}_2$, take $c_1 = -(3 + 2(7)) = -17$ and $c_2 = 3 + 7 = 10$:

$$(0, 3, 7, 0) = -17(0, 1, -1, 0) + 10(0, 2, -1, 0).$$

If S is a subspace of $\mathbf{R}^m$ *generated by a set $\mathcal{T}$* of vectors in S, then by definition $\mathcal{T}$ spans S. The following theorem simplifies the task of determining a subset of $\mathcal{T}$ that also spans S.

Theorem 2   Let S be the subspace of $\mathbf{R}^m$ spanned by the set $\mathcal{T} = \{\mathbf{v}_1, \mathbf{v}_2, \ldots, \mathbf{v}_n\}$. If one of the vectors in $\mathcal{T}$ is a linear combination of the others, then the subset of $\mathcal{T}$ obtained by deleting that vector still spans S.

Proof   For the sake of notational convenience, suppose that $\mathbf{v}_n$ is a linear combination of the other vectors in $\mathcal{T}$, that is, $\mathbf{v}_n = c_1\mathbf{v}_1 + c_2\mathbf{v}_2 + \cdots + c_{n-1}\mathbf{v}_{n-1}$. Let $\mathbf{w}$ be an arbitrary vector in S. Since $\mathcal{T}$ spans S, we have $\mathbf{w} = a_1\mathbf{v}_1 + a_2\mathbf{v}_2 + \cdots + a_n\mathbf{v}_n$ for some scalars $a_1, a_2, \ldots, a_n$. But then

$$\mathbf{w} = a_1\mathbf{v}_1 + a_2\mathbf{v}_2 + \cdots + a_{n-1}\mathbf{v}_{n-1} + a_n(c_1\mathbf{v}_1 + c_2\mathbf{v}_2 + \cdots + c_{n-1}\mathbf{v}_{n-1})$$
$$= (a_1 + a_nc_1)\mathbf{v}_1 + (a_2 + a_nc_2)\mathbf{v}_2 + \cdots + (a_{n-1} + a_nc_{n-1})\mathbf{v}_{n-1}.$$

Thus $\mathbf{w}$ is a linear combination of $\{\mathbf{v}_1, \mathbf{v}_2, \ldots, \mathbf{v}_{n-1}\}$, the subset of $\mathcal{T}$ obtained by deleting $\mathbf{v}_n$, and this subset spans S, as desired.   ∎

Example 4.15

Let S be the subspace of $\mathbf{R}^3$ generated by the set $\mathcal{T} = \{(1, -1, 0), (-2, 1, 2), (-1, -1, 4)\}$. Show that the subset $\mathcal{T}_0 = \{(1, -1, 0), (-2, 1, 2)\}$ spans S.

Applying theorem 2, we show that $(-1, -1, 4)$ is a linear combination of $(1, -1, 0)$ and $(-2, 1, 2)$ by using theorem 5 of section 4.1. We construct the matrix,

$$\begin{bmatrix} 1 & -2 & -1 \\ -1 & 1 & -1 \\ 0 & 2 & 4 \end{bmatrix}$$

and transform it to the row-reduced echelon form,

$$\begin{bmatrix} 1 & 0 & 3 \\ 0 & 1 & 2 \\ 0 & 0 & 0 \end{bmatrix}.$$

Consequently $(-1, -1, 4) = 3(1, -1, 0) + 2(-2, 1, 2)$ (from the third column of the latter matrix). Thus $\mathcal{T}_0$ spans $\mathbf{S}$.

Note: Geometrically we have shown that the three vectors of example 4.15 lie in a plane—the plane generated by $(1, -1, 0)$ and $(-2, 1, 2)$ illustrated in figure 4.4.

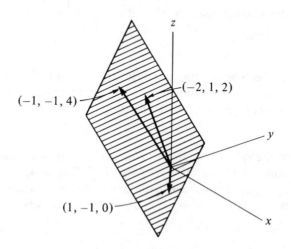

Figure 4.4

4.2 Exercises

In exercises 1 through 6 show that the given sets of vectors are subspaces of $\mathbf{R}^m$.

1. The set of all scalar multiples of the vector $(1, 2, 3)$ (of $\mathbf{R}^3$)

2. The set of all m-vectors whose first component is 0 (of $\mathbf{R}^m$)

3. The set of all linear combinations of the vectors $(1, 0, 1, 0)$ and $(0, 1, 0, 1)$ (of $\mathbf{R}^4$)

4. The set consisting solely of the zero vector, $\{\mathbf{0}\}$ (of $\mathbf{R}^m$)

5. The set of all vectors of the form $(a, b, a - b, a + b)$ (of $\mathbf{R}^4$)

6. The set of all vectors (x, y, z) such that $x + y + z = 0$ (of $\mathbf{R}^3$)

In exercises 7 through 12 show that the given sets of vectors do *not* form subspaces of $\mathbf{R}^m$.

7. The set consisting of the single vector $(1, 1, 1, 1)$ (of $\mathbf{R}^4$)

8. The set of all m-vectors whose first component is 2 (of $\mathbf{R}^m$)

9. The set of all m-vectors *except* the m-vector $\mathbf{0}$ (of $\mathbf{R}^m$)

10. The set of all 4-vectors *except* the vector $(1, 1, 1, 1)$ (of $\mathbf{R}^4$)

11. The set of all 3-vectors the sum of whose components is 1 (of $\mathbf{R}^3$)

12. The set of all solutions of $A\mathbf{x} = \mathbf{b}$, $\mathbf{b} \neq \mathbf{0}$ (of $\mathbf{R}^n$, where A is $m \times n$)

In exercises 13 through 16 determine which sets span $\mathbf{R}^3$.

13. $\{(1, 2, 3), (-1, 0, 1), (0, 1, 2)\}$

14. $\{(-1, 1, 2), (3, 3, 1), (1, 2, 2)\}$

15. $\{(1, 2, -1), (1, 0, 1)\}$

16. $\{(1, 0, 0), (0, 1, 0), (0, 0, 1), (1, 1, 1)\}$

17. Show that $\{\mathbf{e}_2, \mathbf{e}_3\}$ spans the subspace of $\mathbf{R}^4$ consisting of all vectors whose first and last components are zero.

18. Show that $\{(1, 0, -1), (0, 1, -1)\}$ spans the subspace of $\mathbf{R}^3$ consisting of all vectors (x, y, z) such that $x + y + z = 0$.

19. Show that $\{(0, 1, 1)\}$ spans the solution space of $A\mathbf{x} = \mathbf{0}$, where

$$A = \begin{bmatrix} 1 & -1 & 1 \\ 2 & 1 & -1 \\ 0 & -1 & 1 \end{bmatrix}.$$

20. Show that $\{(1, 1, 0), (-2, 0, 1)\}$ spans the solution space of $A\mathbf{x} = \mathbf{0}$, where

$$A = \begin{bmatrix} 1 & -1 & 2 \\ -2 & 2 & -4 \end{bmatrix}.$$

21. Show that, if $\mathbf{v}_1, \mathbf{v}_2, \ldots, \mathbf{v}_n$ are vectors in a subspace, S, of $\mathbf{R}^m$, then $\mathbf{w} = c_1\mathbf{v}_1 + c_2\mathbf{v}_2 + \cdots + c_n\mathbf{v}_n$ is in S for all scalars $c_1, c_2, \ldots, c_n$.

22. Prove theorem 1: let $\mathcal{T} = \{\mathbf{v}_1, \mathbf{v}_2, \ldots, \mathbf{v}_n\}$ be a set of vectors in $\mathbf{R}^m$, and show that the set of all linear combinations of the vectors in $\mathcal{T}$ is a subspace of $\mathbf{R}^m$.

4.3 Basis and Dimension

In section 4.1 we considered linearly independent sets and in section 4.2 sets that spanned a given subspace. Subsets of a subspace that are linearly independent *and* span the subspace are of particular importance in linear algebra. In this section we will investigate the characteristics of such sets.

Basis for a Subspace

Definition Let S be a subspace of $\mathbf{R}^m$. A set $\mathcal{T}$ of vectors in S is a **basis** for S if

i. $\mathcal{T}$ is linearly independent,
ii. $\mathcal{T}$ spans S.

Example 4.16

Show that the set of elementary vectors in $\mathbf{R}^m$, $\mathcal{T} = \{e_1, e_2, \ldots, e_m\}$, is a basis for $\mathbf{R}^m$.

By exercise 25 of section 4.1 $\mathcal{T}$ is linearly independent. Moreover $\mathcal{T}$ spans $\mathbf{R}^m$ (example 4.13). Hence $\mathcal{T}$ is a basis for $\mathbf{R}^m$.

Note: The basis, $\{e_1, e_2, \ldots, e_m\}$ is called the **standard basis** for $\mathbf{R}^m$.

Example 4.17

Show that the set $\mathcal{T} = \{(1, 0, 0), (1, 2, 0), (1, 2, 3)\}$ is a basis for $\mathbf{R}^3$.

We can show that $\mathcal{T}$ is linearly independent and that $\mathcal{T}$ spans $\mathbf{R}^3$ at the same time. Let $\mathbf{w} = (w_1, w_2, w_3)$ be an arbitrary vector in $\mathbf{R}^3$. Then $\mathcal{T}$ spans $\mathbf{R}^3$ if there exist scalars c_1, c_2, c_3 such that

$$c_1(1, 0, 0) + c_2(1, 2, 0) + c_3(1, 2, 3) = (w_1, w_2, w_3).$$

Performing the scalar multiplications and equating corresponding components yields the linear system

$$
\begin{aligned}
c_1 + c_2 + c_3 &= w_1 \\
2c_2 + 2c_3 &= w_2 \\
3c_3 &= w_3,
\end{aligned}
$$

which has the coefficient matrix

$$
A = \begin{bmatrix} 1 & 1 & 1 \\ 0 & 2 & 2 \\ 0 & 0 & 3 \end{bmatrix}.
$$

By theorem 6 of section 2.5 there is a unique solution to this system if the row-reduced echelon form of A is the identity matrix I. Performing the row-reduction we see that this is indeed the case, so $\mathcal{T}$ spans $\mathbf{R}^3$. Moreover, in doing the row-reduction and obtaining the identity matrix, we have also shown that $\mathcal{T}$ is linearly independent (theorem 4 of section 4.1). Consequently $\mathcal{T}$ is a basis for $\mathbf{R}^3$.

Now suppose that we wish to find a basis for the subspace **S** generated by a set, $\mathcal{T}$, of vectors in **S**. By definition $\mathcal{T}$ spans **S**, but $\mathcal{T}$ is not necessarily linearly independent. The following theorem and accompanying procedure show us how to *reduce a spanning set to a basis*.

Theorem 1   Let **S** be the subspace of $\mathbf{R}^m$ generated by the set $\mathcal{T} = \{v_1, v_2, \ldots, v_n\}$. Then there is a subset of $\mathcal{T}$ that is a basis for **S**.

Procedure for reducing a spanning set to a basis   Let A be the $m \times n$ matrix with columns $A^i = \mathbf{v}_i$, and let B be its row-reduced echelon form. If $\mathbf{w}_1, \mathbf{w}_2, \ldots, \mathbf{w}_s$ are the vectors in $\mathcal{T}$ that are transformed into the distinct elementary columns of B, then $\mathcal{T}_0 = \{\mathbf{w}_1, \mathbf{w}_2, \ldots, \mathbf{w}_s\}$ is a basis for **S**.

Proof   By theorem 4 of section 4.1 $\mathcal{T}_0$ is linearly independent, and by theorem 5 of that section the other vectors in $\mathcal{T}$ are linear combinations of those in $\mathcal{T}_0$. But then by theorem 2 of section 4.2, $\mathcal{T}_0$ spans **S** as well. Hence $\mathcal{T}_0$ is a basis for **S**.   ■

Example 4.18

Find a basis for the subspace, **S**, of $\mathbf{R}^4$ generated by $\mathcal{T} = \{\mathbf{v}_1, \mathbf{v}_2, \mathbf{v}_3, \mathbf{v}_4, \mathbf{v}_5\}$, where $\mathbf{v}_1 = (1, 0, -1, 1)$, $\mathbf{v}_2 = (-2, 0, 2, -2)$, $\mathbf{v}_3 = (1, 1, 1, 1)$, $\mathbf{v}_4 = (1, -1, -3, 1)$, and $\mathbf{v}_5 = (1, -1, 1, -1)$.

We apply the procedure, constructing the matrix A with $A^i = \mathbf{v}_i$,

$$A = \begin{bmatrix} 1 & -2 & 1 & 1 & 1 \\ 0 & 0 & 1 & -1 & -1 \\ -1 & 2 & 1 & -3 & 1 \\ 1 & -2 & 1 & 1 & -1 \end{bmatrix},$$

which has the row-reduced echelon form

$$B = \begin{bmatrix} 1 & -2 & 0 & 2 & 0 \\ 0 & 0 & 1 & -1 & 0 \\ 0 & 0 & 0 & 0 & 1 \\ 0 & 0 & 0 & 0 & 0 \end{bmatrix}.$$

We see that $\mathbf{v}_1$, $\mathbf{v}_3$, and $\mathbf{v}_5$ have been transformed into the distinct elementary vectors $\mathbf{e}_1$, $\mathbf{e}_2$, and $\mathbf{e}_3$ ($\mathbf{v}_1$, $\mathbf{v}_3$, and $\mathbf{v}_5$ are the $\mathbf{w}_1$, $\mathbf{w}_2$, and $\mathbf{w}_3$ of the procedure), so $\{\mathbf{v}_1, \mathbf{v}_3, \mathbf{v}_5\}$ is a basis for **S**.

In theorem 1 we learned how to *reduce a spanning set to a basis*. This raises the question, "Is it possible to *extend a linearly independent set to a basis*?" More precisely, if $\mathcal{T}$ is a linearly independent subset of a subspace **S**, "Are there vectors in **S** that can be added to the set $\mathcal{T}$ to obtain a basis for **S**?" The answer is, "Yes," and, if we have a spanning set for **S**, there is a straightforward computational technique to construct this basis. The next theorem and accompanying procedure describe the process.

Theorem 2   Let S be the subspace of $\mathbf{R}^m$ generated by the set $\mathcal{T} = \{v_1, v_2, \ldots, v_n\}$, and let $\mathcal{T}_1 = \{u_1, u_2, \ldots, u_s\}$ be a linearly independent subset of S. Then there is a subset $\mathcal{T}_2$ of $\mathcal{T}$ such that the union of $\mathcal{T}_1$ and $\mathcal{T}_2$ is a basis for S.

Procedure for extending an independent set to a basis   Let A be the $m \times (s + n)$ matrix, with columns $A^i = u_i$ for $i = 1, \ldots, s$ and $A^{s+j} = v_j$ for $j = 1, \ldots, n$. In other words, the first s columns of A are the vectors in $\mathcal{T}_1$, and the last n columns are the vectors in $\mathcal{T}$. Let B be the row-reduced echelon matrix for A, and let $w_1, w_2, \ldots, w_t$ be those vectors in $\mathcal{T}$ that are transformed into distinct elementary columns of B other than $e_1, e_2, \ldots, e_s$. Then $\mathcal{T}_2 = \{w_1, w_2, \ldots, w_t\}$.

Proof   By the procedure of theorem 1 the columns of A that correspond to the distinct elementary columns of B form a basis for S. Since the first s columns of A are linearly independent, they are transformed into e_1, $\ldots, e_s$, the first s columns of B. Thus each element of $\mathcal{T}_1$ is in the basis. Let $\mathcal{T}_2$ be the set of other basis vectors. Then $\mathcal{T}_2$ is a subset of $\mathcal{T}$, since each element comes from the last n columns of A. So the union of $\mathcal{T}_1$ and $\mathcal{T}_2$ is just the set of all basis elements.   ∎

Example 4.19

Let $\mathcal{T}_1 = \{v_1, v_2\}$ be the set of linearly independent vectors in $\mathbf{R}^4$, with $v_1 = (1, 0, -1, 1)$ and $v_2 = (1, 1, 1, -1)$. Extend this set to form a basis for $\mathbf{R}^4$.

Since $\mathbf{R}^4$ is spanned by the set $\mathcal{T} = \{e_1, e_2, e_3, e_4\}$, we may find a subset $\mathcal{T}_2$ of this set $\mathcal{T}$ such that the union of $\mathcal{T}_1$ and $\mathcal{T}_2$ is a basis for $\mathbf{R}^4$. Following the procedure of theorem 2, we construct the matrix

$$\left[\begin{array}{rr|rrrr} 1 & 1 & 1 & 0 & 0 & 0 \\ 0 & 1 & 0 & 1 & 0 & 0 \\ -1 & 1 & 0 & 0 & 1 & 0 \\ 1 & -1 & 0 & 0 & 0 & 1 \end{array}\right]$$

and transform it to row-reduced echelon form

$$\left[\begin{array}{rr|rrrr} 1 & 0 & 0 & 1 & 0 & 1 \\ 0 & 1 & 0 & 1 & 0 & 0 \\ 0 & 0 & 1 & -2 & 0 & -1 \\ 0 & 0 & 0 & 0 & 1 & 1 \end{array}\right].$$

We see that e_1 and e_3 of the set $\mathcal{T}$ have been transformed into distinct elementary columns of the latter matrix, so the set $\{v_1, v_2, e_1, e_3\}$ is a basis for $\mathbf{R}^4$.

We will now state an important theorem concerning bases that will be proved in chapter 7.

Theorem 3   Let S be a subspace of $\mathbf{R}^m$. Then any basis for S contains the same number of vectors as any other basis.

For example, we have shown (example 4.16) that there is a basis for $\mathbf{R}^m$ containing m vectors. As a result theorem 3 tells us that *every* basis for $\mathbf{R}^m$ must contain m vectors.

Dimension of a Subspace

Definition   Let S be a subspace of $\mathbf{R}^m$. The **dimension** of S is equal to the number of vectors in any basis for S.

Note: If we find one basis for a subspace that contains n vectors, and then determine another basis for the same subspace, theorem 3 guarantees that it too will contain n vectors. Consequently there is no problem with the definition—the concept of dimension is "well defined."

Example 4.20

Find the dimension of $\mathbf{R}^2$, $\mathbf{R}^3$, and $\mathbf{R}^m$.
  Bases for these subspaces are given by $\{e_1, e_2\}$, $\{e_1, e_2, e_3\}$, and $\{e_1, e_2, \ldots, e_m\}$, respectively. Thus $\mathbf{R}^2$ has dimension 2, $\mathbf{R}^3$ has dimension 3, and in general $\mathbf{R}^m$ has dimension m.

Example 4.21

Find the dimension of the subspace S of $\mathbf{R}^4$ consisting of all vectors whose first and last components are equal to 0 (see example 4.9).
  In example 4.14 we showed that the vectors $v_1 = (0, 1, -1, 0)$ and $v_2 = (0, 2, -1, 0)$ span S. By transforming the matrix

$$\begin{bmatrix} 0 & 0 \\ 1 & 2 \\ -1 & -1 \\ 0 & 0 \end{bmatrix}$$

to row-reduced echelon form, we can show that they are also linearly independent. Hence $\{v_1, v_2\}$ is a basis for S. Thus S has dimension equal to 2.

One of the more important subspaces of $\mathbf{R}^m$ is the *solution space* of a homogeneous linear system—the set of all solutions of $A\mathbf{x} = \mathbf{0}$ for a given matrix A (see example 4.10). The next theorem and its accompanying procedure describe how we can find a basis for, and hence the dimension of, these subspaces. This is a process we will use several times in the following chapters.

Theorem 4 Let A be an $m \times n$ matrix, and let B be its row-reduced echelon form. Let the rank of A (the number of nonzero rows of B) be equal to r. Then the dimension of the solution space of $A\mathbf{x} = \mathbf{0}$ is $n - r$.

Before we prove theorem 4, we will present a procedure for determining a *basis* of $n - r$ vectors for the solution space and an example illustrating this procedure. The proof of the theorem will simply be a verification that the set given in the procedure is in fact a basis.

Procedure for finding a basis for the solution space of $A\mathbf{x} = \mathbf{0}$

   i. Transform A to its row-reduced echelon form B.
  ii. Interpret B as a system of linear equations (with right side constants all 0), and solve for the variables that correspond to the leading 1's.
 iii. Parameterize the other variables, and express the result so that the entries in a column are coefficients of the same parameter.
 iv. The columns of coefficients form a basis for the solution space of $A\mathbf{x} = \mathbf{0}$.

Example 4.22

Find a basis for and the dimension of the solution space of the homogeneous linear system $A\mathbf{x} = \mathbf{0}$, where

$$A = \begin{bmatrix} 1 & 1 & 5 & 0 & 1 \\ 1 & -1 & 1 & -2 & -1 \\ 1 & 1 & 5 & 1 & 1 \\ 0 & 2 & 4 & 3 & 2 \\ 0 & 1 & 2 & 2 & 1 \end{bmatrix}.$$

Following the procedure just given, we first transform A to its row-reduced echelon form

$$B = \begin{bmatrix} 1 & 0 & 3 & 0 & 0 \\ 0 & 1 & 2 & 0 & 1 \\ 0 & 0 & 0 & 1 & 0 \\ 0 & 0 & 0 & 0 & 0 \\ 0 & 0 & 0 & 0 & 0 \end{bmatrix}.$$

The rank of A is 3, so the dimension of the solution space is $5 - 3 = 2$. Translating B into a system of equations, and solving for x_1, x_2, and x_4 (the variables that correspond to the leading 1's of B), we obtain

$$x_1 = -3x_3,$$
$$x_2 = -2x_3 - x_5,$$
$$x_4 = 0.$$

Next we set $x_3 = t_1$ and $x_5 = t_2$, and write the right-side terms in two columns to get

$$x_1 = -3t_1 \qquad\quad = -3t_1 + 0t_2$$
$$x_2 = -2t_1 - t_2 = -2t_1 - 1t_2$$
$$x_3 = \quad t_1 \qquad\quad = \quad 1t_1 + 0t_2$$
$$x_4 = \quad 0 \qquad\quad = \quad 0t_1 + 0t_2$$
$$x_5 = \qquad\quad t_2 = \quad 0t_1 + 1t_2.$$

Hence the set $\{v_1, v_2\}$, with $v_1 = (-3, -2, 1, 0, 0)$ and $v_2 = (0, -1, 0, 0, 1)$ (the coefficients of the two columns), is a basis for the solution space.

We now prove theorem 4.

Proof of Theorem 4 By theorem 5 of section 2.5 the solution set of $Ax = 0$ contains $(n - r)$ parameters. Recalling the setting (the system of equations 1 in section 2.5 with $d_1 = d_2 = \cdots = d_r = 0$), we assume that the first r columns of A are the ones transformed into a set of distinct elementary columns of B. This gives the system of equations

$$x_1 \qquad\quad + b_{1,r+1}x_{r+1} + \cdots + b_{1,n}x_n = 0$$
$$\quad x_2 \qquad + b_{2,r+1}x_{r+1} + \cdots + b_{2,n}x_n = 0$$
$$\qquad\quad x_r + b_{r,r+1}x_{r+1} + \cdots + b_{r,n}x_n = 0.$$

Solving for $x_1, \ldots, x_r$, and setting $x_{r+1} = t_1$, $x_{r+2} = t_2, \ldots, x_n = t_{n-r}$, we obtain the solution

$$
\begin{aligned}
x_1 \;\;&= -b_{1,r+1}t_1 - b_{1,r+2}t_2 - \cdots - b_{1,n}t_{n-r} \\
x_2 \;\;&= -b_{2,r+1}t_1 - b_{2,r+2}t_2 - \cdots - b_{2,n}t_{n-r} \\
&\;\;\vdots \\
x_r \;\;&= -b_{r,r+1}t_1 - b_{r,r+2}t_2 - \cdots - b_{r,n}t_{n-r} \\
x_{r+1} &= \qquad 1t_1 + \qquad 0t_2 + \cdots + \qquad 0t_{n-r} \\
x_{r+2} &= \qquad 0t_1 + \qquad 1t_2 + \cdots + \qquad 0t_{n-r} \\
&\;\;\vdots \\
x_n \;\;&= \qquad 0t_1 + \qquad 0t_2 + \cdots + \qquad 1t_{n-r}.
\end{aligned}
\tag{1}
$$

Letting $\mathbf{v}_i$ be the column of coefficients of the parameter t_i for each i, we may write the solution given by system (1) in vector form as

$$\mathbf{x} = t_1\mathbf{v}_1 + t_2\mathbf{v}_2 + \cdots + t_{n-r}\mathbf{v}_{n-r} \tag{2}$$

Each vector $\mathbf{v}_i$ is in the solution space (let $t_i = 1$ and $t_j = 0$ for $j \neq i$) and, since any solution $\mathbf{x}$ can be written as a linear combination of $\mathscr{T} = \{\mathbf{v}_1, \ldots, \mathbf{v}_{n-r}\}$, $\mathscr{T}$ spans the solution space. Moreover $\mathscr{T}$ is linearly independent. To see this, note that each vector $\mathbf{v}_i$ has 1 as its $(r + i)$th component and 0 for each of the other last $(n - r)$ components. Set $\mathbf{0}$ equal to a linear combination of the vectors in $\mathscr{T}$, say,

$$\mathbf{0} = c_1\mathbf{v}_1 + c_2\mathbf{v}_2 + \cdots + c_{n-r}\mathbf{v}_{n-r}.$$

Looking only at the $(r + i)$th component, we have

$$0 = c_1(0) + \cdots + c_{i-1}(0) + c_i(1) + c_{i+1}(0) + \cdots + c_{n-r}(0)$$

from which $0 = c_i$. Since this holds for each i, $\mathscr{T}$ is linearly independent. Thus $\mathscr{T}$ spans the solution space and is linearly independent, so it is a basis. ∎

The following theorem gives some useful relationships among the notions of *linear dependence, span, basis,* and *dimension.* We will only prove part i, but the proofs of the other parts are just as easy.

Theorem 5   Let S be a subspace of $\mathbf{R}^m$ whose dimension is n. Let $\mathscr{T} = \{\mathbf{v}_1, \ldots, \mathbf{v}_s\}$ be a subset of S.

 i. If $s > n$, then $\mathscr{T}$ is linearly dependent.
 ii. If $s < n$, then $\mathscr{T}$ does not span S.
 iii. If $s = n$, and $\mathscr{T}$ is linearly independent, then it is a basis for S.
 iv. If $s = n$, and $\mathscr{T}$ spans S, then it is a basis for S.

Proof of i   Suppose $s > n$ and $\mathscr{T}$ is linearly independent. By theorem 2, $\mathscr{T}$ can be extended to a basis for S. But this gives a basis for S with more than n vectors, which contradicts theorem 3. Hence $\mathscr{T}$ must be linearly dependent. ∎

This theorem can sometimes be used to simplify the task of answering questions concerning linear dependence, span, or basis.

Example 4.23

   a. Do the vectors $(0, 1, 2, 3)$, $(1, 3, 2, 6)$, $(-1, 4, 0, 1)$ span $\mathbf{R}^4$?

   Apply part ii of theorem 5. We are given 3 vectors in $\mathbf{R}^4$, which has dimension 4 (example 4.20). Consequently the given vectors do not span $\mathbf{R}^4$.

b. Show that the set $\{(0, 1, 2, 0), (0, 1, 0, 0), (0, 1, 1, 0)\}$ is linearly dependent.

The given vectors all lie in the subspace $\mathbf{S}$ of $\mathbf{R}^4$, consisting of all vectors with first and last components equal to zero. By example 4.21 the dimension of $\mathbf{S}$ is 2. Consequently by part i of theorem 5 the given set is linearly dependent.

c. Show that the set $\{(1, 0, 0), (1, 2, 0), (1, 2, 3)\}$ is a basis for $\mathbf{R}^3$.

By part iii of theorem 5 we need only show that the given set of three vectors is linearly independent (since the dimension of $\mathbf{R}^3$ is three). Let A be the 3×3 matrix with columns formed by the given vectors. Since det $A = 6 \neq 0$, the set is linearly independent by theorem 3 of section 4.1.

The solution technique of part c of example 4.23 is important enough to state as a theorem.

Theorem 6 Let $\mathcal{T} = \{v_1, v_2, \ldots, v_m\}$ be a set of vectors in $\mathbf{R}^m$. Then $\mathcal{T}$ is a basis for $\mathbf{R}^m$ if and only if det $A \neq 0$, where A is the $m \times m$ matrix with $A^i = v_i$.

Proof By theorem 3 of section 4.1 $\mathcal{T}$ is linearly independent if and only if det $A \neq 0$. But since $\mathbf{R}^m$ has dimension m, by part iii of theorem 5, $\mathcal{T}$ is a basis for $\mathbf{R}^m$ if and only if it is linearly independent. Consequently $\mathcal{T}$ is a basis for $\mathbf{R}^m$ if and only if det $A \neq 0$.

Example 4.24

Determine whether or not the set $\mathcal{T} = \{v_1, v_2, v_3, v_4\}$ is a basis for $\mathbf{R}^4$, where $v_1 = (1, 0, 0, 0)$, $v_2 = (-1, 1, 3, 2)$, $v_3 = (1, 2, 3, 1)$, and $v_4 = (1, 1, 2, 1)$.

Here we have four vectors in $\mathbf{R}^4$, so we can apply theorem 6. To do so, we construct the matrix A with $A^i = v_i$,

$$A = \begin{bmatrix} 1 & -1 & 1 & 1 \\ 0 & 1 & 2 & 1 \\ 0 & 3 & 3 & 2 \\ 0 & 2 & 1 & 1 \end{bmatrix}.$$

Since det $A = 0$, $\mathcal{T}$ is not a basis for $\mathbf{R}^4$.

4.3 Exercises

In exercises 1 through 6 determine whether or not the given set forms a basis for the indicated subspace.

1. $\{(1, 2, 3), (-1, 0, 1), (0, 1, 2)\}$ for $\mathbf{R}^3$

2. $\{(-1, 1, 2), (3, 3, 1), (1, 2, 2)\}$ for $\mathbf{R}^3$

3. $\{(1, -1, 0), (0, 1, -1)\}$ for the subspace of $\mathbf{R}^3$ of all (x, y, z) such that $x + y + z = 0$

4. $\{(1, 2, -1, 3), (0, 0, 0, 0)\}$ for the subspace of $\mathbf{R}^4$ of all vectors of the form $(a, b, a - b, a + b)$

5. $\{(1, -1), (-2, 3)\}$ for the subspace of $\mathbf{R}^2$ generated by the set $(1, 2)$ and $(2, 1)$

6. $\{(2, 0, 2), (1, 1, 1)\}$ for the subspace of $\mathbf{R}^3$ generated by $(1, 0, 1)$ and $(0, 1, 0)$

In exercises 7 through 10 find a basis for the subspace generated by the given set of vectors.

7. $\{(1, -1, 1), (2, 0, 1), (1, 1, 0)\}$

8. $\{(1, 0, 1, 0), (0, -1, 1, 2), (2, -1, 3, 2), (1, 1, 1, 1)\}$

9. $\{(-1, 1, 2, -1), (1, 0, -1, 1), (-1, 2, 3, -1), (1, 1, 0, 1)\}$

10. $\{(1, -1, 0), (0, 1, 1), (1, 0, 2)\}$

In exercises 11 through 14 extend the given set to form a basis for $\mathbf{R}^m$.

11. $\{(1, 1, 0), (1, -1, 1)\}$ for $\mathbf{R}^3$

12. $\{(-1, 1, 0, 0), (1, -1, 1, 0), (0, 0, 1, 2)\}$ for $\mathbf{R}^4$

13. $\{(1, 0, 0, 1), (0, 1, 1, 0), (1, 1, 1, 2)\}$ for $\mathbf{R}^4$

14. $\{(-1, -1, 2, 1), (2, 1, -1, -2), (0, -1, 4, 0)\}$ for $\mathbf{R}^4$

In exercises 15 through 18 explain why the given statement is true "by inspection."

15. The set $\{(1, 0, 3), (-1, 1, 0), (1, 2, 4), (0, -1, -2)\}$ is linearly dependent.

16. The set $\{(1, -1, 2), (0, 1, 1)\}$ does not span $\mathbf{R}^3$.

17. If the set $\{\mathbf{v}_1, \mathbf{v}_2, \mathbf{v}_3, \mathbf{v}_4\}$ of vectors in $\mathbf{R}^4$ is linearly independent, then it spans $\mathbf{R}^4$.

18. The set $\{(0, 1, -1, 0), (0, -1, 2, 0)\}$ is linearly independent, and so it spans the subspace of $\mathbf{R}^4$ of all vectors of the form $(0, a, b, 0)$.

In exercises 19 through 24 find a basis for and the dimension of the indicated subspace.

19. The solution space of the homogeneous linear system

$$\begin{array}{rcl} x_1 - 2x_2 + x_3 & = & 0 \\ x_2 - x_3 + x_4 & = & 0 \\ x_1 - x_2 + x_4 & = & 0 \end{array}$$

20. The solution space of

$$\begin{array}{rcl} x_1 - 3x_2 + x_3 - x_5 & = & 0 \\ x_1 - 2x_2 + x_3 - x_4 & = & 0 \\ x_1 - x_2 + x_3 - 2x_4 + x_5 & = & 0 \end{array}$$

21. The subspace of $\mathbf{R}^3$ of all vectors of the form $(x, 2x, z)$

22. The subspace of $\mathbf{R}^4$ of all vectors of the form $(0, x, y, x - y)$

23. The subspace of $\mathbf{R}^4$ generated by the set

$$\{(1, -1, 3, 1), (-2, 0, -2, 0), (0, -1, 2, 1)\}$$

24. The subspace of $\mathbf{R}^3$ generated by the set

$$\{(1, -1, 0), (0, 1, 1), (2, -1, 1), (1, 0, 1)\}$$

25. Let $\mathbf{v}$ be a nonzero vector in $\mathbf{R}^m$. Find the dimension of the subspace S of $\mathbf{R}^m$ consisting of all scalar multiples of $\mathbf{v}$. (Geometrically S is the line through the origin containing $\mathbf{v}$.)

26. Let $\mathbf{v}_1$ and $\mathbf{v}_2$ be two noncollinear vectors in $\mathbf{R}^m$. Find the dimension of the subspace S of $\mathbf{R}^m$ generated by $\mathbf{v}_1$ and $\mathbf{v}_2$. (Geometrically S is the plane containing the given vectors.)

27. Let $\{\mathbf{v}_1, \mathbf{v}_2, \ldots, \mathbf{v}_n\}$ be a basis for a subspace S of $\mathbf{R}^m$. Show that every vector $\mathbf{x}$ in S may be written as a *unique* linear combination of the $\mathbf{v}_i$.

4.4   Rank of a Matrix

In section 2.5 the notion of the *rank* of a matrix was defined as the number of nonzero rows in its row-reduced echelon form. In this section we will introduce the concepts of *column rank* and *row rank* of a matrix and show that these three types of rank are in fact one and the same.

Definition   Let A be an $m \times n$ matrix. The subspace of $\mathbf{R}^m$ generated by the columns of the matrix, $A^1, A^2, \ldots, A^n$, is called the **column space** of A. Its dimension is called the **column rank** of A.

Example 4.25

Let A be given by

$$A = \begin{bmatrix} 1 & -1 & 3 & -1 & 2 \\ 2 & 2 & 1 & -1 & 1 \\ 1 & 0 & 1 & -15 & -1 \\ 1 & 0 & 2 & 4 & 2 \end{bmatrix}.$$

Describe the column space of A, and find a basis for it. Determine the column rank of A.

The column space of A is the subspace of $\mathbf{R}^4$ generated by the vectors $\mathbf{v}_1 = (1, 2, 1, 1)$, $\mathbf{v}_2 = (-1, 2, 0, 0)$, $\mathbf{v}_3 = (3, 1, 1, 2)$, $\mathbf{v}_4 = (-1, -1, -15, 4)$, and $\mathbf{v}_5 = (2, 1, -1, 2)$. To find a basis for this subspace, we follow the procedure of theorem 1 of section 4.3 and transform A into the row-reduced echelon form

$$B = \begin{bmatrix} 1 & 0 & 0 & -34 & 4 \\ 0 & 1 & 0 & 24 & 3 \\ 0 & 0 & 1 & 19 & 3 \\ 0 & 0 & 0 & 0 & 0 \end{bmatrix}.$$

Since the vectors $\mathbf{v}_1$, $\mathbf{v}_2$, and $\mathbf{v}_3$ have been transformed into distinct elementary vectors, the set $\{\mathbf{v}_1, \mathbf{v}_2, \mathbf{v}_3\}$ is a basis for the column space of A. Moreover the dimension (the number of vectors in a basis) of the latter is 3, and so the column rank of A is 3.

Notice that for the matrix of example 4.25 the column rank and the rank (the number of nonzero rows in the row-reduced echelon form) are equal. The following theorem states that this is no coincidence.

Theorem 1   The dimension of the column space of a matrix A is equal to the rank of A.

Proof   Let B be the row-reduced echelon form of A. According to procedure of theorem 1 of section 4.3 we need only show that the number of distinct elementary columns of B is equal to the number of nonzero rows of B. The last statement follows immediately from the definition of row-reduced echelon form. The number of nonzero rows is the same as the number of *leading ones*, and the number of leading ones is in turn the same as the number of distinct elementary columns.   ∎

By considering the rows of a matrix instead of its columns, we can develop a theorem similar to the one just given.

Definition   Let A be an $m \times n$ matrix. The subspace of $\mathbf{R}^n$ generated by the rows of the matrix $A_1, A_2, \ldots, A_m$ is called the **row space** of A. Its dimension is called the **row rank** of A.

Example 4.26

As in example 4.25, let

$$A = \begin{bmatrix} 1 & -1 & 3 & -1 & 2 \\ 2 & 2 & 1 & -1 & 1 \\ 1 & 0 & 1 & -15 & -1 \\ 1 & 0 & 2 & 4 & 2 \end{bmatrix}.$$

Describe the row space of A, and find a basis for it. Determine the row rank of A.

The row space of A is the subspace of $\mathbf{R}^5$ generated by the vectors $\mathbf{w}_1 = (1, -1, 3, -1, 2)$, $\mathbf{w}_2 = (2, 2, 1, -1, 1)$, $\mathbf{w}_3 = (1, 0, 1, -15, -1)$, and $\mathbf{w}_4 = (1, 0, 2, 4, 2)$. To find a basis for it, we again apply theorem 1 of section 4.3. We must first form the matrix whose ith column is $\mathbf{w}_i$. This matrix is simply the transpose of A,

$$A^T = \begin{bmatrix} 1 & 2 & 1 & 1 \\ -1 & 2 & 0 & 0 \\ 3 & 1 & 1 & 2 \\ -1 & -1 & -15 & 4 \\ 2 & 1 & -1 & 2 \end{bmatrix},$$

which has the row-reduced echelon form

$$B = \begin{bmatrix} 1 & 0 & 0 & 2/3 \\ 0 & 1 & 0 & 1/3 \\ 0 & 0 & 1 & -1/3 \\ 0 & 0 & 0 & 0 \\ 0 & 0 & 0 & 0 \end{bmatrix}.$$

Since $\mathbf{w}_1$, $\mathbf{w}_2$, and $\mathbf{w}_3$ are transformed into $\mathbf{e}_1$, $\mathbf{e}_2$, and $\mathbf{e}_3$, the set $\{\mathbf{w}_1, \mathbf{w}_2, \mathbf{w}_3\}$ is a basis for the row space of A, and so the dimension of the latter is 3. Thus the row rank of A is 3.

In example 4.25 we found that the *column rank* of this matrix A is also 3. It is in fact always true that the row rank of a given matrix is the same as its column rank. As an intermediate step toward obtaining this result, we prove the following theorem.

Theorem 2   Let A be an $m \times n$ matrix, and suppose that C is the matrix obtained from A by performing an elementary row operation on the latter. Then the row space of A is the same as the row space of C.

Proof   Let the rows of A be $A_1, A_2, \ldots, A_m$, and let those of C be $C_1, C_2, \ldots, C_m$. Then the row space of A is the subspace $\mathbf{S}_1$ of $\mathbf{R}^n$ generated by $\mathcal{T}_1 = \{A_1, A_2, \ldots, A_m\}$, and the row space of C is the subspace $\mathbf{S}_2$ generated by $\mathcal{T}_2 = \{C_1, C_2, \ldots, C_m\}$. We must show that under each of the three elementary row operations $\mathbf{S}_1 = \mathbf{S}_2$.

i. Suppose C is obtained from A by interchanging two rows. Then $\mathcal{T}_1 = \mathcal{T}_2$, and hence $\mathbf{S}_1 = \mathbf{S}_2$.

ii. Suppose C is obtained from A by multiplication of row i by the non-zero scalar k. Then

$$T_2 = \{A_1, \ldots, A_{i-1}, kA_i, A_{i+1}, \ldots, A_m\}.$$

If $\mathbf{v}$ is in the row space of C, $\mathbf{S}_2$, then there exist scalars $c_1, c_2 \ldots, c_m$ such that

$$\mathbf{v} = c_1 A_1 + \cdots + c_{i-1}A_{i-1} + c_i(kA_i) + c_{i+1}A_{i+1} + \cdots + c_m A_m$$
$$= c_1 A_1 + \cdots + c_{i-1}A_{i-1} + (c_i k)A_i + c_{i+1}A_{i+1} + \cdots + c_m A_m,$$

a linear combination of $\mathcal{T}_1$. Thus $\mathbf{v}$ lies in the row space of A, $\mathbf{S}_1$. Similarly we can show that, if $\mathbf{w}$ is in $\mathbf{S}_1$, it must also be in $\mathbf{S}_2$. Hence $\mathbf{S}_1 = \mathbf{S}_2$.

iii. Suppose C is obtained from A by the addition of k times row j to row i. Then

$$\mathcal{T}_2 = \{A_1, \ldots, A_{i-1}, kA_j + A_i, A_{i+1}, \ldots, A_m\}.$$

By an argument similar to the one in part ii we can show that in this case as well $\mathbf{S}_1 = \mathbf{S}_2$.   ∎

Note: By applying theorem 2 successively a finite number of times, we see that any finite sequence of row operations on a matrix A results in a matrix C that has the same row space as A. In other words, any two matrices that are *row equivalent* have the same row space.

Example 4.27

Show that the matrices

$$A = \begin{bmatrix} 5 & 3 & 4 \\ 2 & 2 & -4 \\ 3 & 2 & 1 \\ -1 & 2 & 1 \end{bmatrix} \quad \text{and} \quad C = \begin{bmatrix} 3 & 1 & 8 \\ 1 & 4 & -3 \\ -2 & -1 & -3 \\ 1 & 1 & -2 \end{bmatrix}$$

have the same row space and find a basis for this subspace of $\mathbf{R}^3$.

By performing the computations, we see that both A and C have the row-reduced echelon form

$$B = \begin{bmatrix} 1 & 0 & 0 \\ 0 & 1 & 0 \\ 0 & 0 & 1 \\ 0 & 0 & 0 \end{bmatrix}.$$

Thus A and C are row equivalent to the same matrix, so they are row equivalent to one another and have the same row space.

This row space is also that of B—the one generated by $\{(1, 0, 0), (0, 1, 0), (0, 0, 1), (0, 0, 0)\}$. A basis for this subspace is clearly $\{(1, 0, 0), (0, 1, 0), (0, 0, 1)\}$.

Note: Examples 4.26 and 4.27 demonstrate two essentially different methods of producing bases for the row space of a matrix. The first provides a basis that consists of a *subset* of the given rows. The second provides a basis that ordinarily is *not* a subset of the original rows. However, the elements of the latter are in some sense simpler than those of the former.

We will now give the main result of this section.

Theorem 3   Let A be a matrix. The row rank, the column rank, and the rank of A are all equal.

Proof   By theorem 1 the column rank of A is equal to the rank of A. Thus it suffices to show that the row rank of A is equal to the rank of A.

Let B be the row-reduced echelon form of A. According to theorem 2 and the definition of rank, we need only show that the nonzero rows of B form a basis for the row space of B. Since these nonzero rows span the row space, it suffices to prove that the nonzero rows of B form a linearly independent set. We leave the verification of this last statement as exercise 13.   ∎

4.4 Exercises

In exercises 1 through 4 verify in each case that the row rank is equal to the column rank by explicitly finding the dimensions of the row space and the column space of the given matrix.

1. $\begin{bmatrix} 1 & 2 & 0 \\ 0 & 0 & 1 \\ 0 & 0 & 0 \end{bmatrix}$

2. $\begin{bmatrix} 1 & 2 & 1 \\ 2 & 1 & -1 \end{bmatrix}$

3. $\begin{bmatrix} 1 & -1 & 3 \\ 0 & 1 & 1 \\ 1 & 1 & 0 \\ 2 & -1 & 1 \end{bmatrix}$

4. $\begin{bmatrix} 1 & -1 & 1 & 0 \\ 1 & 1 & 0 & 0 \\ 1 & 0 & 0 & 1 \end{bmatrix}$

5–8. Find a basis for the column space of each matrix of exercises 1 through 4.

9–12. Find a basis for the row space of each matrix of exercises 1 through 4.

13. Prove that the set of nonzero rows of the row-reduced echelon form of a matrix is linearly independent.

The method of finding a basis for the row space of a matrix A given by example 4.26 has the advantage of introducing no new mathematical tools, simply finding a basis for the column space of A^T. There is an alternate approach that parallels the development based on the idea of *column operations*. Exercises 14 through 22 explore this idea.

14. Define the three **elementary column operations** on a matrix (see section 2.2).

15. Define the notion of a matrix being in **column-reduced echelon form.**

16. For a matrix A prove that a column operation on A has exactly the same effect as the corresponding row operation on A^T followed by transposing the resulting matrix.

17. If A is a matrix, and B is its column-reduced echelon form, prove that the rows of A that correspond to a distinct set of elementary rows of B form a basis for the row space of A.

18–21. Find a basis for the row space of each matrix in exercises 1 through 4 using exercise 17.

22. Recall that, to solve the matrix equation $A\mathbf{x} = \mathbf{b}$, we augment A by $\mathbf{b}$, obtaining $[A|\mathbf{b}]$, and row-reduce. We can use this procedure to solve the equation $\mathbf{x}^T A = \mathbf{b}^T$ by solving $A^T \mathbf{x} = \mathbf{b}$ since $(\mathbf{x}^T A)^T = A^T \mathbf{x}$. Devise a technique (using column operations) to solve the problem $\mathbf{x}^T A = \mathbf{b}^T$ without using the transpose of A.

4.5   Orthonormal Bases

The standard basis for $\mathbf{R}^m$, $\{\mathbf{e}_1, \mathbf{e}_2, \ldots, \mathbf{e}_m\}$, has several computational advantages over an arbitrary one. The standard basis vectors are pairwise orthogonal, each has unit length, and it is a simple matter to write an arbitrary vector in $\mathbf{R}^m$ as a linear combination of these basis vectors. Unfortunately subspaces of $\mathbf{R}^m$ do not ordinarily have bases that are subsets of $\{\mathbf{e}_1, \mathbf{e}_2, \ldots, \mathbf{e}_m\}$. Nevertheless we can construct bases for them— *orthonormal* ones—that do have the nice properties of the standard basis. In this section we will investigate the properties of orthonormal bases and present a method for replacing an arbitrary basis by an orthonormal one.

Definition   An **orthogonal set** of vectors in $\mathbf{R}^m$ is one in which every pair of vectors is orthogonal. An **orthonormal set** of vectors is an orthogonal one in which every vector has length equal to 1.

Note: Let $\mathscr{S} = \{\mathbf{v}_1, \mathbf{v}_2, \ldots, \mathbf{v}_n\}$ be a set of vectors in $\mathbf{R}^m$. Using the dot product and norm notation of section 1.4,

i. $\mathscr{S}$ is *orthogonal* if and only if $\mathbf{v}_i \cdot \mathbf{v}_j = 0$ whenever $i \neq j$,

ii. $\mathscr{S}$ is *orthonormal* if and only if $\mathbf{v}_i \cdot \mathbf{v}_j = 0$ whenever $i \neq j$ *and* $\|\mathbf{v}_i\| = 1$ for each i. (The last equation is equivalent to $\mathbf{v}_i \cdot \mathbf{v}_i = 1$.)

Example 4.28

Show that $\mathscr{T} = \{\mathbf{v}_1, \mathbf{v}_2, \mathbf{v}_3\}$ is an orthonormal set in $\mathbf{R}^3$, where

$$\mathbf{v}_1 = (1/3, 2/3, -2/3),$$
$$\mathbf{v}_2 = (0, 1/\sqrt{2}, 1/\sqrt{2}),$$
and
$$\mathbf{v}_3 = (-4/3\sqrt{2}, 1/3\sqrt{2}, -1/3\sqrt{2}).$$

We first check that $\mathscr{T}$ is orthogonal by verifying that the dot products $\mathbf{v}_1 \cdot \mathbf{v}_2$, $\mathbf{v}_1 \cdot \mathbf{v}_3$, and $\mathbf{v}_2 \cdot \mathbf{v}_3$ are all equal to 0. We then check the "length = 1" condition by verifying that $\|\mathbf{v}_1\| = \|\mathbf{v}_2\| = \|\mathbf{v}_3\| = 1$. Thus $\mathscr{T}$ is an orthonormal set.

The following theorem (which was exercise 28 of section 4.1) gives an important property of orthogonal sets that do not contain the zero vector.

Theorem 1   Let $\mathscr{T} = \{\mathbf{v}_1, \mathbf{v}_2, \ldots, \mathbf{v}_n\}$ be an orthogonal set of nonzero vectors in $\mathbf{R}^m$. Then $\mathscr{T}$ is linearly independent.

Proof   Consider the equation $c_1\mathbf{v}_1 + c_2\mathbf{v}_2 + \cdots + c_n\mathbf{v}_n = \mathbf{0}$. We will show that all c_i must be zero, which by theorem 1 of section 4.1 demonstrates the linear independence of $\mathscr{T}$. To accomplish this, form the dot product of both sides of this equation with each vector $\mathbf{v}_j$. For each j this yields the equation

$$c_1(\mathbf{v}_1 \cdot \mathbf{v}_j) + c_2(\mathbf{v}_2 \cdot \mathbf{v}_j) + \cdots + c_j(\mathbf{v}_j \cdot \mathbf{v}_j) + \cdots + c_n(\mathbf{v}_n \cdot \mathbf{v}_j) = \mathbf{0} \cdot \mathbf{v}_j.$$

But, because the set $\mathscr{T}$ is orthogonal, all dot products in this equation *except* $\mathbf{v}_j \cdot \mathbf{v}_j$ are equal to zero. Thus the equation reduces to

$$c_j(\mathbf{v}_j \cdot \mathbf{v}_j) = 0.$$

Now by hypothesis $\mathbf{v}_j \neq \mathbf{0}$ for $j = 1, 2, \ldots, n$, so $\mathbf{v}_j \cdot \mathbf{v}_j \neq 0$, and the last equation implies that $c_j = 0$ ($j = 1, 2, \ldots, n$), as desired.   ∎

Since $\mathbf{0}$ cannot be a member of an *orthonormal* set ($\|\mathbf{0}\| = 0 \neq 1$), theorem 1 has an immediate corollary.

Corollary   An orthonormal set of vectors is linearly independent.

As a consequence of this theorem and corollary, if an orthonormal set, or an orthogonal set of nonzero vectors, spans a subspace of $\mathbf{R}^m$, it is a basis for this subspace.

Definition   Let S be a subspace of $\mathbf{R}^m$. If a set $\mathcal{T}$ of vectors is a basis for S, and

  i. if $\mathcal{T}$ is *orthogonal*, it is called an **orthogonal basis** for S,
  ii. if $\mathcal{T}$ is *orthonormal*, it is called an **orthonormal basis** for S.

For example, it is easy to check that the standard basis for $\mathbf{R}^m$, $\{\mathbf{e}_1, \mathbf{e}_2, \ldots, \mathbf{e}_m\}$, is an orthonormal set (exercise 17) and that consequently it is an orthonormal basis for $\mathbf{R}^m$. However, the standard basis is certainly not the *only* orthonormal one for $\mathbf{R}^m$. For example, the set $\mathcal{T}$ of example 4.28 is an orthonormal set containing three vectors. By theorem 1 and part iii of theorem 5 of section 4.3 it is an orthonormal basis for $\mathbf{R}^3$.

The next theorem provides one of the major reasons why orthonormal bases are so desirable. It is a very simple matter to write an arbitrary vector in a given subspace as a linear combination of the vectors in an orthonormal basis for that subspace.

Theorem 2   Let $\mathcal{T} = \{\mathbf{v}_1, \mathbf{v}_2, \ldots, \mathbf{v}_n\}$ be an orthonormal basis for a subspace S of $\mathbf{R}^m$. Let $\mathbf{x}$ be an arbitrary vector in S. Then $\mathbf{x} = c_1\mathbf{v}_1 + c_2\mathbf{v}_2 + \cdots + c_n\mathbf{v}_n$, where $c_j = \mathbf{x} \cdot \mathbf{v}_j$ $(j = 1, 2, \ldots, n)$.

Proof   Since $\mathcal{T}$ is a basis for S, we know there exist scalars $c_1, c_2, \ldots, c_n$ such that $\mathbf{x} = c_1\mathbf{v}_1 + c_2\mathbf{v}_2 + \cdots + c_n\mathbf{v}_n$. Moreover

$$\mathbf{x} \cdot \mathbf{v}_j = (c_1\mathbf{v}_1 + c_2\mathbf{v}_2 + \cdots + c_j\mathbf{v}_j + \cdots + c_n\mathbf{v}_n) \cdot \mathbf{v}_j$$
$$= c_1\mathbf{v}_1 \cdot \mathbf{v}_j + c_2\mathbf{v}_2 \cdot \mathbf{v}_j + \cdots + c_j\mathbf{v}_j \cdot \mathbf{v}_j + \cdots + c_n\mathbf{v}_n \cdot \mathbf{v}_j$$
$$= c_j\mathbf{v}_j \cdot \mathbf{v}_j$$

due to the orthogonality of $\mathcal{T}$. But $\mathbf{v}_j \cdot \mathbf{v}_j = 1$ since $\mathcal{T}$ is orthonormal, so $\mathbf{x} \cdot \mathbf{v}_j = c_j$, as desired.   ∎

Example 4.29

Express $\mathbf{x} = (1, 2, 3)$ as a linear combination of the vectors

$$\mathbf{v}_1 = (1/2, 2/3, -2/3),$$

and
$$\mathbf{v}_2 = (0, 1/\sqrt{2}, 1/\sqrt{2}),$$

$$\mathbf{v}_3 = (-4/3\sqrt{2}, 1/3\sqrt{2}, -1/3\sqrt{2}).$$

In example 4.28 we showed that $\{\mathbf{v}_1, \mathbf{v}_2, \mathbf{v}_3\}$ is an orthonormal set. By theorem 1 this set is linearly independent, and by part iii of theorem 5 of section 4.3 it is a basis for $\mathbf{R}^3$. We may therefore apply theorem 2 (with $\mathbf{S} = \mathbf{R}^3$) to obtain $\mathbf{x} = c_1\mathbf{v}_1 + c_2\mathbf{v}_2 + c_3\mathbf{v}_3$, with

$$c_1 = \mathbf{x} \cdot \mathbf{v}_1 = -1/3,$$

and
$$c_2 = \mathbf{x} \cdot \mathbf{v}_2 = 5/\sqrt{2},$$

$$c_3 = \mathbf{x} \cdot \mathbf{v}_3 = -5/3\sqrt{2}.$$

In the preceding discussion, we have defined *orthonormal basis* and given some of its properties. Now we will describe how to construct one from a known, but nonorthonormal, basis.

If the known basis is *orthogonal*, all we need do to produce an orthonormal one is to *normalize* each vector in the given basis, in other words, divide each vector by its length. This preserves the orthogonality of the given set and at the same time creates vectors of length 1 (see section 1.4).

Example 4.30

Find an orthonormal basis for the subspace $\mathbf{S}$ of $\mathbf{R}^4$ generated by the orthogonal set $\mathscr{T} = \{\mathbf{v}_1, \mathbf{v}_2, \mathbf{v}_3\}$, where

$$\mathbf{v}_1 = (0, 2, -1, 1), \quad \mathbf{v}_2 = (0, 0, 1, 1), \quad \mathbf{v}_3 = (-2, 1, 1, -1).$$

Since $\mathscr{T}$ is orthogonal, it is linearly independent, and since $\mathscr{T}$ generates $\mathbf{S}$, it spans $\mathbf{S}$. Hence $\mathscr{T}$ is an *orthogonal* basis for $\mathbf{S}$. To obtain an ortho*normal* basis for $\mathbf{S}$, all we need do is normalize each vector in $\mathscr{T}$: form the set

$$\mathscr{T}' = \{(1/\|\mathbf{v}_1\|)\mathbf{v}_1, (1/\|\mathbf{v}_2\|)\mathbf{v}_2, (1/\|\mathbf{v}_3\|)\mathbf{v}_3\}$$
$$= \{(0, 2/\sqrt{6}, -1/\sqrt{6}, 1/\sqrt{6}), (0, 0, 1/\sqrt{2}, 1/\sqrt{2}),$$
$$(-2/\sqrt{7}, 1/\sqrt{7}, 1/\sqrt{7}, -1/\sqrt{7})\}.$$

Then $\mathscr{T}'$ is an orthonormal basis for $\mathbf{S}$.

When the known basis is not orthogonal, the process of finding an orthonormal basis is more complicated but still algorithmic. To motivate the procedure, we consider the special case of two vectors.

Let $\mathbf{u}$ and $\mathbf{v}$ be nonzero vectors in $\mathbf{R}^m$, and let $\mathbf{x}$ be the vector along $\mathbf{v}$ such that the points $\mathbf{0}$, $\mathbf{x}$, and $\mathbf{u}$ form a right triangle. The vector $\mathbf{x}$ is called the **orthogonal projection** of $\mathbf{u}$ onto $\mathbf{v}$ (figure 4.5).

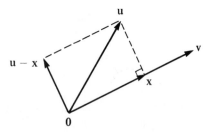

Figure 4.5

Theorem 3 Let $\mathbf{u}$ and $\mathbf{v}$ be nonzero vectors in $\mathbf{R}^m$. The orthogonal projection of $\mathbf{u}$ onto $\mathbf{v}$ is given by

$$\mathbf{x} = \left(\frac{\mathbf{u} \cdot \mathbf{v}}{\mathbf{v} \cdot \mathbf{v}}\right)\mathbf{v}.$$

Proof Since $\mathbf{u} - \mathbf{x}$ is equivalent to $\overrightarrow{\mathbf{x}\mathbf{u}}$, we need only check that $\mathbf{u} - \mathbf{x}$ is orthogonal to $\mathbf{v}$:

$$(\mathbf{u} - \mathbf{x}) \cdot \mathbf{v} = \left(\mathbf{u} - \left(\frac{\mathbf{u} \cdot \mathbf{v}}{\mathbf{v} \cdot \mathbf{v}}\right)\mathbf{v}\right) \cdot \mathbf{v} = \mathbf{u} \cdot \mathbf{v} - \left(\frac{\mathbf{u} \cdot \mathbf{v}}{\mathbf{v} \cdot \mathbf{v}}\right)\mathbf{v} \cdot \mathbf{v} = 0. \quad \blacksquare$$

Notice that the sets $\{\mathbf{v}, \mathbf{u}\}$ and $\{\mathbf{v}, \mathbf{u} - \mathbf{x}\}$ span the same set because $\mathbf{u}$ is a linear combination of $\mathbf{v}$ and $\mathbf{u} - \mathbf{x}$, while $\mathbf{u} - \mathbf{x}$ is a linear combination of $\mathbf{v}$ and $\mathbf{u}$. Specifically we have

$$\mathbf{u} = \left(\frac{\mathbf{u} \cdot \mathbf{v}}{\mathbf{v} \cdot \mathbf{v}}\right)\mathbf{v} + (1)(\mathbf{u} - \mathbf{x})$$

and

$$\mathbf{u} - \mathbf{x} = (1)\mathbf{u} + (-1)\left(\frac{\mathbf{u} \cdot \mathbf{v}}{\mathbf{v} \cdot \mathbf{v}}\right)\mathbf{v}.$$

It follows that, if $\{\mathbf{v}, \mathbf{u}\}$ is a basis for a subspace of $\mathbf{R}^m$, then $\{\mathbf{v}, \mathbf{u} - \mathbf{x}\}$ is an *orthogonal* basis for the same subspace.

This process of replacing a basis by an orthogonal one for the same subspace can be generalized to any finite number of vectors by the procedure described next. The proof of its validity is simply an extension of the argument given for the case of two vectors. It will also be demonstrated more generally in chapter 7.

Procedure for constructing an orthogonal basis (Gram-Schmidt Process)
Let $\mathcal{T} = \{\mathbf{u}_1, \mathbf{u}_2, \ldots, \mathbf{u}_n\}$ be a basis for a subspace S of $\mathbf{R}^m$. Let

$\mathcal{T}' = \{v_1, v_2, \ldots, v_n\}$ be defined as follows:

$$v_1 = u_1,$$

$$v_2 = u_2 - \left(\frac{u_2 \cdot v_1}{v_1 \cdot v_1}\right)v_1,$$

$$v_3 = u_3 - \left(\frac{u_3 \cdot v_1}{v_1 \cdot v_1}\right)v_1 - \left(\frac{u_3 \cdot v_2}{v_2 \cdot v_2}\right)v_2,$$

$$\vdots$$

$$v_n = u_n - \left(\frac{u_n \cdot v_1}{v_1 \cdot v_1}\right)v_1 - \left(\frac{u_n \cdot v_2}{v_2 \cdot v_2}\right)v_2 - \cdots - \left(\frac{u_n \cdot v_{n-1}}{v_{n-1} \cdot v_{n-1}}\right)v_{n-1}.$$

Then the set $\mathcal{T}'$ is an *orthogonal* basis for **S**. An *orthonormal* basis for **S** is given by $\mathcal{T}'' = \{w_1, w_2, \ldots, w_n\}$, where $w_i = (1/\|v_i\|)v_i$ for each $i = 1, \ldots, n$.

Example 4.31

Let **S** be the subspace of $\mathbf{R}^5$ generated by the set $\mathcal{T} = \{u_1, u_2, u_3\}$, where $u_1 = (-1, -1, 1, 0, 0)$, $u_2 = (0, -1, 0, 0, 1)$, and $u_3 = (1, -1, 0, 1, 0)$. Find an orthonormal basis for **S**.

We apply the Gram-Schmidt process to $\mathcal{T}$.

$$v_1 = u_1 = (-1, -1, 1, 0, 0),$$

$$v_2 = u_2 - \left(\frac{u_2 \cdot v_1}{v_1 \cdot v_1}\right)v_1$$

$$= (0, -1, 0, 0, 1) - \left(\frac{1}{3}\right)(-1, -1, 1, 0, 0)$$

$$= (1/3, -2/3, -1/3, 0, 1),$$

$$v_3 = u_3 - \left(\frac{u_3 \cdot v_1}{v_1 \cdot v_1}\right)v_1 - \left(\frac{u_3 \cdot v_2}{v_2 \cdot v_2}\right)v_2$$

$$= (1, -1, 0, 1, 0) - \left(\frac{0}{3}\right)(-1, -1, 1, 0, 0)$$

$$- \left(\frac{1}{5/3}\right)(1/3, -2/3, -1/3, 0, 1)$$

$$= (4/5, -3/5, 1/5, 1, -3/5).$$

The set $\mathcal{T}' = \{v_1, v_2, v_3\}$ is an orthogonal basis for **S**. Normalizing each vector, we obtain an orthonormal basis $\mathcal{T}'' = \{w_1, w_2, w_3\}$, where

$$w_1 = (-\sqrt{3}/3, -\sqrt{3}/3, \sqrt{3}/3, 0, 0),$$
$$w_2 = (\sqrt{15}/15, -2\sqrt{15}/15, -\sqrt{15}/15, 0, \sqrt{15}/5),$$
$$w_3 = (2\sqrt{15}/15, -\sqrt{15}/10, \sqrt{15}/30, \sqrt{15}/6, -\sqrt{15}/10).$$

4.5 Exercises

In exercises 1 through 6 determine which sets are orthogonal bases and which are orthonormal bases for the indicated subspaces of $\mathbf{R}^m$.

1. $\{(1/\sqrt{5}, 2/\sqrt{5}, 0), (-2/\sqrt{5}, 1/\sqrt{5}, 0), (0, 0, 1)\}$ for $\mathbf{R}^3$
2. $\{(0, 1, 0), (1, 0, 1), (0, 0, 1)\}$ for $\mathbf{R}^3$
3. $\{(1, 2, 2), (2, 1, -2), (1, -2, 2)\}$ for $\mathbf{R}^3$
4. $\{(0, \sin\theta, \cos\theta, 0), (0, \cos\theta, -\sin\theta, 0)\}$, where θ is any real number for the subspace of $\mathbf{R}^4$ of all vectors whose first and last components are zero.

5–6. $\{(\sqrt{2}/2, 0, \sqrt{2}/2)\}$ for the solution space of $A\mathbf{x} = \mathbf{0}$, where

$$5. \ A = \begin{bmatrix} 1 & 0 & -1 \\ 0 & 1 & 0 \\ 1 & 1 & -1 \end{bmatrix} \qquad 6. \ A = \begin{bmatrix} 1 & 0 & -1 \\ 1 & 1 & -1 \\ 2 & 1 & -2 \end{bmatrix}$$

In exercises 7 and 8 express the given vector $\mathbf{x}$ as a linear combination of the vectors in the orthonormal basis $\{(1/3, -2/3, 2/3), (-2/3, 1/3, 2/3), (2/3, 2/3, 1/3)\}$.

7. $\mathbf{x} = (1, 2, 3)$ 8. $\mathbf{x} = (-1, 0, 1)$

In exercises 9 and 10 the given set is an *orthogonal* basis for a certain subspace of $\mathbf{R}^4$. Find an ortho*normal* basis for this subspace.

9. $\{(4, 5, 0, -2), (-2, 2, 0, 1), (1, 0, 0, 2)\}$
10. $\{(1, -1, 1, -1), (1, 1, 1, 1), (1, 0, -1, 0)\}$

In exercises 11 through 16 the given set is a basis for a certain subspace of $\mathbf{R}^m$. Use the Gram-Schmidt process to find an orthonormal basis for this subspace.

11. $\{(1, -1, 0), (0, 1, -1)\}$
12. $\{(1, 0, -1), (1, 0, 1)\}$
13. $\{(-1, 1, 0, 0), (1, -1, 1, 0), (0, 0, 1, 2)\}$
14. $\{(-1, -1, 2, 1), (2, 1, -1, -2), (1, 1, 0, 1)\}$
15. $\{(1, 0, 0, 0), (1, 1, 0, 0), (1, 1, 1, 0), (1, 1, 1, 1)\}$
16. $\{(1, 1, 1, 0), (1, -1, 1, -1), (1, 0, 1, 1), (0, 1, 1, 1)\}$

17. Show that the standard basis for $\mathbf{R}^m$, $\{\mathbf{e}_1, \mathbf{e}_2, \ldots, \mathbf{e}_m\}$, is an orthonormal set.

Linear Transformations

5

To this point we have been studying facts about $\mathbf{R}^m$ for a fixed value of m. In this chapter we will explore *linear transformations* between $\mathbf{R}^n$ and $\mathbf{R}^m$ where m may or may not equal n. It will be seen that matrices play an important role in describing linear transformations and that we may develop an algebra of linear transformations analogous to the algebra of matrices.

5.1 Definition of a Linear Transformation

A **function** or **map** from one set to another is an association of each element of the first set with a unique element of the second. If $\mathscr{A}$ and $\mathscr{B}$ are sets, the notation

$$f: \mathscr{A} \to \mathscr{B}$$

denotes a function from $\mathscr{A}$ to $\mathscr{B}$. For a in $\mathscr{A}$ the unique element of $\mathscr{B}$ associated with a is called the **image** of a under f and is denoted by $f(a)$.

For our purposes, the sets $\mathscr{A}$ and $\mathscr{B}$ will be the spaces $\mathbf{R}^n$ and $\mathbf{R}^m$ for some choice of n and m. We will not be interested in all functions from $\mathbf{R}^n$ to $\mathbf{R}^m$ but only those that are *linear* in the following sense.

Definition   A **linear transformation** from $\mathbf{R}^n$ to $\mathbf{R}^m$ is a function $T: \mathbf{R}^n \to \mathbf{R}^m$ that satisfies the following two conditions. For each $\mathbf{u}$ and $\mathbf{v}$ in $\mathbf{R}^n$ and a in $\mathbf{R}$,

   i. $T(a\mathbf{u}) = aT(\mathbf{u})$ (scalars factor out),
  ii. $T(\mathbf{u} + \mathbf{v}) = T(\mathbf{u}) + T(\mathbf{v})$ (T is additive).

Example 5.1

Let $T : \mathbf{R}^2 \to \mathbf{R}^3$ be described by

$$T(x, y) = (x + y, x, 2x - y).$$

Show that T is a linear transformation from $\mathbf{R}^2$ to $\mathbf{R}^3$.

For $\mathbf{u} = (x_1, y_1)$ and $\mathbf{v} = (x_2, y_2)$ in $\mathbf{R}^2$ and a in $\mathbf{R}$ we check conditions i and ii.

171

i. $T(a\mathbf{u}) = T(a(x_1, y_1))$
$= T((ax_1, ay_1))$
$= (ax_1 + ay_1, ax_1, 2(ax_1) - ay_1)$
$= a(x_1 + y_1, x_1, 2x_1 - y_1)$
$= aT((x_1, y_1)) = aT(\mathbf{u})$.

ii. $T(\mathbf{u} + \mathbf{v}) = T((x_1, y_1) + (x_2, y_2))$
$= T((x_1 + x_2, y_1 + y_2))$
$= ((x_1 + x_2) + (y_1 + y_2), x_1 + x_2, \; 2(x_1 + x_2)$
$\quad - (y_1 + y_2))$
$= ((x_1 + y_1) + (x_2 + y_2), x_1 + x_2, (2x_1 - y_1)$
$\quad + (2x_2 - y_2))$
$= (x_1 + y_1, x_1, 2x_1 - y_1) + (x_2 + y_2, x_2, 2x_2 - y_2)$
$= T(\mathbf{u}) + T(\mathbf{v})$.

Example 5.2

Let $T : \mathbf{R}^n \to \mathbf{R}^m$ be described by $T(\mathbf{u}) = \mathbf{0}$, for all $\mathbf{u}$ in $\mathbf{R}^n$. Show that T is a linear transformation.

i. For a in $\mathbf{R}$ and $\mathbf{u}$ in $\mathbf{R}^n$, $T(a\mathbf{u}) = \mathbf{0} = a\mathbf{0} = aT(\mathbf{u})$.
ii. For $\mathbf{u}, \mathbf{v}$ in $\mathbf{R}^n$, $T(\mathbf{u} + \mathbf{v}) = \mathbf{0} = \mathbf{0} + \mathbf{0} = T(\mathbf{u}) + T(\mathbf{v})$.

Thus T is a linear transformation.

Definition   The transformation of example 5.2 is called the **zero transformation** from $\mathbf{R}^n$ to $\mathbf{R}^m$ and will be denoted by 0.

Example 5.3

Let $T : \mathbf{R}^2 \to \mathbf{R}^2$ be defined by $T(\mathbf{u}) = \mathbf{u} + (1, 2)$, for all $\mathbf{u}$ in $\mathbf{R}^2$. Show that T is *not* a linear transformation.

To show that condition i is violated by this function, we let $a = 0$ and $\mathbf{u} = (0, 0)$. Then

$$T(a\mathbf{u}) = T((0, 0)) = (0, 0) + (1, 2) = (1, 2),$$

but

$$aT(\mathbf{u}) = 0T((0, 0)) = 0((0, 0) + (1, 2)) = 0(1, 2) = (0, 0).$$

Thus it is not always true that $T(a\mathbf{u}) = aT(\mathbf{u})$, and condition i is violated: T is not linear. (It can be shown that condition ii does not hold either.)

Notice that the function described in example 5.3 is a *translation* of the plane $\mathbf{R}^2$ by the vector $(1, 2)$: for each point $\mathbf{u}$, the line segment $\overrightarrow{\mathbf{u}T(\mathbf{u})}$ is equivalent to the vector $(1, 2)$ as shown in figure 5.1. In particular the origin $\mathbf{0}$ gets mapped by T to the point $(1, 2)$, $T(\mathbf{0}) = (1, 2)$.

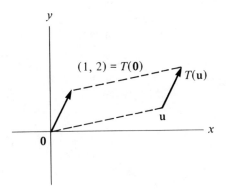

Figure 5.1

According to statement a of the following theorem this function could not be linear.

Theorem 1   Let $T : \mathbf{R}^n \to \mathbf{R}^m$ be a linear transformation. Then

a. $T(\mathbf{0}) = \mathbf{0}$,
b. $T(-\mathbf{u}) = -T(\mathbf{u})$ for each $\mathbf{u}$ in $\mathbf{R}^n$,
c. $T(\mathbf{u} - \mathbf{v}) = T(\mathbf{u}) - T(\mathbf{v})$ for each $\mathbf{u}$ and $\mathbf{v}$ in $\mathbf{R}^n$.

Note: The vector $\mathbf{0}$ of $T(\mathbf{0})$ is an element of $\mathbf{R}^n$, and the vector $\mathbf{0}$ on the right side of the equation given in theorem 1a is an element of $\mathbf{R}^m$. We use the same notation for each, but you should always be aware of where the vector is located.

Proof of 1a   Since $\mathbf{0} = \mathbf{0} + \mathbf{0}$ (in $\mathbf{R}^n$),

$$T(\mathbf{0}) = T(\mathbf{0} + \mathbf{0}),$$
$$T(\mathbf{0}) = T(\mathbf{0}) + T(\mathbf{0}) \quad (T \text{ is additive}).$$

Since $T(\mathbf{0})$ is in $\mathbf{R}^m$, using $\mathbf{0}$ in $\mathbf{R}^m$, we have

$$T(\mathbf{0}) + \mathbf{0} = T(\mathbf{0}) + T(\mathbf{0}).$$

Adding $-T(\mathbf{0})$ to both sides, we have

$$\mathbf{0} = T(\mathbf{0}).$$

Proof of 1b   By 1a, and the fact that $\mathbf{0} = \mathbf{u} + (-\mathbf{u})$, we have

$$\mathbf{0} = T(\mathbf{0}) = T(\mathbf{u} + -\mathbf{u}) = T(\mathbf{u}) + T(-\mathbf{u}).$$

That is,
$$0 = T(\mathbf{u}) + T(-\mathbf{u}).$$
Adding $-T(\mathbf{u})$ to both sides, we have
$$-T(\mathbf{u}) = T(-\mathbf{u}).$$
The proof of 1c is exercise 16 at the end of this section. ∎

Example 5.4

Let A be the matrix
$$A = \begin{bmatrix} 2 & 0 & 1 \\ 1 & 1 & -1 \end{bmatrix},$$
and let $\mathbf{x}$ be an element of $\mathbf{R}^3$. Let $T(\mathbf{x}) = A\mathbf{x}$, where $A\mathbf{x}$ denotes matrix multiplication. Show that T is a linear transformation.

We let $\mathbf{x} = (x, y, z)$ and compute $T(\mathbf{x})$,

$$T(\mathbf{x}) = A\mathbf{x} = \begin{bmatrix} 2 & 0 & 1 \\ 1 & 1 & -1 \end{bmatrix} \begin{bmatrix} x \\ y \\ z \end{bmatrix}$$

$$= \begin{bmatrix} 2x + z \\ x + y - z \end{bmatrix} = (2x + z, x + y - z).$$

Thus we have the rule $T(x, y, z) = (2x + z, x + y - z)$, and we could follow the procedure of example 5.1 to verify that T is linear. (You should do it for the practice.)

Example 5.4 is a special case of the following theorem.

Theorem 2   Let A be an $m \times n$ matrix. Then $T : \mathbf{R}^n \to \mathbf{R}^m$ defined by $T(\mathbf{x}) = A\mathbf{x}$ is a linear transformation.

Proof   First note that $\mathbf{x}$ is to be written as a column vector so that the matrix multiplication will be defined: A is $m \times n$, and $\mathbf{x}$ is $n \times 1$. Thus the product is defined, and the result is an $m \times 1$ matrix. The latter may be viewed as an element of $\mathbf{R}^m$.

To check property i of the definition of linear transformation, let a be a scalar, and let $\mathbf{x}$ be an element of $\mathbf{R}^n$. Then

$$T(a\mathbf{x}) = A(a\mathbf{x})$$
$$= aA\mathbf{x} \quad \text{(theorem 1n of section 2.3)}$$
$$= aT(\mathbf{x}).$$

For property ii let $\mathbf{x}$ and $\mathbf{y}$ be elements of $\mathbf{R}^n$. Then

$$T(\mathbf{x} + \mathbf{y}) = A(\mathbf{x} + \mathbf{y})$$
$$= A\mathbf{x} + A\mathbf{y} \quad \text{(theorem 1g of section 2.3)}$$
$$= T(\mathbf{x}) + T(\mathbf{y}).$$

Thus T is a linear transformation.   ∎

Examples of linear transformations given by theorem 2 are of the utmost importance. In fact it will be shown in section 5.2 that in some sense all linear transformations are of this type. For the time being such functions provide an infinite supply of linear transformations for which we need not check the definition; that has already been done in theorem 2. We continue with some interesting special cases.

Example 5.5

Let θ be any real number, and view θ as the radian measure of an angle. Let T_θ be given by $T_\theta(\mathbf{x}) = A_\theta \mathbf{x}$, where

$$A_\theta = \begin{bmatrix} \cos\theta & -\sin\theta \\ \sin\theta & \cos\theta \end{bmatrix}.$$

Then T_θ is the **rotation by** θ. In other words, T_θ is the linear transformation that rotates each vector of $\mathbf{R}^2$ through an angle of θ radians.

To see that this is so, we need only show that for all $\mathbf{x}$ in $\mathbf{R}^2$, $\mathbf{x}$ and $T_\theta(\mathbf{x})$ have the same length and that the angle between these vectors is θ (figure 5.2).

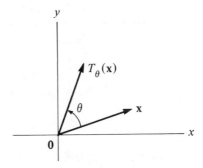

Figure 5.2

To show that $\mathbf{x}$ and $T_\theta(\mathbf{x})$ have the same length, let $\mathbf{x} = (x, y)$. Then $T_\theta(\mathbf{x}) = (x\cos\theta - y\sin\theta, \ x\sin\theta + y\cos\theta)$ so that

$$\|T_\theta(\mathbf{x})\| = \|(x\cos\theta - y\sin\theta, \ x\sin\theta + y\cos\theta)\|$$

$$= \sqrt{(x \cos \theta - y \sin \theta)^2 + (x \sin \theta + y \cos \theta)^2}$$
$$= \sqrt{x^2 \cos^2 \theta - 2xy \cos \theta \sin \theta + y^2 \sin^2 \theta + x^2 \sin^2 \theta + 2xy \sin \theta \cos \theta + y^2 \cos^2 \theta}$$
$$= \sqrt{x^2(\cos^2 \theta + \sin^2 \theta) + y^2(\sin^2 \theta + \cos^2 \theta)}$$
$$= \sqrt{x^2 + y^2} \quad (\text{since } \sin^2 \theta + \cos^2 \theta = 1)$$
$$= \|\mathbf{x}\|.$$

By theorem 2 of section 1.2, the cosine of the angle ψ between x and $T_\theta(\mathbf{x})$ is given by

$$\cos \psi = \frac{\mathbf{x} \cdot T_\theta(\mathbf{x})}{\|\mathbf{x}\| \, \|T_\theta(\mathbf{x})\|} = \frac{x^2 \cos \theta - xy \sin \theta + xy \sin \theta + y^2 \cos \theta}{\|\mathbf{x}\| \, \|\mathbf{x}\|}$$

$$= \frac{(x^2 + y^2)}{\|\mathbf{x}\|^2} \cos \theta$$

$$= \frac{\|\mathbf{x}\|^2}{\|\mathbf{x}\|^2} \cos \theta = \cos \theta.$$

Thus $\cos \psi = \cos \theta$. If $0 \leq \theta \leq \pi$, this implies that $\psi = \theta$. Since the angle ψ between the two vectors is restricted so that $0 \leq \psi \leq \pi$, the angles will *not* be equal unless θ also satisfies the condition $0 \leq \theta \leq \pi$, a condition we have not imposed. It can be shown, however, that θ and ψ still determine the same direction from the vector $\mathbf{x}$. Figure 5.3 demonstrates the situation for $\pi < \theta < 2\pi$. In this case, $\psi = 2\pi - \theta$.

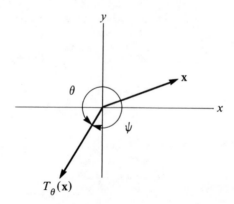

Figure 5.3

As a specific case of example 5.5, let $\theta = \pi/6$ radians. Then $\cos \pi/6 = \sqrt{3}/2$ and $\sin \pi/6 = 1/2$ so that

$$A = \begin{bmatrix} \sqrt{3}/2 & -1/2 \\ 1/2 & \sqrt{3}/2 \end{bmatrix}.$$

This matrix determines the linear transformation that rotates each vector in $\mathbf{R}^2$ by $\pi/6$ and leaves its length unchanged.

Another special case of example 5.5 is $\theta = 0$ radians, no rotation at all. Since $\cos 0 = 1$, and $\sin 0 = 0$, we have the 2×2 identity matrix,

$$A = \begin{bmatrix} 1 & 0 \\ 0 & 1 \end{bmatrix}$$

It should be no surprise that multiplication by $A = I$ leaves each vector unchanged. In fact the function "leave each vector unchanged" is always a linear transformation.

Definition   For any n, define $I : \mathbf{R}^n \to \mathbf{R}^n$ by $I(\mathbf{x}) = \mathbf{x}$ for each $\mathbf{x}$ in $\mathbf{R}^n$. Then I is a linear transformation (see exercise 4) that is called the **identity transformation** on $\mathbf{R}^n$. It is always true that $I(\mathbf{x}) = I\mathbf{x}$, where I is the $n \times n$ identity matrix.

Example 5.6

Let c be a positive scalar, and let $A = cI$, where I is the $n \times n$ identity matrix. Then T_c given by $T_c(\mathbf{x}) = A\mathbf{x}$ is the **dilation of $\mathbf{R}^n$ by c** if $c > 1$ and is the **contraction by c** if $0 < c < 1$. In other words, T_c is the linear transformation that "stretches" or "shrinks" each vector of $\mathbf{R}^n$ by a factor of c but leaves its direction unchanged. Of course, if $c = 1$, T_c is the identity transformation. Figures 5.4 and 5.5 indicate the result when $c = 3$ and $c = 1/2$, respectively.

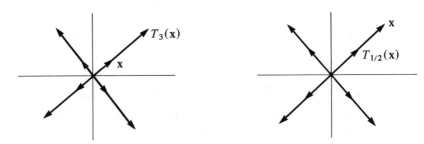

Figure 5.4   **Figure 5.5**

To verify the assertions, note that $T_c(\mathbf{x}) = c\mathbf{x}$. Thus T_c is just multiplication by the scalar c, and by section 1.4 both the length and the direction are as stated.

Example 5.7

Let θ be any real number, and view θ as the radian measure of an angle. Let T^θ be given by $T^\theta(\mathbf{x}) = A\mathbf{x}$, where

$$A = \begin{bmatrix} \cos 2\theta & \sin 2\theta \\ \sin 2\theta & -\cos 2\theta \end{bmatrix}.$$

Then T^θ is the **reflection** in the line l through the origin that forms angle θ with the positive x-axis: for each $\mathbf{x}$ in $\mathbf{R}^2$, $T^\theta(\mathbf{x})$ is situated so that the perpendicular bisector of the segment from the point $\mathbf{x}$ to $T^\theta(\mathbf{x})$ is the line l (figure 5.6). We leave the verification of this statement as exercise 21 at the end of the section.

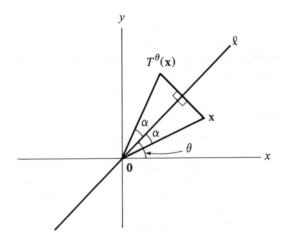

Figure 5.6

Note: The linear transformations of examples 5.5, 5.6, and 5.7 are all **linear operators**, in other words, they are all linear transformations that map the space back into itself. They are in fact much more special than that; they are **similarities**, they preserve the shape of geometric figures. Examples 5.5 and 5.7 are also **isometries**, not only do they preserve shape but size (distance between any two points) as well.

In the preceding examples we checked the effect or "action" that a linear transformation $T : \mathbf{R}^n \to \mathbf{R}^m$ has on an arbitrary vector $\mathbf{x}$ in $\mathbf{R}^n$. Determining the effect of a linear transformation can also be done by checking its action on a specific, finite set of vectors. The next theorem makes this precise.

Definition   Let $S : \mathbf{R}^n \to \mathbf{R}^m$ and $T : \mathbf{R}^n \to \mathbf{R}^m$ be linear transformations. Then $S = T$ if $S(\mathbf{x}) = T(\mathbf{x})$ for all $\mathbf{x}$ in $\mathbf{R}^n$.

Theorem 3   A linear transformation is determined by its action on a basis. That is, if $S : \mathbf{R}^n \to \mathbf{R}^m$ and $T : \mathbf{R}^n \to \mathbf{R}^m$ are linear transformations, and $\mathcal{B} = \{\mathbf{b}_1, \mathbf{b}_2, \ldots, \mathbf{b}_n\}$ is a basis for $\mathbf{R}^n$ and $S(\mathbf{b}_i) = T(\mathbf{b}_i)$ for each i, then $S = T$.

Proof   Since $\mathcal{B}$ is a basis for $\mathbf{R}^n$, for each $\mathbf{x}$ in $\mathbf{R}^n$ there exist scalars $c_1, \ldots, c_n$ with $\mathbf{x} = c_1\mathbf{b}_1 + c_2\mathbf{b}_2 + \cdots + c_n\mathbf{b}_n$.
Then

$$
\begin{aligned}
S(\mathbf{x}) &= S(c_1\mathbf{b}_1 + c_2\mathbf{b}_2 + \cdots + c_n\mathbf{b}_n) \\
&= cS(\mathbf{b}_1) + c_2S(\mathbf{b}_2) + \cdots + c_nS(\mathbf{b}_n) \quad \text{(exercise 23)} \\
&= cT(\mathbf{b}_1) + c_2T(\mathbf{b}_2) + \cdots + c_nT(\mathbf{b}_n) \quad \text{(hypothesis)} \\
&= T(c_1\mathbf{b}_1 + c_2\mathbf{b}_2 + \cdots + c_n\mathbf{b}_n) \quad \text{(exercise 23)} \\
&= T(\mathbf{x}).
\end{aligned}
$$

Thus $S = T$.   ∎

Example 5.8

Describe the action of the linear transformation $T : \mathbf{R}^2 \to \mathbf{R}^2$ given by $T(x, y) = (2x - y, x + y)$.

Theorem 3 applies only to *linear transformations* so it must be checked that T is a linear transformation. This detail is exercise 1 at the end of the section. We proceed to check the action of T on a basis for $\mathbf{R}^2$.

For the basis we choose the most convenient one, the standard basis $\{\mathbf{e}_1, \mathbf{e}_2\}$. We have

$$T(\mathbf{e}_1) = T(1, 0) = (2, 1) \quad \text{and} \quad T(\mathbf{e}_2) = T(0, 1) = (-1, 1).$$

Next we sketch $\{\mathbf{e}_1, \mathbf{e}_2\}$ and $\{T(\mathbf{e}_1), T(\mathbf{e}_2)\}$ on separate copies of the space $\mathbf{R}^2$ (figures 5.7 and 5.8).

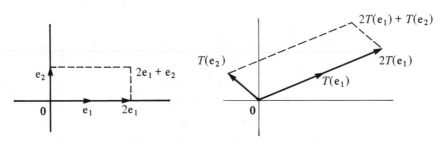

Figure 5.7   **Figure 5.8**

Here T is seen to rotate and stretch the vectors, but unevenly: $\mathbf{e}_1$ is not rotated as much as $\mathbf{e}_2$, and $\mathbf{e}_2$ is not stretched as much as $\mathbf{e}_1$. The importance of theorem 3 is that it shows us that the effect of T on all other vectors is determined by its effect on these two. To be precise, for any $x = (x, y)$ in $\mathbf{R}^2$, since $x = xe_1 + ye_2$, we have $T(x) = xT(e_1) + yT(e_2)$. For example, consider the point $(2, 1)$. Since $(2, 1) = 2e_1 + e_2$, the point $(2, 1)$ completes the parallelogram determined by $2e_1$ and e_2. Then $T(2, 1)$ must be the point that completes the parallelogram determined by $2T(e_1)$ and $T(e_2)$ (figure 5.8).

5.1 Exercises

In exercises 1 through 4 show that each function is a linear transformation.

1. $S(x, y) = (2x - y, x + y)$ $(S : \mathbf{R}^2 \rightarrow \mathbf{R}^2)$

2. $T(x, y) = (2x, x + y, x - 2y)$ $(T : \mathbf{R}^2 \rightarrow \mathbf{R}^3)$

3. $U(x, y, z) = x + y + z$ $(U : \mathbf{R}^3 \rightarrow \mathbf{R})$

4. $I(x_1, \ldots, x_n) = (x_1, \ldots, x_n)$ $(I : \mathbf{R}^n \rightarrow \mathbf{R}^n)$

In exercises 5 through 8 determine whether or not the given function is a linear transformation, and justify your answer.

5. $F(x, y, z) = (0, 2x + y)$ $(F : \mathbf{R}^3 \rightarrow \mathbf{R}^2)$

6. $G(x, y, z) = (xy, y, x - z)$ $(G : \mathbf{R}^3 \rightarrow \mathbf{R}^3)$

7. $H(x, y) = \sqrt{x^2 + y^2}$ $(H : \mathbf{R}^2 \rightarrow \mathbf{R})$

8. $K(x, y) = (x, \sin y, 2x + y)$ $(K : \mathbf{R}^2 \rightarrow \mathbf{R}^3)$

In exercises 9 through 12 let $T : \mathbf{R}^n \rightarrow \mathbf{R}^m$ be the linear transformation given by $T(x) = Ax$, multiplication by the matrix A. Find n and m and $T(x_0)$ for each given vector x_0.

9. $A = \begin{bmatrix} 2 & 3 & 1 \\ 1 & -1 & 0 \end{bmatrix}$, $x_0 = (1, 4, 2)$ **10.** $A = \begin{bmatrix} -1 & 1 & 2 \\ 2 & 4 & -1 \\ 1 & 0 & 1 \end{bmatrix}$, $x_0 = (1, 4, 2)$

11. $A = \begin{bmatrix} 1 & 2 \\ 0 & 1 \\ 1 & 0 \end{bmatrix}$, $x_0 = (2, -3)$ **12.** $A = \begin{bmatrix} 0 & 1 & 0 \\ 1 & 0 & 0 \\ 0 & 0 & 1 \end{bmatrix}$, $x_0 = (x, y, z)$

13. The function $\Pi_i : \mathbf{R}^n \rightarrow \mathbf{R}$ defined by $\Pi_i(x_1, \ldots, x_n) = x_i$ is called the ith *projection map*. Prove that each projection map is a linear transformation.

14. If A is an $m \times n$ matrix, prove that the function $T : \mathbf{R}^m \rightarrow \mathbf{R}^n$ defined by $T(x) = xA$ is a linear transformation, where x is viewed as a matrix with one row. (*Hint:* follow theorem 2.)

15. For a linear transformation $T : \mathbf{R}^n \rightarrow \mathbf{R}^m$, the **negative** of T, $-T : \mathbf{R}^n \rightarrow \mathbf{R}^m$, is defined by $(-T)(x) = -(T(x))$. Prove that $-T$ is a linear transformation.

16. Prove theorem 1c.

17. Find the matrix for the linear transformation $T_{3\pi/4}$, the rotation by $3\pi/4$.

18. Find the matrix for the linear transformation $T_{-\pi/2}$, the rotation by $-\pi/2$.

19. Find the matrix for the linear transformation $T^{\pi/4}$, the reflection in the line $x = y$.

20. Find the matrix for the linear transformation T^0, the reflection in the x-axis.

21. Complete the verification of example 5.7.

22. Let $F : \mathbf{R}^n \to \mathbf{R}^m$ be a function. Prove that F is a linear transformation if and only if $F(a\mathbf{u} + \mathbf{v}) = aF(\mathbf{u}) + F(\mathbf{v})$ for each $\mathbf{u}$ and $\mathbf{v}$ in $\mathbf{R}^n$ and a in $\mathbf{R}$.

23. Let $T : \mathbf{R}^n \to \mathbf{R}^m$ be a linear transformation. Prove that for any linear combination $c_1\mathbf{v}_1 + \cdots + c_k\mathbf{v}_k$ in $\mathbf{R}^n$, $T(c_1\mathbf{v}_1 + \cdots + c_k\mathbf{v}_k) = c_1 T(\mathbf{v}_1) + \cdots + c_k T(\mathbf{v}_k)$.

24. Let $T : \mathbf{R}^n \to \mathbf{R}^m$ be a linear transformation. If $\mathscr{S}$ is a subset of $\mathbf{R}^n$, then $T[\mathscr{S}]$ is the set of all images $T(\mathbf{s})$ for $\mathbf{s}$ in $\mathscr{S}$. If l is a line in $\mathbf{R}^n$, show that $T[l]$ is either a line or a point in $\mathbf{R}^m$. (*Hint:* use exercise 23 and the point-parallel form for a line.)

5.2 Algebra of Linear Transformations

In the first section of this chapter matrices were used to provide examples of linear transformations, while other linear transformations were presented by different means. In this section we will show that *every* linear transformation may be viewed as multiplication by an associated matrix. Furthermore we will introduce an algebra of linear transformations that corresponds exactly to the algebra of the matrices associated with them.

Matrix of a Linear Transformation

We begin with a theorem that gives the procedure for finding the matrix just described.

Theorem 1   Let $T : \mathbf{R}^n \to \mathbf{R}^m$ be a linear transformation, and let A be the $m \times n$ matrix with the ith column given by $A^i = T(\mathbf{e}_i)$. Then A is the unique matrix for which $A\mathbf{x} = T(\mathbf{x})$ for all $\mathbf{x}$ in $\mathbf{R}^n$.

Proof   Since $A^i = (a_{1i}, \ldots, a_{mi}) = T(\mathbf{e}_i)$ for each $i = 1, \ldots, n$, we have the following computation:

$$A e_i = \begin{bmatrix} a_{11} & \cdots & a_{1i} & \cdots & a_{1n} \\ a_{21} & \cdots & a_{2i} & \cdots & a_{2n} \\ \vdots & & \vdots & & \vdots \\ a_{m1} & \cdots & a_{mi} & \cdots & a_{mn} \end{bmatrix} \begin{bmatrix} 0 \\ \vdots \\ 0 \\ 1 \\ 0 \\ \vdots \\ 0 \end{bmatrix}$$

$$= \begin{bmatrix} a_{1i} \\ a_{2i} \\ \vdots \\ a_{mi} \end{bmatrix} = (a_{1i}, \ldots, a_{mi}) = T(e_i).$$

Thus multiplication on the left by A agrees with the linear transformation T on a basis for $\mathbf{R}^n$ (the standard one) and is itself a linear transformation (theorem 2 of section 5.1). By theorem 3 of section 5.1 $A\mathbf{x} = T(\mathbf{x})$ for all $\mathbf{x}$ in $\mathbf{R}^n$.

To show uniqueness, suppose B is a matrix such that $B\mathbf{x} = T(\mathbf{x})$ for all $\mathbf{x}$ in $\mathbf{R}^n$. Then in particular $Be_i = T(e_i)$ for $i = 1, \ldots, n$. But $Be_i = B^i$ and $T(e_i) = A^i$, which means the ith column of B is the ith column of A for each i. Thus $B = A$. ∎

Note: The matrix A of theorem 1 is called the matrix **associated with** T or the matrix that **represents** T.

Example 5.9

Let $T : \mathbf{R}^2 \rightarrow \mathbf{R}^3$ be the linear transformation of example 5.1. For each $\mathbf{x} = (x, y)$ in $\mathbf{R}^2$, $T(x, y) = (x + y, x, 2x - y)$. Find the matrix that represents T.

We have already shown that T is linear, so we proceed with the computation of the associated matrix A:

$$A^1 = T(e_1) = T(1, 0) = (1, 1, 2),$$
$$A^2 = T(e_2) = T(0, 1) = (1, 0, -1).$$

Thus the matrix A is given by

$$A = \begin{bmatrix} 1 & 1 \\ 1 & 0 \\ 2 & -1 \end{bmatrix}.$$

By theorem 1 we do not need to verify that this matrix A represents T. To emphasize the point, however, we do it anyway. For any $\mathbf{x} = (x, y)$ in $\mathbf{R}^2$

$$Ax = \begin{bmatrix} 1 & 1 \\ 1 & 0 \\ 2 & -1 \end{bmatrix} \begin{bmatrix} x \\ y \end{bmatrix} = \begin{bmatrix} x + y \\ x \\ 2x - y \end{bmatrix}.$$

The resulting vector is exactly $T(x, y)$.

Note: Before applying theorem 1, it is essential to check that the function T is indeed a linear transformation. Following the given procedure will always produce a matrix, even for nonlinear functions. However, unless the function is linear, the resulting matrix will not represent the function (see exercise 19).

Example 5.10

Find the matrix associated with the identity transformation $I : \mathbf{R}^n \to \mathbf{R}^n$.
By definition $I(\mathbf{x}) = \mathbf{x}$ for each $\mathbf{x}$ in $\mathbf{R}^n$. Then in particular $I(\mathbf{e}_i) = \mathbf{e}_i = A^i$ for each i, and thus the matrix A associated with I is the matrix with ith column equal to $\mathbf{e}_i$ for each i. Thus, $A = I$, the $n \times n$ identity matrix.

Note: The identity *transformation*, $I : \mathbf{R}^n \to \mathbf{R}^n$, is represented by the $n \times n$ identity *matrix*, I. Because of their close relationship, using the same symbol for both will present no difficulties. Similarly the zero transformation $0 : \mathbf{R}^n \to \mathbf{R}^m$ is represented by the $m \times n$ zero matrix, 0. Again the slight ambiguity is no problem.

Composition of Linear Transformations

In general, if $f : \mathcal{A} \to \mathcal{B}$ and $g : \mathcal{B} \to \mathcal{C}$ are functions, the **composition** of f and g is the function $g \circ f : \mathcal{A} \to \mathcal{C}$ defined by $(g \circ f)(a) = g(f(a))$ for all a in $\mathcal{A}$. That is, first apply f to obtain $f(a)$, and then apply g to the result. As usual we will only be concerned with functions that are linear transformations.

Example 5.11

Let $S : \mathbf{R}^2 \to \mathbf{R}^3$ be defined by $S(x, y) = (x - y, x + y, 2y)$, and let $T : \mathbf{R}^3 \to \mathbf{R}^3$ be defined by $T(x, y, z) = (x, 2x - y, x + y + z)$. Determine $T \circ S$.
Before computing the general result let us find $(T \circ S)(1, 2)$. First, $S(1, 2) = (-1, 3, 4)$. Now apply T to this result,

$$(T \circ S)(1, 2) = T(S(1, 2)) = T(-1, 3, 4) = (-1, -5, 6).$$

The general result is computed in exactly the same way but, unless we change variable names, it is easy to become confused. Let us rewrite T as $T(s, t, u) = (s, 2s - t, s + t + u)$. Now to evaluate T at $S(x, y) = (x - y, x + y, 2y)$, we will have $s = x - y$, $t = x + y$, and $u = 2y$:

$$(T \circ S)(x, y) = T(S(x, y)) = T(x - y, x + y, 2y)$$
$$= (x - y, 2(x - y) - (x + y), (x - y) + (x + y) + 2y)$$
$$= (x - y, x - 3y, 2x + 2y).$$

The resulting description of $T \circ S$ is thus

$$(T \circ S)(x, y) = (x - y, x - 3y, 2x + 2y).$$

In example 5.11 you can check to see that S, T, and $T \circ S$ are all linear transformations. No mention was made of the composition in reverse order, $S \circ T$, since the image of any vector $T(\mathbf{x})$ is in $\mathbf{R}^3$, and S is only defined for vectors in $\mathbf{R}^2$. These observations are stated more generally in the following theorem.

Theorem 2 Let $S : \mathbf{R}^n \to \mathbf{R}^m$ and $T : \mathbf{R}^k \to \mathbf{R}^l$ be linear transformations. Then the composition $T \circ S : \mathbf{R}^n \to \mathbf{R}^l$ is defined if and only if $m = k$. In this case $T \circ S$ is a linear transformation.

Proof That m must equal k in order to compose the functions is clear from the definition of composition. It remains to be shown that, when $m = k$, the resulting composition is a linear transformation.

For any $\mathbf{u}$ in $\mathbf{R}^n$ and a in $\mathbf{R}$

$$\begin{aligned}(T \circ S)(a\mathbf{u}) &= T(S(a\mathbf{u})) && \text{(definition of composition)} \\ &= T(aS(\mathbf{u})) && \text{(S is linear)} \\ &= aT(S(\mathbf{u})) && \text{(T is linear)} \\ &= a(T \circ S)(\mathbf{u}) && \text{(definition of composition).}\end{aligned}$$

That $(T \circ S)(\mathbf{u} + \mathbf{v}) = (T \circ S)(\mathbf{u}) + (T \circ S)(\mathbf{v})$ for all $\mathbf{u}$ and $\mathbf{v}$ is done in a similar manner and is left as exercise 13. ∎

Since the composition of two linear transformations is a linear transformation, and since linear transformations can be represented by matrices (theorem 1), it is natural to inquire about the nature of the matrix that represents the composition. The next theorem describes the situation.

Theorem 3 Let $S : \mathbf{R}^n \to \mathbf{R}^m$ and $T : \mathbf{R}^m \to \mathbf{R}^l$ be linear transformations, and let

A be the matrix that represents S and B the matrix that represents T. Then the matrix product BA represents the composition $T \circ S$.

Proof Since B is $l \times m$, and A is $m \times n$, the matrix product BA is defined and is an $l \times n$ matrix as required ($T \circ S$ is a linear transformation from $\mathbf{R}^n$ to $\mathbf{R}^l$). For any $\mathbf{x}$ in $\mathbf{R}^n$

$$
\begin{aligned}
(BA)\mathbf{x} &= (B)(A\mathbf{x}) & \text{(matrix multiplication is associative)} \\
&= (B)(S(\mathbf{x})) & (A \text{ represents } S) \\
&= T(S(\mathbf{x})) & (B \text{ represents } T) \\
&= (T \circ S)(\mathbf{x}) & \text{(definition of composition)}.
\end{aligned}
$$

Thus multiplication on the left by BA has exactly the same effect as mapping by the linear transformation $T \circ S$ and BA represents $T \circ S$. ∎

Example 5.12

Let S and T be as in example 5.11. Find the matrices A, B, and C that represent S, T, and $T \circ S$, respectively, and verify that $C = BA$.
From example 5.11

$$
\begin{aligned}
S(x, y) &= (x - y, x + y, 2y), \\
T(x, y, z) &= (x, 2x - y, x + y + z), \\
(T \circ S)(x, y) &= (x - y, x - 3y, 2x + 2y).
\end{aligned}
$$

Following the procedure described in theorem 1 for each of these linear transformations,

$$
A = \begin{bmatrix} 1 & -1 \\ 1 & 1 \\ 0 & 2 \end{bmatrix}, \quad B = \begin{bmatrix} 1 & 0 & 0 \\ 2 & -1 & 0 \\ 1 & 1 & 1 \end{bmatrix}, \quad \text{and} \quad C = \begin{bmatrix} 1 & -1 \\ 1 & -3 \\ 2 & 2 \end{bmatrix}.
$$

The essence of theorem 3 is that it is unnecessary to first evaluate $T \circ S$ and then compute its corresponding matrix; exactly the same result may be obtained by computing A and B and then $C = BA$. Performing this multiplication, we do indeed have

$$
BA = \begin{bmatrix} 1 & 0 & 0 \\ 2 & -1 & 0 \\ 1 & 1 & 1 \end{bmatrix} \begin{bmatrix} 1 & -1 \\ 1 & 1 \\ 0 & 2 \end{bmatrix} = \begin{bmatrix} 1 & -1 \\ 1 & -3 \\ 2 & 2 \end{bmatrix} = C.
$$

Example 5.13

Show that the composition of a rotation of $\mathbf{R}^2$ by an angle θ followed by a rotation by an angle ψ is a rotation by the angle $\theta + \psi$. (Obviously this is true geometrically; what is intended is an analytical argument.)

From example 5.5 we have that the rotation by θ is represented by the matrix

$$A_\theta = \begin{bmatrix} \cos\theta & -\sin\theta \\ \sin\theta & \cos\theta \end{bmatrix},$$

and the rotation by ψ is represented by the matrix

$$A_\psi = \begin{bmatrix} \cos\psi & -\sin\psi \\ \sin\psi & \cos\psi \end{bmatrix}.$$

By theorem 3 the matrix that represents the composition is the product $A_\psi A_\theta$. Computing the product, we have

$$A_\psi A_\theta = \begin{bmatrix} \cos\psi\cos\theta - \sin\psi\sin\theta & -\cos\psi\sin\theta - \sin\psi\cos\theta \\ \sin\psi\cos\theta + \cos\psi\sin\theta & -\sin\psi\sin\theta + \cos\psi\cos\theta \end{bmatrix}$$

$$= \begin{bmatrix} \cos(\theta+\psi) & -\sin(\theta+\psi) \\ \sin(\theta+\psi) & \cos(\theta+\psi) \end{bmatrix} = A_{\theta+\psi}.$$

It is left as an exercise to show that the composition of two *reflections* is also a *rotation*.

Algebra of Linear Transformations

Since composition of linear transformations and multiplication of the associated matrices are so closely related by theorem 3, it is common to speak of the **product** of two linear transformations instead of their composition. In this vein, for $S : \mathbf{R}^n \to \mathbf{R}^m$ and $T : \mathbf{R}^m \to \mathbf{R}^l$, the composition $T \circ S$ is often abbreviated to TS.

In general a linear transformation cannot be composed with itself. In other words, for $T : \mathbf{R}^n \to \mathbf{R}^m$, $TT = T \circ T$ is not defined unless $n = m$. If $n = m$, the composition is defined and can be repeated as often as we wish. The convenient notation $T^2, T^3, \ldots$ symbolizes $T \circ T, T \circ (T \circ T), \ldots$. If A is the matrix that represents T, then A is a square matrix of order n, and repeated application of theorem 3 implies that A^k corresponds to T^k.

The following definition and theorem summarize this discussion. Notice that the theorem is simply a restatement of theorem 3, using the new terminology.

Definition   Let $S : \mathbf{R}^n \to \mathbf{R}^m$, $T : \mathbf{R}^m \to \mathbf{R}^l$, and $L : \mathbf{R}^n \to \mathbf{R}^n$ be linear transformations.

  i. The **product** TS is the linear transformation $T \circ S$.
  ii. The **powers** of L are the linear transformations $L^0 = I$, $L^1 = L$, $L^2 = LL, \ldots, L^{k+1} = L(L^k)$ for all $k = 2, 3, \ldots$.

Theorem 4   Let $S : \mathbf{R}^n \to \mathbf{R}^m$, $T : \mathbf{R}^m \to \mathbf{R}^l$, and $L : \mathbf{R}^n \to \mathbf{R}^n$ be linear transformations, and let their associated matrices be A, B, and C, respectively.

i. The matrix that represents the product TS is the matrix product BA.
ii. The matrix that represents the power L^k is the matrix power C^k, $k = 0, 1, 2, \ldots$.

Thus we have a "multiplication" for linear transformations that has properties very much like those of matrix multiplication. Sums, differences, and negatives of linear transformations and products of linear transformations by scalars can also be defined in a very natural way.

Definition   Let S and T be linear transformations from $\mathbf{R}^n$ to $\mathbf{R}^m$, and let c be a scalar.

i. The **sum** $S + T$ is the function, given by $(S + T)(\mathbf{x}) = S(\mathbf{x}) + T(\mathbf{x})$ for all $\mathbf{x}$ in $\mathbf{R}^n$.
ii. The **difference** $S - T$ is the function, given by $(S - T)(\mathbf{x}) = S(\mathbf{x}) - T(\mathbf{x})$ for all $\mathbf{x}$ in $\mathbf{R}^n$.
iii. The **negative** $-S$ is the function given by $(-S)(\mathbf{x}) = -S(\mathbf{x})$ for all $\mathbf{x}$ in $\mathbf{R}^n$.
iv. The **scalar product** cS is the function given by $(cS)(\mathbf{x}) = (c)(S(\mathbf{x}))$ for all $\mathbf{x}$ in $\mathbf{R}^n$.

Example 5.14

Let S and T be linear transformations from $\mathbf{R}^3$ to $\mathbf{R}^2$ defined by $S(x, y, z) = (x + y - z, 3y + z)$ and $T(x, y, z) = (2x + z, y + z)$. Describe the transformation $3S - 2T$.

For any $\mathbf{x} = (x, y, z)$ we follow the definition:

$$
\begin{aligned}
(3S - 2T)(x, y, z) &= (3S)(x, y, z) - (2T)(x, y, z) \\
&= 3(S(x, y, z)) - 2(T(x, y, z)) \\
&= 3(x + y - z, 3y + z) - 2(2x + z, y + z) \\
&= (3x + 3y - 3z, 9y + 3z) - (4x + 2z, 2y + 2z) \\
&= (-x + 3y - 5z, 7y + z).
\end{aligned}
$$

It would be easy to verify directly that the resulting function $3S - 2T$ of example 5.14 is a linear transformation, but the following theorem makes this unnecessary.

Theorem 5   Let S and T be linear transformations from $\mathbf{R}^n$ to $\mathbf{R}^m$, and let c be a scalar. Then $S + T$, $S - T$, $-S$, and cS are all linear transformations from $\mathbf{R}^n$ to $\mathbf{R}^m$.

Proof We prove only the case of $S + T$ and leave the remaining ones as an exercise. We must check the two defining properties of linear transformation. If $\mathbf{x}$ and $\mathbf{y}$ are vectors in $\mathbf{R}^n$, then

$$
\begin{aligned}
(S + T)(\mathbf{x} + \mathbf{y}) &= S(\mathbf{x} + \mathbf{y}) + T(\mathbf{x} + \mathbf{y}) && \text{(definition of } S + T) \\
&= S(\mathbf{x}) + S(\mathbf{y}) + T(\mathbf{x}) + T(\mathbf{y}) && \text{(linearity of } S \text{ and } T) \\
&= S(\mathbf{x}) + T(\mathbf{x}) + S(\mathbf{y}) + T(\mathbf{y}) && \text{(vector algebra)} \\
&= (S + T)(\mathbf{x}) + (S + T)(\mathbf{y}) && \text{(definition of } S + T).
\end{aligned}
$$

If c is any scalar,

$$
\begin{aligned}
(S + T)(c\mathbf{x}) &= S(c\mathbf{x}) + T(c\mathbf{x}) && \text{(definition of } S + T) \\
&= cS(\mathbf{x}) + cT(\mathbf{x}) && \text{(linearity of } S \text{ and } T) \\
&= c(S(\mathbf{x}) + T(\mathbf{x})) && \text{(vector algebra)} \\
&= c((S + T)(\mathbf{x})) && \text{(definition of } S + T).
\end{aligned}
$$

Thus the sum of two linear transformations is a linear transformation. ∎

Since the sum, difference, negative, and scalar product functions are linear transformations, there is a matrix that represents each of them. These associated matrices are compatible with the corresponding operation in the following sense.

Theorem 6 Let S and T be linear transformations from $\mathbf{R}^n$ to $\mathbf{R}^m$, and let their associated matrices be A and B, respectively. Then the associated matrix of the linear transformation

  i. $S + T$ is $A + B$,
  ii. $S - T$ is $A - B$,
  iii. $-S$ is $-A$,
  iv. cS is cA.

Proof Again we prove only the case of $S + T$ and leave the remaining ones as an exercise. For any $\mathbf{x}$ in $\mathbf{R}^n$

$$
\begin{aligned}
(A + B)\mathbf{x} &= A\mathbf{x} + B\mathbf{x} && \text{(theorem 1g of section 2.3)} \\
&= S(\mathbf{x}) + T(\mathbf{x}) && (A \text{ and } B \text{ represent } S \text{ and } T) \\
&= (S + T)(\mathbf{x}) && \text{(definition of } S + T).
\end{aligned}
$$

Therefore the matrix $A + B$ represents the linear transformation $S + T$. ∎

Example 5.15

Let S and T be the linear transformations of example 5.14. Let A, B, and C be the matrices that represent S, T, and $3S - 2T$, respectively. Verify directly that $C = 3A - 2B$.

Following theorem 1, we have from example 5.14 that

$$A = \begin{bmatrix} 1 & 1 & -1 \\ 0 & 3 & 1 \end{bmatrix}, \quad B = \begin{bmatrix} 2 & 0 & 1 \\ 0 & 1 & 1 \end{bmatrix}, \quad \text{and} \quad C = \begin{bmatrix} -1 & 3 & -5 \\ 0 & 7 & 1 \end{bmatrix}.$$

Finally we have

$$3A - 2B = 3\begin{bmatrix} 1 & 1 & -1 \\ 0 & 3 & 1 \end{bmatrix} - 2\begin{bmatrix} 2 & 0 & 1 \\ 0 & 1 & 1 \end{bmatrix} = \begin{bmatrix} -1 & 3 & -5 \\ 0 & 7 & 1 \end{bmatrix} = C,$$

as required.

Example 5.16

Let $T : \mathbf{R}^3 \to \mathbf{R}^3$ be defined by $T(x, y, z) = (x + 2y - z, 3x - z, y + z)$. Describe the transformation $2T^2 - T + 3I$.

We use the algebra of the corresponding matrices to simplify the computation. Let A represent T, and note that the identity matrix I of order 3 represents I. Then $2A^2 - A + 3I$ represents $2T^2 - T + 3I$. By theorem 1, A is given by

$$A = \begin{bmatrix} 1 & 2 & -1 \\ 3 & 0 & -1 \\ 0 & 1 & 1 \end{bmatrix},$$

and hence

$$2A^2 - A + 3I = 2\begin{bmatrix} 1 & 2 & -1 \\ 3 & 0 & -1 \\ 0 & 1 & 1 \end{bmatrix}^2 - \begin{bmatrix} 1 & 2 & -1 \\ 3 & 0 & -1 \\ 0 & 1 & 1 \end{bmatrix} + 3\begin{bmatrix} 1 & 0 & 0 \\ 0 & 1 & 0 \\ 0 & 0 & 1 \end{bmatrix}$$

$$= \begin{bmatrix} 14 & 2 & -8 \\ 6 & 10 & -8 \\ 6 & 2 & 0 \end{bmatrix} - \begin{bmatrix} 1 & 2 & -1 \\ 3 & 0 & -1 \\ 0 & 1 & 1 \end{bmatrix} + \begin{bmatrix} 3 & 0 & 0 \\ 0 & 3 & 0 \\ 0 & 0 & 3 \end{bmatrix}$$

$$= \begin{bmatrix} 16 & 0 & -7 \\ 3 & 13 & -7 \\ 6 & 1 & 2 \end{bmatrix}.$$

Finally,

$$(2T^2 - T + 3I)(x, y, z) = \begin{bmatrix} 16 & 0 & -7 \\ 3 & 13 & -7 \\ 6 & 1 & 2 \end{bmatrix}\begin{bmatrix} x \\ y \\ z \end{bmatrix} = \begin{bmatrix} 16x & -7z \\ 3x + 13y - 7z \\ 6x + y + 2z \end{bmatrix},$$

so that $(2T^2 - T + 3I)(x, y, z) = (16x - 7z, 3x + 13y - 7z, 6x + y + 2z)$.

We conclude this section with those properties of linear transformations that correspond to the analogous ones for matrices (theorem 1 of section 2.3).

Theorem 7   Let S, T, and U be linear transformations and a and b scalars. Assume that the linear transformations are such that each of the following operations is defined. Then

    a. $S + T = T + S$ (commutative law for addition),
    b. $S + (T + U) = (S + T) + U$ (associative law for addition),
    c. $S + 0 = S$,
    d. $S + (-S) = 0$,
    e. $S(TU) = (ST)U$ (associative law for multiplication),
    f. $SI = S$, $IS = S$,
    g. $\left.\begin{array}{l} S(T + U) = ST + SU \\ (S + T)U = SU + TU \end{array}\right\}$ (distributive laws),
    h. $a(S + T) = aS + aT$,
    i. $(a + b)S = aS + bS$,
    j. $(ab)S = a(bS)$,
    k. $1S = S$,
    l. $S0 = 0$, $0S = 0$,
    m. $a0 = 0$,
    n. $a(ST) = (aS)T$.

Proof   We prove parts a, e, g, and j and leave the others as an exercise. To show two linear transformations are equal, we show that they are equal as functions, that they agree at any vector for which they are defined. In these arguments, when we say "for each $\mathbf{x}$," we really mean "for each $\mathbf{x}$ for which the functions are defined."

Proof of 1a   For each $\mathbf{x}$

$$
\begin{aligned}
(S + T)(\mathbf{x}) &= S(\mathbf{x}) + T(\mathbf{x}) && \text{(definition of sum)} \\
&= T(\mathbf{x}) + S(\mathbf{x}) && \text{(theorem 2a of section 1.4)} \\
&= (T + S)(\mathbf{x}) && \text{(definition of sum).}
\end{aligned}
$$

Therefore $S + T = T + S$.

Proof of 1e   For each $\mathbf{x}$

$$
\begin{aligned}
S(TU)(\mathbf{x}) &= S(TU(\mathbf{x})) && \text{(definition of product)} \\
&= S(T(U(\mathbf{x}))) && \text{(definition of product)} \\
&= ST(U(\mathbf{x})) = (ST)U(\mathbf{x}) && \text{(definition of product).}
\end{aligned}
$$

Therefore $S(TU) = (ST)U$.

Proof of 1g For each **x**

$$
\begin{aligned}
(S(T + U))(\mathbf{x}) &= S((T + U)(\mathbf{x})) && \text{(definition of product)} \\
&= S(T(\mathbf{x}) + U(\mathbf{x})) && \text{(definition of sum)} \\
&= S(T(\mathbf{x})) + S(U(\mathbf{x})) && (S \text{ is a linear transformation)} \\
&= ST(\mathbf{x}) + SU(\mathbf{x}) && \text{(definition of product)} \\
&= (ST + SU)(\mathbf{x}) && \text{(definition of sum)}.
\end{aligned}
$$

Therefore $S(T + U) = ST + SU$. (The other case of 1g is similarly handled.)

Proof of 1j For each **x**

$$
\begin{aligned}
(ab)S(\mathbf{x}) &= (ab)(S(\mathbf{x})) && \text{(definition of multiplication by scalar)} \\
&= a(bS(\mathbf{x})) && \text{(theorem 2e of section 1.4)} \\
&= (a(bS))(\mathbf{x}) && \text{(definition of multiplication by scalar)}.
\end{aligned}
$$

Therefore $(ab)S = a(bS)$.   ■

5.2 Exercises

In exercises 1 through 5 compute the matrix that represents the given linear transformation.

1. $K(x, y, z) = (x, x + y, x + y + z)$
2. $L(x, y, z) = (2x - y, x + 2y)$
3. $S(x, y, z) = (z, y, x)$
4. $T(x, y) = (2x + y, x + y, x - y, x - 2y)$
5. $\Pi_i : \mathbf{R}^n \to \mathbf{R}$ defined by $\Pi_i(x_1, \ldots, x_n) = x_i$

In exercises 6 through 11 find the matrix of the indicated linear transformation. The transformations K, L, S, and T are defined in exercises 1 through 4.

6. $LK (= L \circ K)$ 7. TL
8. S^2 9. $K + S$
10. $3K - 2S$ 11. $2S^3 - S^2 + 3S - 4I$

12. Show that the composition of one reflection T^θ followed by another T^ψ is the rotation by the angle $2(\psi - \theta)$.

13. Finish the proof of theorem 2 by showing that, if S and T are linear transformations such that $T \circ S$ is defined, then $(T \circ S)(\mathbf{u} + \mathbf{v}) = (T \circ S)\mathbf{u} + (T \circ S)\mathbf{v}$.

14. Complete the proof of theorem 5.

15. Complete the proof of theorem 6.

16. Complete the proof of theorem 7.

17. Show that the composition of a contraction/dilation by a contraction/dilation is a contraction/dilation.

18. If $S : \mathbf{R}^n \to \mathbf{R}^n$ is a linear transformation, and $T_c : \mathbf{R}^n \to \mathbf{R}^n$ is a contraction/dilation, prove that $ST_c = T_c S$. In other words, show that a contraction/dilation *commutes* with any linear operator.

19. Let $f : \mathbf{R}^2 \to \mathbf{R}^2$ be the *nonlinear* function defined by $f(x, y) = (xy, x^2)$. Let A be the 2×2 matrix constructed as in theorem 1. Show that A does *not* represent f. In other words, show that for some $\mathbf{x}$ in $\mathbf{R}^2$, $f(\mathbf{x}) \neq A\mathbf{x}$.

5.3 Kernel and Image

In this section we will investigate further properties of linear transformations as well as two very important subspaces associated with linear transformations.

Kernel of a Linear Transformation

If T is a linear transformation, theorem 1 of section 5.1 showed us that $T(\mathbf{0}) = \mathbf{0}$. There may or may not be other vectors $\mathbf{x}$ such that $T(\mathbf{x}) = \mathbf{0}$.

Definition Let $T : \mathbf{R}^n \to \mathbf{R}^m$ be a linear transformation. The set of all vectors $\mathbf{x}$ in $\mathbf{R}^n$ that satisfy $T(\mathbf{x}) = \mathbf{0}$ is called the **kernel** (or **nullspace**) of T and is denoted by $\mathrm{Ker}(T)$.

Example 5.17

Compute $\mathrm{Ker}(T)$ for $T : \mathbf{R}^2 \to \mathbf{R}^3$ given by $T(x, y) = (x + y, x, 2x - y)$.
That the transformation T is linear has been shown in example 5.1. To find $\mathrm{Ker}(T)$, we must solve the equation $T(x, y) = (0, 0, 0)$ for all possible ordered pairs (x, y). That is, we solve

$$(x + y, x, 2x - y) = (0, 0, 0).$$

Equating components, we have

$$x + y = 0$$
$$x \quad\quad = 0$$
$$2x - y = 0.$$

This homogeneous system of linear equations can easily be solved by inspection. The second equation implies that $x = 0$ which in turn implies that $y = 0$ from the first equation. Thus $\mathrm{Ker}(T) = \{\mathbf{0}\}$, the set that consists solely of the zero vector.

Example 5.18

Let 0 be the zero transformation from $\mathbf{R}^n$ to $\mathbf{R}^m$, that is, $0(\mathbf{x}) = \mathbf{0}$ for each $\mathbf{x}$ in $\mathbf{R}^n$. Compute $\text{Ker}(T)$.

That 0 is a linear transformation was verified in example 5.2. In this case it is clear that $\text{Ker}(0) = \mathbf{R}^n$ since every $\mathbf{x}$ in $\mathbf{R}^n$ gets mapped into $\mathbf{0}$.

Example 5.19

Describe $\text{Ker}(T)$, where $T : \mathbf{R}^3 \to \mathbf{R}^2$ is the linear transformation given by $T(\mathbf{x}) = A\mathbf{x}$ and

$$A = \begin{bmatrix} 2 & 0 & 1 \\ 1 & 1 & -1 \end{bmatrix}.$$

Since T is given by matrix multiplication, it is linear by theorem 2 of section 5.1. To compute $\text{Ker}(T)$ we let $\mathbf{x} = (x, y, z)$, determine $T(\mathbf{x})$, and set it equal to zero.

$$T(\mathbf{x}) = \begin{bmatrix} 2 & 0 & 1 \\ 1 & 1 & -1 \end{bmatrix} \begin{bmatrix} x \\ y \\ z \end{bmatrix} = \begin{bmatrix} 2x & + z \\ x + y & - z \end{bmatrix} = \begin{bmatrix} 0 \\ 0 \end{bmatrix}.$$

Equating components, we have

$$2x \quad + z = 0$$
$$x + y - z = 0.$$

We solve the resulting homogeneous system by the methods of chapter 2, forming the augmented matrix and transforming it to row-reduced echelon form:

$$\begin{bmatrix} 2 & 0 & 1 & | & 0 \\ 1 & 1 & -1 & | & 0 \end{bmatrix} \to \begin{bmatrix} 1 & 0 & 1/2 & | & 0 \\ 0 & 1 & -3/2 & | & 0 \end{bmatrix}.$$

Thus

$$x = (-1/2)t,$$
$$y = (3/2)t,$$
$$z = t,$$

and we have a one-parameter family of solutions. Expressing this result in vector form,

$$\mathbf{x} = t(-1/2, 3/2, 1),$$

we see that $\text{Ker}(T)$ is the line through the origin determined by the vector $(-1/2, 3/2, 1)$ (or more conveniently, by $(-1, 3, 2)$).

In each of these three examples, the kernel has been a subspace of $\mathbf{R}^n$. The following theorem shows that the kernel of a linear transformation is *always* a subspace. The dimension of the kernel is called the **nullity** of the linear transformation.

Theorem 1   Let $T : \mathbf{R}^n \to \mathbf{R}^m$ be a linear transformation. Then the kernel of T is a subspace of $\mathbf{R}^n$.

Proof   That $\text{Ker}(T)$ is a *subset* of $\mathbf{R}^n$ is clear from the definition.
For $\mathbf{u}$ and $\mathbf{v}$ in $\text{Ker}(T)$, we compute $T(\mathbf{u} + \mathbf{v})$,

$$T(\mathbf{u} + \mathbf{v}) = T(\mathbf{u}) + T(\mathbf{v}) = \mathbf{0} + \mathbf{0} = \mathbf{0}.$$

Thus $\mathbf{u} + \mathbf{v}$ is in $\text{Ker}(T)$ as well.
For $\mathbf{u}$ in $\text{Ker}(T)$ and c any scalar,

$$T(c\mathbf{u}) = cT(\mathbf{u}) = c\mathbf{0} = \mathbf{0}.$$

Thus $c\mathbf{u}$ is also in $\text{Ker}(T)$.
These two properties confirm that the kernel of a linear transformation is a subspace of $\mathbf{R}^n$.   ∎

If the linear transformation is given by a matrix as in example 5.19, computation of the kernel is particularly easy. Notice that in this example, we obtained the kernel by augmenting the matrix A by a column of zeros and row-reducing. That this is always the case is simply a statement of the fact that $T(\mathbf{x}) = \mathbf{0}$ if and only if $A\mathbf{x} = \mathbf{0}$. The latter equation may be solved by row-reducing the augmented matrix $[A|\mathbf{0}]$, and a basis for the kernel may be produced by following the procedure of theorem 4 of section 4.3. This observation is stated as the next theorem. Recall that the *rank* of a matrix is the number of nonzero rows in its row-reduced echelon form.

Theorem 2   Let $T : \mathbf{R}^n \to \mathbf{R}^m$ be a linear transformation, and let A be the matrix that represents T. Then the kernel of T is the solution space of the vector equation $A\mathbf{x} = \mathbf{0}$. Thus $\dim(\text{Ker}(T))$, the nullity of T, is $n - r$ where r is the rank of the matrix A.

Procedure for computing a basis for the kernel of a transformation   Solve $A\mathbf{x} = \mathbf{0}$ by transforming the matrix A to row-reduced echelon form, and then follow the procedure of theorem 4 of section 4.3. The result is a basis for $\text{Ker}(T)$.

Example 5.20

Let $T : \mathbf{R}^5 \to \mathbf{R}^4$ be given by $T(\mathbf{x}) = A\mathbf{x}$, where A is the matrix

$$A = \begin{bmatrix} 1 & 2 & 0 & 1 & 0 \\ 2 & 4 & 1 & 0 & 0 \\ 0 & 0 & 1 & -2 & 1 \\ 1 & 2 & 1 & -1 & 1 \end{bmatrix}.$$

Find the nullity of T and a basis for $\mathrm{Ker}(T)$.
The row-reduced echelon form of A is the matrix

$$B = \begin{bmatrix} 1 & 2 & 0 & 1 & 0 \\ 0 & 0 & 1 & -2 & 0 \\ 0 & 0 & 0 & 0 & 1 \\ 0 & 0 & 0 & 0 & 0 \end{bmatrix}. \qquad x_5 = 0$$

Since B has three nonzero rows, the rank of A is $r = 3$. Since $n = 5$ in this example, the nullity is $n - r = 2$. Following the procedure of theorem 4 of section 4.3, we determine that a basis for $\mathrm{Ker}(T)$ is $\{(-2, 1, 0, 0, 0), (-1, 0, 2, 1, 0)\}$.

A function $f : \mathscr{A} \to \mathscr{B}$ is **one-to-one** if, whenever a_1 and a_2 satisfy $f(a_1) = f(a_2)$, then $a_1 = a_2$. In other words, no two elements of $\mathscr{A}$ are mapped to the same element of $\mathscr{B}$. It is particularly easy to check to see if a *linear transformation* is one-to-one.

Theorem 3 Let $T : \mathbf{R}^n \to \mathbf{R}^m$ be a linear transformation. Then T is one-to-one if and only if $\mathrm{Ker}(T) = \{\mathbf{0}\}$. $\qquad$ Thus

Proof If T is one-to-one, then $T(\mathbf{x}) = \mathbf{0}$ can have only one solution, namely $\mathbf{x} = \mathbf{0}$. That is, $\mathrm{Ker}(T) = \{\mathbf{0}\}$. Conversely suppose $\mathrm{Ker}(T) = \{\mathbf{0}\}$ and $T(\mathbf{u}) = T(\mathbf{v})$. We need to show that $\mathbf{u} = \mathbf{v}$. Since $T(\mathbf{u}) = T(\mathbf{v})$,

$$T(\mathbf{u}) - T(\mathbf{v}) = \mathbf{0},$$

and therefore $T(\mathbf{u} - \mathbf{v}) = \mathbf{0}$ (theorem 1c of section 5.1). Thus $\mathbf{u} - \mathbf{v}$ is in $\mathrm{Ker}(T)$. But then $\mathbf{u} - \mathbf{v} = \mathbf{0}$ since $\mathbf{0}$ is the *only* element of $\mathrm{Ker}(T)$. Hence $\mathbf{u} = \mathbf{v}$, and T is one-to-one. $\qquad\blacksquare$

From the computations given in example 5.17, we see that the linear transformation $T(x, y) = (x + y, x, 2x - y)$ is one-to-one. On the other hand, the linear transformation $T(\mathbf{x}) = A\mathbf{x}$ given in example 5.20 is not.

Image of a Linear Transformation

Let $T : \mathbf{R}^n \to \mathbf{R}^m$ be a linear transformation. The set of all vectors $T(\mathbf{x})$ for $\mathbf{x}$ in $\mathbf{R}^n$ is called the **image** of T and is denoted by $\text{Im}(T)$.

Example 5.21

Describe $\text{Im}(0)$, where 0 is the zero transformation from $\mathbf{R}^n$ to $\mathbf{R}^m$.
Since $0(\mathbf{x}) = \mathbf{0}$ for all $\mathbf{x}$ in $\mathbf{R}^n$, $\text{Im}(0) = \{\mathbf{0}\}$.

Example 5.22

Describe $\text{Im}(T)$, where $T : \mathbf{R} \to \mathbf{R}^2$ is given by $T(x) = (x, 2x)$ for each x in $\mathbf{R}$.
Letting $T(x) = (x, y)$, we have $y = 2x$ for each x in the real numbers $\mathbf{R}$. Thus $\text{Im}(T)$ is the set of points on the line in $\mathbf{R}^2$ that passes through the origin and has slope 2.

Just as the kernel of a linear transformation $T : \mathbf{R}^n \to \mathbf{R}^m$ is a subspace of $\mathbf{R}^n$ (theorem 1), the image of T is a subspace of $\mathbf{R}^m$. The dimension of the image is called the **rank** of T.

Theorem 4   Let $T : \mathbf{R}^n \to \mathbf{R}^m$ be a linear transformation. Then the image of T is a subspace of $\mathbf{R}^m$.

Proof   That $\text{Im}(T)$ is a subset of $\mathbf{R}^m$ is clear from the definition.
For $\mathbf{u}$ and $\mathbf{v}$ in $\text{Im}(T)$, we must show that $\mathbf{u} + \mathbf{v}$ is also in $\text{Im}(T)$. Since $\mathbf{u}$ is in $\text{Im}(T)$, there is some $\mathbf{x}$ in $\mathbf{R}^n$ such that $T(\mathbf{x}) = \mathbf{u}$. Similarly there is some $\mathbf{y}$ in $\mathbf{R}^n$ such that $T(\mathbf{y}) = \mathbf{v}$. Let $\mathbf{z} = \mathbf{x} + \mathbf{y}$. Then

$$T(\mathbf{z}) = T(\mathbf{x} + \mathbf{y}) = T(\mathbf{x}) + T(\mathbf{y}) = \mathbf{u} + \mathbf{v},$$

and $\mathbf{u} + \mathbf{v}$ is the image of $\mathbf{z}$ under T. That is, $\mathbf{u} + \mathbf{v}$ is in $\text{Im}(T)$.
For any scalar c, $T(c\mathbf{x}) = cT(\mathbf{x}) = c\mathbf{u}$. Thus $c\mathbf{u}$ is the image of $c\mathbf{x}$ under T. That is, $c\mathbf{u}$ is in $\text{Im}(T)$.
These two properties confirm that the image of a linear transformation is a subspace of $\mathbf{R}^m$.   ■

Example 5.23

Describe $\text{Im}(T)$, where $T : \mathbf{R}^3 \to \mathbf{R}^3$ is given by $T(\mathbf{x}) = A\mathbf{x}$ and

$$A = \begin{bmatrix} 2 & 1 & 1 \\ 1 & -1 & 0 \\ 1 & 2 & 1 \end{bmatrix}.$$

Letting $\mathbf{x} = (x, y, z)$, we compute $T(\mathbf{x})$,

$$T(\mathbf{x}) = A\mathbf{x} = \begin{bmatrix} 2 & 1 & 1 \\ 1 & -1 & 0 \\ 1 & 2 & 1 \end{bmatrix} \begin{bmatrix} x \\ y \\ z \end{bmatrix} = \begin{bmatrix} 2x + y + z \\ x - y \\ x + 2y + z \end{bmatrix}$$

$$= x \begin{bmatrix} 2 \\ 1 \\ 1 \end{bmatrix} + y \begin{bmatrix} 1 \\ -1 \\ 2 \end{bmatrix} + z \begin{bmatrix} 1 \\ 0 \\ 1 \end{bmatrix}.$$

Thus we see that for any $\mathbf{x}$ in $\mathbf{R}^3$, $T(\mathbf{x})$ is a linear combination of the vectors $(2, 1, 1)$, $(1, -1, 2)$, and $(1, 0, 1)$. Furthermore any given linear combination of these,

$$\mathbf{y} = a(2, 1, 1) + b(1, -1, 2) + c(1, 0, 1),$$

can be seen to be an image of some element $\mathbf{x}$ in $\mathbf{R}^3$ under T by letting $\mathbf{x} = (a, b, c)$:

$$T(a, b, c) = a(2, 1, 1) + b(1, -1, 2) + c(1, 0, 1).$$

Therefore the set of vectors $\{(2, 1, 1), (1, -1, 2), (1, 0, 1)\}$ generates $\text{Im}(T)$. Knowing a spanning set for $\text{Im}(T)$, we can produce a basis for it following theorem 1 of section 4.3. We form the matrix with the spanning set as columns and row-reduce,

$$\begin{bmatrix} 2 & 1 & 1 \\ 1 & -1 & 0 \\ 1 & 2 & 1 \end{bmatrix} \rightarrow \begin{bmatrix} 1 & 0 & 1/3 \\ 0 & 1 & 1/3 \\ 0 & 0 & 0 \end{bmatrix}.$$

Thus a basis for $\text{Im}(T)$ is given by the first two columns, $\{(2, 1, 1), (1, -1, 2)\}$. From a geometric point of view $\text{Im}(T)$ is a plane in $\mathbf{R}^3$.

You probably noticed that the matrix of columns we needed to row-reduce in example 5.23 was exactly the same matrix A as was used to define the linear transformation T at the beginning of example 5.23. The following theorem shows that this is no coincidence. Recall that the *column space* of a matrix A is the set of all linear combinations of its columns (see section 4.4).

Theorem 5 Let $T : \mathbf{R}^n \rightarrow \mathbf{R}^m$ be a linear transformation, and let A be the matrix which represents T. Then $\text{Im}(T)$ is the column space of A, and $\dim(\text{Im}(T))$, the rank of the transformation T, is equal to the rank of the matrix A.

Procedure for computing a basis for the image of a transformation Transform the matrix A to its row-reduced echelon matrix form, B. A basis for $\text{Im}(T)$ consists of those columns of A transformed into the distinct elementary columns of B.

Proof Let $\mathbf{x} = (x_1, \ldots, x_n)$. Then we have

$$T(\mathbf{x}) = A\mathbf{x} = \begin{bmatrix} a_{11} & a_{12} \cdots a_{1n} \\ a_{21} & a_{22} \cdots a_{2n} \\ \vdots & \vdots & \vdots \\ a_{m1} & a_{m2} \cdots a_{mn} \end{bmatrix} \begin{bmatrix} x_1 \\ x_2 \\ \vdots \\ x_n \end{bmatrix}$$

$$= \begin{bmatrix} a_{11}x_1 + a_{12}x_2 + \cdots + a_{1n}x_n \\ a_{21}x_1 + a_{22}x_2 + \cdots + a_{2n}x_n \\ \vdots & \vdots & \vdots \\ a_{m1}x_1 + a_{m2}x_2 + \cdots + a_{mn}x_n \end{bmatrix}$$

$$= x_1 \begin{bmatrix} a_{11} \\ a_{21} \\ \vdots \\ a_{m1} \end{bmatrix} + x_2 \begin{bmatrix} a_{12} \\ a_{22} \\ \vdots \\ a_{m2} \end{bmatrix} + \cdots + x_n \begin{bmatrix} a_{1n} \\ a_{2n} \\ \vdots \\ a_{mn} \end{bmatrix}$$

$$= x_1 A^1 + x_2 A^2 + \cdots + x_n A^n,$$

where A^i denotes the ith column of A. In other words, for each $\mathbf{x}$ in $\mathbf{R}^n$, $T(\mathbf{x})$ is a linear combination of the columns of A. Conversely, if $\mathbf{y}$ in $\mathbf{R}^m$ is a linear combination of the columns of A,

$$\mathbf{y} = c_1 A^1 + c_2 A^2 + \cdots + c_n A^n,$$

then $\mathbf{y} = T(\mathbf{c})$, where $\mathbf{c} = (c_1, \ldots, c_n)$. That is, $\mathbf{y}$ is in $\text{Im}(T)$ if and only if $\mathbf{y}$ is a linear combination of the columns of A. ∎

The last procedure described is essentially that of theorem 1 of section 4.3. It is appropriate here as a consequence of theorem 5.

Example 5.24

For the matrix A let $T : \mathbf{R}^5 \to \mathbf{R}^4$ be defined by $T(\mathbf{x}) = A\mathbf{x}$ for each $\mathbf{x}$ in $\mathbf{R}^5$. Find the rank of T and a basis for $\text{Im}(T)$.

$$A = \begin{bmatrix} 1 & 2 & 0 & 1 & 0 \\ 2 & 4 & 1 & 0 & 0 \\ 0 & 0 & 1 & -2 & 1 \\ 1 & 2 & 1 & -1 & 1 \end{bmatrix}.$$

We row-reduce A to obtain

$$B = \begin{bmatrix} 1 & 2 & 0 & 1 & 0 \\ 0 & 0 & 1 & -2 & 0 \\ 0 & 0 & 0 & 0 & 1 \\ 0 & 0 & 0 & 0 & 0 \end{bmatrix}.$$

Since B has three columns that contain leading ones (the first, third, and fifth), the rank of T is 3. A basis for $\text{Im}(T)$ is given by the columns of A that correspond to these, $\{A^1, A^3, A^5\}$. That is, $\{(1, 2, 0, 1), (0, 1, 1, 1), (0, 0, 1, 1)\}$ is a basis for $\text{Im}(T)$.

A function $f : \mathscr{A} \to \mathscr{B}$ is **onto** if $\text{Im}(f)$, the set of all images $f(a)$, is the set $\mathscr{B}$. The following theorem gives an easy test to see whether or not a *linear transformation* is onto.

Theorem 6 Let $T : \mathbf{R}^n \to \mathbf{R}^m$ be a linear transformation, and let A be the matrix that represents T. Then T is onto if and only if the rank of A is equal to m.

Proof Since $\text{Im}(T)$ is a subspace of $\mathbf{R}^m$, we have that $\text{Im}(T) = \mathbf{R}^m$ if and only if $\dim(\text{Im}(T)) = m$. By theorem 5, $\dim(\text{Im}(T))$ is equal to the rank of A, so we have the desired result. ∎

As examples of theorem 6 note that the linear transformation of example 5.19 is onto since $\dim(\text{Im}(T)) = 3$, which is the rank of the matrix, but that of example 5.20 is not since $\dim(\text{Im}(T)) = 2$, which is less than 4.

Combining theorem 2 with theorem 5 yields an interesting result. Assume that A is the $m \times n$ matrix that represents the linear transformation T, and let r be the rank of A. As a consequence of theorem 2 the dimension of the kernel (the nullity of A) is $n - r$. As a consequence of theorem 5 the dimension of the image (the rank of A) is r. Combining these facts, we obtain the following theorem.

Theorem 7 Let $T : \mathbf{R}^n \to \mathbf{R}^m$ be a linear transformation. Then the sum of the dimension of the kernel of T and the dimension of the image of T is equal to n. That is,

$$\dim(\text{Ker}(T)) + \dim(\text{Im}(T)) = n$$

or

$$\text{nullity} + \text{rank} = n.$$

An interesting special case of theorems 3 and 6 is for a linear transformation from $\mathbf{R}^n$ into *itself*. The following theorem explains the result.

Theorem 8   Let $T : \mathbf{R}^n \to \mathbf{R}^n$ be a linear transformation. Then T is one-to-one if and only if it is onto.

Proof   From theorem 7 we have

$$\dim(\text{Ker}(T)) + \dim(\text{Im}(T)) = n.$$

By theorem 3, T is one-to-one if and only if $\text{Ker}(T) = \{0\}$; that is, if and only if $\dim(\text{Ker}(T)) = 0$. But then the equation above implies that $\dim(\text{Im}(T)) = n$; that is, T is onto $\mathbf{R}^n$ by theorem 6.   ∎

Let A be the matrix that represents a linear transformation T from $\mathbf{R}^n$ into itself. Then A is an order n square matrix. Since T is one-to-one if and only if it is onto, by theorems 3 and 6 it is one-to-one if and only if the columns of A are a basis for $\mathbf{R}^n$. By theorem 6 of section 4.3 and theorem 5 of section 3.2 this occurs if and only if A is *invertible*. In this situation A^{-1} also gives rise to a linear transformation T^{-1} on $\mathbf{R}^n$ defined by $T^{-1}(\mathbf{x}) = A^{-1}\mathbf{x}$, for each $\mathbf{x}$. The linear transformation T^{-1} is called the **inverse** of T, and we say that T is **invertible**. It is easy to confirm that $T^{-1}T = TT^{-1} = I$ since for each $\mathbf{x}$ in $\mathbf{R}^n$

$$T^{-1}T(\mathbf{x}) = T^{-1}(T(\mathbf{x})) = T^{-1}(A\mathbf{x}) = A^{-1}A\mathbf{x} = I\mathbf{x} = I(\mathbf{x}),$$

and similarly $TT^{-1}(\mathbf{x}) = I(\mathbf{x})$.

The following theorem is an immediate consequence of the preceding remarks and is left as an exercise.

Theorem 9   Let $T : \mathbf{R}^n \to \mathbf{R}^n$ be a linear transformation. The following are equivalent:

i. T is one-to-one,
ii. T is onto,
iii. T is invertible,
iv. $\det A \neq 0$, where A is the matrix that represents T.

Example 5.25

Let $T : \mathbf{R}^3 \to \mathbf{R}^3$ be defined by $T(x, y, z) = (2x - y, \; x + y + z, \; 2y - z)$ for each $\mathbf{x} = (x, y, z)$ in $\mathbf{R}^3$. Show that T is invertible, and compute T^{-1}.

We first compute the matrix A that represents T,

$$A = \begin{bmatrix} 2 & -1 & 0 \\ 1 & 1 & 1 \\ 0 & 2 & -1 \end{bmatrix}.$$

Since det $A = -7 \neq 0$, A is invertible. Its inverse is

$$A^{-1} = \frac{1}{7} \begin{bmatrix} 3 & 1 & 1 \\ -1 & 2 & 2 \\ -2 & 4 & -3 \end{bmatrix}.$$

Thus T^{-1} is given by

$$T^{-1}(\mathbf{x}) = A^{-1}\mathbf{x} = \frac{1}{7} \begin{bmatrix} 3 & 1 & 1 \\ -1 & 2 & 2 \\ -2 & 4 & -3 \end{bmatrix} \begin{bmatrix} x \\ y \\ z \end{bmatrix} = \begin{bmatrix} (3x + y + z)/7 \\ (-x + 2y + 2z)/7 \\ (-2x + 4y - 3z)/7 \end{bmatrix}$$

or

$$T^{-1}(x, y, z) = ((3x + y + z)/7, (-x + 2y + 2z)/7, (-2x + 4y - 3z)/7).$$

5.3 Exercises

In exercises 1 through 4 determine whether or not $\mathbf{v}_1$ or $\mathbf{v}_2$ is in the kernel of the given linear transformation.

1. $T : \mathbf{R}^4 \to \mathbf{R}^3$ given by $T(\mathbf{x}) = A\mathbf{x}$, where

$$A = \begin{bmatrix} 1 & 2 & -1 & 1 \\ 1 & 0 & 1 & 1 \\ 2 & -4 & 6 & 2 \end{bmatrix} \quad \text{and} \quad \mathbf{v}_1 = (-2, 0, 0, 2), \quad \mathbf{v}_2 = (-2, 2, 2, 0)$$

2. $T : \mathbf{R}^3 \to \mathbf{R}^4$ given by $T(\mathbf{x}) = A\mathbf{x}$, where

$$A = \begin{bmatrix} 1 & 1 & 0 \\ 2 & 0 & 3 \\ -1 & 1 & 1 \\ 1 & 1 & 2 \end{bmatrix} \quad \text{and} \quad \mathbf{v}_1 = (-3, 3, 2), \quad \mathbf{v}_2 = (0, 0, 0)$$

3. $T : \mathbf{R}^3 \to \mathbf{R}^2$ given by $T(x, y, z) = (3x + y - z, x + 2y + z)$ and $\mathbf{v}_1 = (3, -4, 5)$, $\mathbf{v}_2 = (1, 4, 7)$

4. $T : \mathbf{R}^4 \to \mathbf{R}$ given by $T(x_1, x_2, x_3, x_4) = x_1 + 2x_2 + 3x_3 + 4x_4$ and $\mathbf{v}_1 = (1, 2, 1, -2)$, $\mathbf{v}_2 = (2, 3, 0, -2)$.

In exercises 5 through 8 determine whether or not $\mathbf{w}_1$ or $\mathbf{w}_2$ is in the image of the corresponding linear transformation of exercises 1 through 4.

5. $\mathbf{w}_1 = (1, 3, 1)$, $\mathbf{w}_2 = (-1, -1, -2)$

6. $\mathbf{w}_1 = (0, 0, -8, -4)$, $\mathbf{w}_2 = (-1, 13, 0, 7)$

7. $\mathbf{w}_1 = (1, 1)$, $\mathbf{w}_2 = (-3, 1)$

8. $\mathbf{w}_1 = -13$, $\mathbf{w}_2 = 0$

9–12. For each of the linear transformations T in exercises 1 through 4 find the nullity of T, and give a basis for $\text{Ker}(T)$. Which of the transformations T are one-to-one?

13–16. For each of the linear transformations T of exercises 1 through 4 find the rank of T, and give a basis for $\text{Im}(T)$. Which of the transformations T are onto?

17. If $T : \mathbf{R}^n \to \mathbf{R}^m$ is a linear transformation, prove that $\dim(\text{Im}(T)) \leq n$.

18. If $T : \mathbf{R}^n \to \mathbf{R}^m$ is a linear transformation, and $n < m$, prove that T is not onto.

19. If $T : \mathbf{R}^n \to \mathbf{R}^m$ is a linear transformation, and $n > m$, prove that T is not one-to-one.

20. If $T : \mathbf{R}^n \to \mathbf{R}^m$ is a linear transformation that is one-to-one, and $\mathscr{B}$ is a basis for $\mathbf{R}^n$, prove that $T[\mathscr{B}]$, the set of all images $T(\mathbf{b})$ for $\mathbf{b}$ in $\mathscr{B}$, is a basis for $\text{Im}(T)$.

In exercises 21 through 27 determine whether or not the given linear transformation is invertible. If it is invertible, compute its inverse.

21. $T : \mathbf{R}^3 \to \mathbf{R}^3$ given by $T(x, y, z) = (x + z, x - y + z, y + 2z)$

22. $T : \mathbf{R}^2 \to \mathbf{R}^2$ given by $T(x, y) = (3x + 2y, -6x - 4y)$

23. $T : \mathbf{R}^3 \to \mathbf{R}^3$ given by $T(\mathbf{x}) = A\mathbf{x}$, where

$$A = \begin{bmatrix} 3 & 1 & 2 \\ 5 & -3 & 4 \\ 1 & -2 & 1 \end{bmatrix}$$

24. $T : \mathbf{R}^4 \to \mathbf{R}^4$ given by $T(\mathbf{x}) = A\mathbf{x}$, where

$$A = \begin{bmatrix} -1 & 2 & 1 & 0 \\ -1 & 1 & 0 & -1 \\ 2 & -1 & 0 & 4 \\ 1 & -2 & 0 & 0 \end{bmatrix}$$

25. $T_\theta : \mathbf{R}^2 \to \mathbf{R}^2$, the rotation of the plane given by example 5.5

26. $T^\theta : \mathbf{R}^2 \to \mathbf{R}^2$, the reflection of the plane given by example 5.7

27. $T_c : \mathbf{R}^n \to \mathbf{R}^n$, the contraction/dilation given by example 5.6.

5.4   Change of Basis

The matrix associated with a linear transformation, $T : \mathbf{R}^n \to \mathbf{R}^m$, was computed by using the standard basis for $\mathbf{R}^n$ (see section 5.2). There are times (for example, see sections 6.2 and 8.4) when it is more convenient to use a different basis for $\mathbf{R}^n$. In this section we will demonstrate how to construct matrices of linear transformations using bases other than the standard one and then show how all such matrices are related.

Recall from section 4.3 (exercise 27) that given a basis $\mathscr{B}$ of $\mathbf{R}^n$ and an x in $\mathbf{R}^n$, there exist unique coefficients that can be used to express x as a linear combination of elements of $\mathscr{B}$. We will identify this vector of coefficients as follows.

Definition   Let $\mathscr{B} = \{\mathbf{b}_1, \ldots, \mathbf{b}_n\}$ be a basis for $\mathbf{R}^n$ and x be in $\mathbf{R}^n$. The unique vector

$$\mathbf{y}_{\mathscr{B}} = (y_1, \ldots, y_n)_{\mathscr{B}} \quad \text{such that} \quad x = y_1\mathbf{b}_1 + \cdots + y_n\mathbf{b}_n$$

is called the **coordinate vector** of x with respect to the basis $\mathscr{B}$ (in the order listed).

Example 5.26

Find the coordinate vector of $x = (2, -1, 1)$ with respect to the basis $\mathscr{B} = \{(2, 1, 3), (1, 0, 1), (-1, 1, 1)\}$ of $\mathbf{R}^3$.

This is simply new terminology for the old technique of producing coefficients to show that x is in the span of $\mathscr{B}$. We form the matrix with columns given by the vectors in $\mathscr{B}$, adjoin x, and transform the augmented matrix to row-reduced echelon form:

$$\begin{bmatrix} 2 & 1 & -1 & | & 2 \\ 1 & 0 & 1 & | & -1 \\ 3 & 1 & 1 & | & 1 \end{bmatrix} \to \begin{bmatrix} 1 & 0 & 0 & | & -1 \\ 0 & 1 & 0 & | & 4 \\ 0 & 0 & 1 & | & 0 \end{bmatrix}.$$

The last column $\mathbf{y}_{\mathscr{B}} = (-1, 4, 0)_{\mathscr{B}}$ is the coordinate vector of x with respect to $\mathscr{B}$ since by theorem 5 of section 4.1, $(2, -1, 1) = (-1)(2, 1, 3) + 4(1, 0, 1) + 0(-1, 1, 1)$.

The components of any vector x in $\mathbf{R}^n$ are its coordinates with respect to the standard basis $\{\mathbf{e}_1, \ldots, \mathbf{e}_n\}$ since $x = x_1\mathbf{e}_1 + \cdots + x_n\mathbf{e}_n$. When we find the coordinate vector for x with respect to some other basis, we are in effect making a **change of basis**. For example, let $n = 2$. Then any vector $x = (x, y)$ can be written as $x = x\mathbf{e}_1 + y\mathbf{e}_2$; the ordered pair (x, y) expresses x in terms of the standard basis. Now, if $\mathscr{B} = \{\mathbf{b}_1, \mathbf{b}_2\}$ is some other basis, there exist unique scalars, s and t such that $x = s\mathbf{b}_1 + t\mathbf{b}_2$

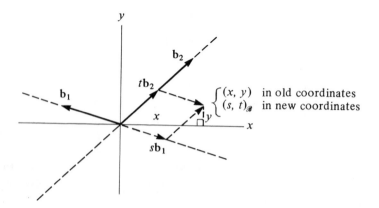

Figure 5.9

(figure 5.9). The coordinate vector $(s, t)_{\mathscr{B}}$ of $\mathbf{x}$ with respect to $\mathscr{B}$ represents the same point as $\mathbf{x}$ (albeit with different coordinates) as long as it is understood that $(s, t)_{\mathscr{B}}$ means the point obtained by adding the vectors $s\mathbf{b}_1$ and $t\mathbf{b}_2$. When this convention is made, the usual coordinate axes are replaced by the lines determined by $\mathbf{b}_1$ and $\mathbf{b}_2$, and each point is expressed in terms of these. For instance, $(1, 0)_{\mathscr{B}}$ is the point $\mathbf{b}_1$ and $(1, 1)_{\mathscr{B}}$ is the point $\mathbf{b}_1 + \mathbf{b}_2$ in the new coordinates. The following theorem shows that the conversion of coordinates from the standard basis to some other basis can be effected by multiplication by a matrix, instead of by solving for the coordinate vector as was done in example 5.26.

Theorem 1   Let $\mathscr{B} = \{\mathbf{b}_1, \ldots, \mathbf{b}_n\}$ be a basis for $\mathbf{R}^n$, and let P be the matrix with ith column $P^i = \mathbf{b}_i$. Then P is invertible, and P^{-1} is the *change of basis* matrix. In other words, for each $\mathbf{x}$ in $\mathbf{R}^n$, $P^{-1}\mathbf{x}$ is the coordinate vector of $\mathbf{x}$ with respect to $\mathscr{B}$, $y_{\mathscr{B}} = P^{-1}\mathbf{x}$.

Proof   The matrix P is invertible since it is square of order n with n independent columns. Since a linear transformation is determined by its action on a basis (theorem 3 of section 5.1), we only need to confirm that multiplication by P^{-1} "acts right" on a basis. Note that each basis element $\mathbf{b}_i$ has coordinate vector $\mathbf{e}_i$ with respect to $\mathscr{B}$, since

$$\mathbf{b}_i = 0\mathbf{b}_1 + \cdots + 0\mathbf{b}_{i-1} + 1\mathbf{b}_i + 0\mathbf{b}_{i+1} + \cdots + 0\mathbf{b}_n.$$

Thus the transformation must carry $\mathbf{b}_i$ into $\mathbf{e}_i$ for each i. But this is exactly the effect of multiplication on the left by P^{-1}, since $P^{-1}P = I$, or by columns, $P^{-1}P^i = P^{-1}\mathbf{b}_i = I^i = \mathbf{e}_i$.   ∎

Example 5.27

For $\mathcal{B} = \{(2, 1, 3), (1, 0, 1), (-1, 1, 1)\}$, the basis of $\mathbf{R}^3$ given in example 5.26, compute the change of basis matrix, and use it to find the coordinate vector of $\mathbf{x} = (2, -1, 1)$.
Following theorem 1, we let

$$P = \begin{bmatrix} 2 & 1 & -1 \\ 1 & 0 & 1 \\ 3 & 1 & 1 \end{bmatrix}.$$

Then the change of basis matrix is

$$P^{-1} = \begin{bmatrix} 1 & 2 & -1 \\ -2 & -5 & 3 \\ -1 & -1 & 1 \end{bmatrix}.$$

Finally, the required coordinate vector is

$$P^{-1}(2, -1, 1) = \begin{bmatrix} 1 & 2 & -1 \\ -2 & -5 & 3 \\ -1 & -1 & 1 \end{bmatrix}\begin{bmatrix} 2 \\ -1 \\ 1 \end{bmatrix} = \begin{bmatrix} -1 \\ 4 \\ 0 \end{bmatrix}_{\mathcal{B}} = (-1, 4, 0)_{\mathcal{B}}.$$

This of course agrees with the result of example 5.1.

If P^{-1} is a change of basis matrix for $\mathbf{R}^n$, then any vector $\mathbf{x}$ is represented by $P^{-1}\mathbf{x}$ with respect to the new basis. In other words, $\mathbf{x}$ and $P^{-1}\mathbf{x}$ each represent the same point but express it in terms of different bases. If $T : \mathbf{R}^n \to \mathbf{R}^n$ is a linear transformation, the point represented by $T(\mathbf{x})$ can also be represented in terms of the new basis as $P^{-1}T(\mathbf{x})$.

Example 5.28

Let $T : \mathbf{R}^3 \to \mathbf{R}^3$ be defined by $T(x, y, z) = (x + y, z, 2x - y)$. Find the coordinate vector of $T(2, -1, 1)$ with respect to the basis

$$\mathcal{B} = \{(2, 1, 3), (1, 0, 1), (-1, 1, 1)\}.$$

From example 5.27 we have the change of basis matrix P^{-1}, so the result is

$$P^{-1}T(2, -1, 1) = \begin{bmatrix} 1 & 2 & -1 \\ -2 & -5 & 3 \\ -1 & -1 & 1 \end{bmatrix}\begin{bmatrix} 1 \\ 1 \\ 5 \end{bmatrix} = \begin{bmatrix} -2 \\ 8 \\ 3 \end{bmatrix}_{\mathcal{B}}.$$

That is, $P^{-1}T(2, -1, 1) = (-2, 8, 3)_{\mathcal{B}}.$

In example 5.27 we computed the coordinates of the vector $(2, -1, 1)$ with respect to a basis $\mathscr{B}$, and in example 5.28 we computed the coordinates of the vector $T(2, -1, 1)$ with respect to the same basis. However, to make the computation we essentially went back to the expression for the point in terms of the standard basis. The next theorem shows that it is possible to go directly from the coordinate vector of $\mathbf{x}$ to the coordinate vector of $T(\mathbf{x})$ by means of multiplication by the appropriate matrix.

Definition   The matrix $A_{\mathscr{B}}$ **represents** T with **respect to the basis** $\mathscr{B}$ if

$$P^{-1}(T(\mathbf{x})) = A_{\mathscr{B}}(P^{-1}\mathbf{x})$$

for every $\mathbf{x}$ in $\mathbf{R}^n$, where P^{-1} is the change of basis matrix for $\mathscr{B}$. In other words, multiplication of the coordinate vector of $\mathbf{x}$ with respect to $\mathscr{B}$ by $A_{\mathscr{B}}$ gives the coordinate vector of $T(\mathbf{x})$ with respect to $\mathscr{B}$.

Note: Let A be the matrix associated with a linear transformation $T: \mathbf{R}^n \to \mathbf{R}^n$. Then A is the matrix that represents T with respect to the *standard* basis, since $P^{-1} = P = I$ is the change of basis matrix for the standard basis.

Theorem 2   Let $T: \mathbf{R}^n \to \mathbf{R}^n$ be a linear transformation with associated matrix A, and let $\mathscr{B}$ be a basis for $\mathbf{R}^n$. Let P^{-1} be the change of basis matrix for $\mathscr{B}$. Then $P^{-1}AP$ is the matrix that represents T with respect to the basis $\mathscr{B}$, $A_{\mathscr{B}} = P^{-1}AP$.

Proof   For each $\mathbf{x}$ in $\mathbf{R}^n$, $P^{-1}\mathbf{x}$ is the coordinate vector of the same point with respect to the basis $\mathscr{B}$. Checking the action of $P^{-1}AP$ on any such vector, we have

$$(P^{-1}AP)(P^{-1}\mathbf{x}) = P^{-1}AI\mathbf{x} = P^{-1}A\mathbf{x} = P^{-1}T(\mathbf{x}),$$

since A is the matrix associated with T. But this last vector, $P^{-1}T(\mathbf{x})$, is the coordinate vector of $T(\mathbf{x})$ with respect to the basis $\mathscr{B}$.   ∎

Example 5.29

Let $T: \mathbf{R}^3 \to \mathbf{R}^3$ be given by $T(x, y, z) = (x + y, z, 2x - y)$, and let $\mathscr{B} = \{(2, 1, 3), (1, 0, 1), (-1, 1, 1)\}$. Calculate $A_{\mathscr{B}}$ for T, and use it to find the coordinate vector for the transformed point with coordinate vector $(-1, 4, 0)_{\mathscr{B}}$.

To use theorem 2, we need the matrix A associated with T,

$$A = \begin{bmatrix} 1 & 1 & 0 \\ 0 & 0 & 1 \\ 2 & -1 & 0 \end{bmatrix}.$$

From example 5.27 we have P and P^{-1}, so applying theorem 2, we have

$$A_{\mathscr{B}} = P^{-1}AP = \begin{bmatrix} 1 & 2 & -1 \\ -2 & -5 & 3 \\ -1 & -1 & 1 \end{bmatrix} \begin{bmatrix} 1 & 1 & 0 \\ 0 & 0 & 1 \\ 2 & -1 & 0 \end{bmatrix} \begin{bmatrix} 2 & 1 & -1 \\ 1 & 0 & 1 \\ 3 & 1 & 1 \end{bmatrix}$$

$$= \begin{bmatrix} 6 & 1 & 5 \\ -12 & -1 & -14 \\ -3 & 0 & -4 \end{bmatrix}.$$

Finally, we use $A_{\mathscr{B}}$ to compute the coordinate vector for the transformed point with coordinate vector $(-1, 4, 0)_{\mathscr{B}}$,

$$\begin{bmatrix} 6 & 1 & 5 \\ -12 & -1 & -14 \\ -3 & 0 & -4 \end{bmatrix} \begin{bmatrix} -1 \\ 4 \\ 0 \end{bmatrix}_{\mathscr{B}} = \begin{bmatrix} -2 \\ 8 \\ 3 \end{bmatrix}_{\mathscr{B}}.$$

As it must, the result agrees with example 5.28.

The relationship between the matrices $P^{-1}AP$ and A is important enough to warrant the following definition.

Definition   Let A and B be square matrices of order n. We say that B is **similar** to A if there exists an invertible matrix P such that

$$B = P^{-1}AP.$$

Several remarks are in order. For one, P must also be of order n in order that the multiplication be defined. For another, $B = P^{-1}AP$ implies that $(P^{-1})^{-1}BP^{-1} = A$ so that, if B is similar to A, then A is also similar to B (using P^{-1}). For a third, the matrix P is not unique. To see this, let $A = B = I$, the identity matrix of order n. Then *any* invertible matrix of order n will serve for P, since $I = P^{-1}IP$, no matter what P is.

By theorem 2, if a matrix A is associated with a linear transformation $T : \mathbf{R}^n \rightarrow \mathbf{R}^n$, if $\mathscr{B}$ is a basis for $\mathbf{R}^n$, and if B is the matrix that represents T with respect to $\mathscr{B}$, then B is similar to A. It is natural to ask if the converse is true. That is, let B be similar to A, and let T be the linear transformation $T(\mathbf{x}) = A\mathbf{x}$. Is there a basis $\mathscr{B}$ such that B represents T with respect to $\mathscr{B}$? The answer is yes.

Theorem 3   Let A and B be square matrices of order n. Suppose B is similar to A, and let $T : \mathbf{R}^n \to \mathbf{R}^n$ be the linear transformation given by $T(\mathbf{x}) = A\mathbf{x}$, for each $\mathbf{x}$ in $\mathbf{R}^n$. Then if P is an invertible matrix such that $B = P^{-1}AP$, the set $\mathscr{B}$ of columns $\mathbf{b}_i = P^i$, $i = 1, \ldots, n$, is a basis for $\mathbf{R}^n$ such that B represents T with respect to $\mathscr{B}A_{\mathscr{B}} = B$.

Proof   First note that $\mathscr{B}$ is a basis, since the columns of P are linearly independent (theorem 6 of section 2.5 together with theorem 4 of section 4.1), and there are n of them.

Since P is the matrix with columns $\mathbf{b}_i = P^i$, by theorem 1, $P^{-1}\mathbf{x}$ is the coordinate vector of $\mathbf{x}$ with respect to $\mathscr{B}$.

Finally by theorem 2, $B = P^{-1}AP$ represents T with respect to $\mathscr{B}$. ∎

Example 5.30

Let $T : \mathbf{R}^3 \to \mathbf{R}^3$ be given by multiplication by the matrix A below. Using the given matrices P and B, show that B is similar to A, and find a basis for $\mathbf{R}^3$ with respect to which B represents the linear transformation T.

$$A = \begin{bmatrix} 2 & 1 & 3 \\ 1 & -2 & 1 \\ 0 & 2 & 1 \end{bmatrix}, \quad P = \begin{bmatrix} 1 & 0 & 1 \\ 2 & 1 & 2 \\ 0 & 2 & 1 \end{bmatrix}, \quad B = \begin{bmatrix} -22 & -25 & -30 \\ -11 & -14 & -16 \\ 26 & 32 & 37 \end{bmatrix}.$$

First compute P^{-1} to be

$$P^{-1} = \begin{bmatrix} -3 & 2 & -1 \\ -2 & 1 & 0 \\ 4 & -2 & 1 \end{bmatrix}.$$

Then direct multiplication confirms that $B = P^{-1}AP$ so that B is similar to A. Finally, theorem 3 assures us that the columns of P provide a basis for $\mathbf{R}^3$ with the required property,

$$\mathscr{B} = \{(1, 2, 0), (0, 1, 2), (1, 2, 1)\}.$$

5.4   Exercises

In exercises 1 through 4 verify that the given set $\mathscr{B}$ is a basis for $\mathbf{R}^n$, and find the coordinate vector of $\mathbf{v}$ with respect to $\mathscr{B}$.

1. $\mathscr{B} = \{(1, 2), (1, -2)\}$, $\mathbf{v} = (-1, 3)$, $(n = 2)$
2. $\mathscr{B} = \{(1, 2, 1), (-1, 1, 2), (1, 2, 3)\}$, $\mathbf{v} = (2, 1, 0)$, $(n = 3)$
3. $\mathscr{B} = \{(1, 2, 1, -1), (0, 2, 1, 0), (1, 1, 0, 1), (1, 3, 0, 0)\}$, $\mathbf{v} = (-1, 1, -1, 1)$, $(n = 4)$
4. $\mathscr{B} = \{(-1, 1, -1, 1), (2, 1, 0, 1), (1, 0, 0, 2), (0, 0, 3, 2)\}$, $\mathbf{v} = (2, 1, 2, 3)$, $(n = 4)$

5–8. Compute the change of basis matrix for each of the bases in exercises 1 through 4, and use it to find the coordinate vector of $\mathbf{v}$ with respect to $\mathscr{B}$.

9. Let $T : \mathbf{R}^2 \rightarrow \mathbf{R}^2$ be given by $T(\mathbf{x}) = A\mathbf{x}$, where A is the matrix

$$A = \begin{bmatrix} 1 & 1 \\ 0 & -1 \end{bmatrix}.$$

Find the matrix $A_{\mathscr{B}}$ that represents T with respect to the basis $\mathscr{B}$ of exercise 1.

10. Let $T : \mathbf{R}^3 \rightarrow \mathbf{R}^3$ be given by $T(\mathbf{x}) = A\mathbf{x}$, where A is the matrix

$$A = \begin{bmatrix} 1 & 1 & 3 \\ 2 & -1 & 1 \\ 1 & 2 & 0 \end{bmatrix}.$$

Find the matrix $A_{\mathscr{B}}$ that represents T with respect to the basis $\mathscr{B}$ of exercise 2.

11. Let $T : \mathbf{R}^3 \rightarrow \mathbf{R}^3$ be given by $T(\mathbf{x}) = T(x, y, z) = (2x - y, x + y + z, y - z)$. Find the matrix that represents T with respect to the basis $\mathscr{B}$ of exercise 2.

12. Let $T : \mathbf{R}^4 \rightarrow \mathbf{R}^4$ be given by $T(\mathbf{x}) = T(w, x, y, z) = (w + y - z, 2w - y, w + 2x + y + z, y - z)$. Find the matrix that represents T with respect to the basis $\mathscr{B}$ of exercise 3.

13. If A is similar to B, prove that $\det A = \det B$.

14. Let A and B be square matrices of the same order with $\det A = \det B$. Show that it is not necessarily true that A is similar to B.

15. If A is similar to B, prove that A^2 is similar to B^2.

16. If A is similar to B, prove that rank $A = $ rank B. (*Hint:* use theorem 3.)

Eigenvalues and Eigenvectors

6

In this chapter we will discuss the topic of *eigenvalues* and *eigenvectors* of linear transformations and matrices. These entities have many far-reaching applications in science and mathematics. One of these applications will be given in section 6.2 and some others in chapter 8.

6.1 Definitions and Examples

From a geometric point of view a linear transformation T, of $\mathbf{R}^n$ into itself, maps one vector into another by altering the magnitude and/or direction of the former. The *eigenvectors* of T are those nonzero vectors, $\mathbf{x}$, for which $T(\mathbf{x})$ and $\mathbf{x}$ are collinear. In other words, eigenvectors are characterized by the property that $T(\mathbf{x})$ is a scalar multiple of $\mathbf{x}$. The scalar involved is called the *eigenvalue* of T associated with the eigenvector $\mathbf{x}$. In more precise terms we have the following definition.

Definition Let $T : \mathbf{R}^n \to \mathbf{R}^n$ be a linear transformation. A scalar λ is called an **eigenvalue** of T if there is a *nonzero* vector $\mathbf{x}$ in $\mathbf{R}^n$ such that $T(\mathbf{x}) = \lambda \mathbf{x}$. The vector $\mathbf{x}$ is said to be an **eigenvector** corresponding to the eigenvalue λ.

Eigenvalues are sometimes referred to as *characteristic values, proper values,* or *latent values.* Similarly other names for eigenvectors are *characteristic vectors, proper vectors,* and *latent vectors.* In the first two sections of this chapter we will consider only eigenvalues that are real numbers. However, in section 6.3 we will allow eigenvalues and the components of eigenvectors to be complex numbers.

Example 6.1

Let $T : \mathbf{R}^3 \to \mathbf{R}^3$ be the linear transformation given by

$$T(x, y, z) = (0, y + z, 3y - z).$$

Show that $\lambda_1 = 0$ and $\lambda_2 = 2$ are eigenvalues of T with corresponding eigenvectors $\mathbf{x}_1 = (1, 0, 0)$ and $\mathbf{x}_2 = (0, 1, 1)$, respectively.

We must verify that $T(\mathbf{x}_1) = \lambda_1\mathbf{x}_1$ and $T(\mathbf{x}_2) = \lambda_2\mathbf{x}_2$. Now

$$T(\mathbf{x}_1) = T(1, 0, 0) = (0, 0, 0) = 0(1, 0, 0) = \lambda_1\mathbf{x}_1$$

and

$$T(\mathbf{x}_2) = T(0, 1, 1) = (0, 2, 2) = 2(0, 1, 1) = \lambda_2\mathbf{x}_2,$$

as desired.

Note: By definition an eigen*vector* must be nonzero. A linear transformation may have a zero eigen*value*, however, as can be seen from example 6.1.

It is easy to verify that for the linear transformation of example 6.1 any nonzero vector of the form $(s, 0, 0)$ is an eigenvector corresponding to $\lambda_1 = 0$, while any nonzero vector of the form $(0, t, t)$ is an eigenvector for $\lambda_2 = 2$. In general, if T is any linear transformation with eigenvalue λ_0 and corresponding eigenvector $\mathbf{x}_0$, then all vectors of the form $c\mathbf{x}_0, c \neq 0$, are eigenvectors of T corresponding to λ_0. This statement follows from the fact that

$$T(c\mathbf{x}_0) = cT(\mathbf{x}_0) = c(\lambda\mathbf{x}_0) = \lambda(c\mathbf{x}_0).$$

Example 6.2

Let $T : \mathbf{R}^2 \rightarrow \mathbf{R}^2$ be the linear transformation that maps every vector into its reflection in the x-axis (see example 5.7). Find the eigenvalues and eigenvectors of T.

By the nature of the given transformation the only nonzero vectors collinear with their images under T are those that lie along the coordinate axes. Thus all eigenvectors of T are of the form $c_1(1, 0)$ or $c_2(0, 1)$, where c_1 and c_2 are nonzero scalars. Moreover for any vector $\mathbf{x}$ lying on the x-axis, $T(\mathbf{x}) = \mathbf{x}$, while if $\mathbf{x}$ lies on the y-axis, we have $T(\mathbf{x}) = -\mathbf{x}$. So the eigenvectors of the form $c_1(1, 0)$ satisfy $T(\mathbf{x}) = \mathbf{x}$, while those of the form $c_2(0, 1)$ satisfy $T(\mathbf{x}) = -\mathbf{x}$. Comparing these equations to the defining one for an eigenvalue, $T(\mathbf{x}) = \lambda\mathbf{x}$, we see that the eigenvalue corresponding to $c_1(1, 0)$ is $\lambda_1 = 1$ and that corresponding to $c_2(0, 1)$ is $\lambda_2 = -1$.

Note: Let $T : \mathbf{R}^n \rightarrow \mathbf{R}^n$ be a linear transformation. For a scalar λ and a vector $\mathbf{x}$ we may rewrite the equation $T(\mathbf{x}) = \lambda\mathbf{x}$ as $T(\mathbf{x}) = \lambda I(\mathbf{x})$ or $(T - \lambda I)(\mathbf{x}) = \mathbf{0}$, where I is the identity transformation. This last equation tells us that a nonzero vector $\mathbf{x}$ is an eigenvector of T if and only if $\mathbf{x}$ lies in the kernel of the linear transformation $T - \lambda I$. From theorem 1 of section 5.3 we know the kernel of this linear transformation is a subspace of $\mathbf{R}^n$. Thus the set of all eigenvectors of T that correspond to λ, together with the zero vector, forms a subspace of $\mathbf{R}^n$, $\mathrm{Ker}(T - \lambda I)$. This subspace is called the **eigenspace** of λ. For example, the eigenspaces of $\lambda_1 = 1$ and $\lambda_2 = -1$ of example 6.2 are $\{c_1(1, 0)\}$ and $\{c_2(0, 1)\}$, respectively.

In section 5.2 we demonstrated that any linear transformation T : $\mathbf{R}^n \to \mathbf{R}^n$ may be represented as multiplication by an $n \times n$ matrix A, that is, for any $\mathbf{x}$ in $\mathbf{R}^n$, $T(\mathbf{x}) = A\mathbf{x}$. Consequently it is natural to define the eigenvalues and eigenvectors of a square matrix in the following way.

Definition Let A be an $n \times n$ matrix. A scalar λ is called an **eigenvalue** of A if there is a *nonzero* vector $\mathbf{x}$ in $\mathbf{R}^n$ such that $A\mathbf{x} = \lambda\mathbf{x}$. The vector $\mathbf{x}$ is said to be an **eigenvector** corresponding to the eigenvalue λ.

Notice that if a square matrix A represents a linear transformation T, then λ is an eigenvalue of A if and only if it is an eigenvalue of T. Moreover the corresponding eigenvectors for both A and T are the same.

We will now develop a systematic procedure for determining the eigenvalues and eigenvectors of an $n \times n$ matrix A. First suppose we know that λ_0 is an eigenvalue of A. How do we find corresponding eigenvectors $\mathbf{x}$? By definition $A\mathbf{x} = \lambda_0\mathbf{x}$, which may be rewritten as $A\mathbf{x} = \lambda_0 I\mathbf{x}$ or $(A - \lambda_0 I)\mathbf{x} = \mathbf{0}$, where I is the $n \times n$ identity matrix. Thus $\mathbf{x}$ is an eigenvector corresponding to λ_0 if and only if it is a nonzero solution of the homogeneous linear system $(A - \lambda_0 I)\mathbf{x} = \mathbf{0}$.

Example 6.3

The matrix

$$A = \begin{bmatrix} 1 & 1 \\ 6 & 2 \end{bmatrix}$$

has eigenvalues $\lambda_1 = -1$ and $\lambda_2 = 4$. Find the corresponding eigenvectors.

All eigenvectors corresponding to $\lambda_1 = -1$ will be nonzero solutions of the homogeneous linear system $(A - \lambda_1 I)\mathbf{x} = \mathbf{0}$ or $(A + I)\mathbf{x} = \mathbf{0}$, where

$$A + I = \begin{bmatrix} 2 & 1 \\ 6 & 3 \end{bmatrix}.$$

The solution of this system is found to be all vectors of the form $(s, -2s)$. So all eigenvectors corresponding to $\lambda_1 = -1$ are of the form $(s, -2s)$ where $s \neq 0$.

Similarly, to find the eigenvectors corresponding to $\lambda_2 = 4$, we solve the system $(A - 4I)\mathbf{x} = \mathbf{0}$. This yields eigenvectors of the form $(t, 3t)$, $t \neq 0$.

Note: The **eigenspace** of A corresponding to λ_0 is just the *solution space* of the linear homogeneous system $(A - \lambda_0 I)\mathbf{x} = \mathbf{0}$ and consists of all eigenvectors of A together with the zero vector. Thus the eigenspaces of λ_1 and λ_2 of example 6.3 are $\{(s, -2s)\}$ and $\{(t, 3t)\}$, respectively.

In example 6.3 we saw how to find eigenvectors corresponding to a particular known eigenvalue. The next question is how to find the eigenvalues of an $n \times n$ matrix A in the first place.

Consider the equation $(A - \lambda I)\mathbf{x} = \mathbf{0}$. We seek scalars λ for which there exists *some nonzero* $\mathbf{x}$ satisfying the equation. By theorem 6 of section 2.5 there exists a nonzero solution, $\mathbf{x}$, of the linear homogeneous system $(A - \lambda I)\mathbf{x} = \mathbf{0}$ if and only if the matrix $A - \lambda I$ is noninvertible. So the problem of finding the eigenvalues of A reduces to that of determining all scalars λ for which $A - \lambda I$ is noninvertible. But by theorem 5 of section 3.2, $A - \lambda I$ is noninvertible if and only if $\det(A - \lambda I) = 0$. Thus λ is an eigenvalue of A if and only if $\det(A - \lambda I) = 0$.

Example 6.4

Find the eigenvalues of

$$A = \begin{bmatrix} 3 & -1 \\ 2 & 0 \end{bmatrix}.$$

We must determine all scalars λ for which $\det(A - \lambda I) = 0$. Now

$$A - \lambda I = \begin{bmatrix} 3 & -1 \\ 2 & 0 \end{bmatrix} - \lambda \begin{bmatrix} 1 & 0 \\ 0 & 1 \end{bmatrix} = \begin{bmatrix} 3 - \lambda & -1 \\ 2 & -\lambda \end{bmatrix},$$

so that $\det(A - \lambda I) = (3 - \lambda)(-\lambda) + 2 = \lambda^2 - 3\lambda + 2$. Thus the eigenvalues of A are those scalars λ that satisfy $\lambda^2 - 3\lambda + 2 = 0$. Solving this quadratic equation, we obtain $\lambda_1 = 1$ and $\lambda_2 = 2$.

The discussion that preceded examples 6.3 and 6.4 can be formalized to provide the proof of the following theorem.

Theorem 1   Let A be an $n \times n$ matrix, and let I be the identity matrix of order n. Then

a. The eigenvalues of A are those scalars λ that satisfy the equation $\det(A - \lambda I) = 0$,

b. The eigenvectors of A corresponding to an eigenvalue λ_0 are the non-zero solutions $\mathbf{x}$ of the homogeneous linear system $(A - \lambda_0 I)\mathbf{x} = \mathbf{0}$.

To facilitate the implementation of theorem 1, we restate it as the following procedure.

Finding eigenvalues and corresponding eigenvectors of a square matrix A

  i. Form the matrix $A - \lambda I$ by subtracting λ from each diagonal entry of A.
 ii. Solve the equation $\det(A - \lambda I) = 0$. The real solutions are the eigenvalues of A.
iii. For each eigenvalue λ_0, form the matrix $A - \lambda_0 I$.
 iv. Solve the homogeneous linear system $(A - \lambda_0 I)\mathbf{x} = \mathbf{0}$. The nonzero solutions are the eigenvectors of A corresponding to λ_0 (procedure of theorem 4 of section 4.3).

Definition The *eigenspace* of a matrix A and an eigenvalue λ_0 is the set of *all* solutions of $(A - \lambda_0 I)\mathbf{x} = \mathbf{0}$.

From theorem 4 of section 4.3 the dimension of the eigenspace for λ_0 is $n - r$, where r is the rank of $A - \lambda_0 I$. The associated procedure provides a basis for the eigenspace.

Example 6.5

Find the eigenvalues and corresponding eigenvectors, as well as a basis for each eigenspace, of the matrix

$$A = \begin{bmatrix} 4 & 1 & -3 \\ 0 & 0 & 2 \\ 0 & 0 & -3 \end{bmatrix}.$$

Following the procedure for finding the eigenvalues of A, we compute

$$\det(A - \lambda I) = \det \begin{bmatrix} 4 - \lambda & 1 & -3 \\ 0 & -\lambda & 2 \\ 0 & 0 & -3 - \lambda \end{bmatrix} = (4 - \lambda)(-\lambda)(-3 - \lambda).$$

We now set this expression equal to zero, obtaining the equation $(4 - \lambda)(-\lambda)(-3 - \lambda) = 0$. The eigenvalues of A are the roots of this equation:

$$\lambda_1 = 4, \quad \lambda_2 = 0, \quad \text{and} \quad \lambda_3 = -3.$$

To find the eigenvectors corresponding to $\lambda_1 = 4$, we form the matrix

$$A - 4I = \begin{bmatrix} 0 & 1 & -3 \\ 0 & -4 & 2 \\ 0 & 0 & -7 \end{bmatrix}$$

and solve the homogeneous linear system $(A - 4I)\mathbf{x} = \mathbf{0}$. Doing so, we see that all eigenvectors corresponding to $\lambda_1 = 4$ take the form $(t, 0, 0)$, where $t \neq 0$, and that a basis for the eigenspace of $\lambda_1 = 4$ is $\{(1, 0, 0)\}$.

To find the eigenspaces of $\lambda_2 = 0$ and $\lambda_3 = -3$, we work in a similar manner with the matrices $A - 0I$ ($= A$) and $A + 3I$, respectively. This yields eigenvectors of the form $(r, -4r, 0)$, $r \neq 0$, for λ_2 and $(11s, -14s, 21s)$, $s \neq 0$, for λ_3. The respective eigenspaces have bases given by $\{(1, -4, 0)\}$ and $\{(11, -14, 21)\}$.

Note: The equation $\det(A - \lambda I) = 0$, whose roots are the eigenvalues of A, is called the **characteristic equation** of A. If A is $n \times n$, the expression $\det(A - \lambda I)$ is a polynomial of degree equal to n (see exercise 20 of section 3.1) known as the **characteristic polynomial** for A. For example, the characteristic polynomials of examples 6.4 and 6.5 are, respectively, $\lambda^2 - 3\lambda + 2$ (of degree 2) and $(4 - \lambda)(-\lambda)(-3 - \lambda) = -\lambda^3 + \lambda^2 + 12\lambda$ (of degree 3). Since an nth degree polynomial equation has at most n roots, we see that an $n \times n$ matrix has at most n eigenvalues. As the next example illustrates, a matrix may have fewer than n eigenvalues.

Example 6.6

Find the eigenvalues and bases for the corresponding eigenspaces of

$$A = \begin{bmatrix} -1 & 1 & 1 & 0 \\ 0 & 0 & 0 & -1 \\ 0 & 1 & 0 & 0 \\ 0 & 1 & 1 & -1 \end{bmatrix}.$$

Here the characteristic polynomial is of fourth degree and is computed (by expanding down the first column of $A - \lambda I$) to be

$$\det(A - \lambda I) = (-1 - \lambda)\det\begin{bmatrix} -\lambda & 0 & -1 \\ 1 & -\lambda & 0 \\ 1 & 1 & -1 - \lambda \end{bmatrix}$$

$$= (-1 - \lambda)[-\lambda^3 - \lambda^2 - \lambda - 1].$$

Thus one of the eigenvalues of A is $\lambda = -1$ (from the first factor), and the others are the real roots of the cubic equation $\lambda^3 + \lambda^2 + \lambda + 1 = 0$ (from the second factor). Since

$$\lambda^3 + \lambda^2 + \lambda + 1 = \lambda^2(\lambda + 1) + (\lambda + 1) = (\lambda^2 + 1)(\lambda + 1),$$

we see that -1 also satisfies this equation. We can find the remaining eigenvalues of A by solving the quadratic equation $\lambda^2 + 1 = 0$. Since the latter has no real roots (try, for example, the quadratic formula), A has only one eigenvalue, $\lambda_1 = -1$.

To find the eigenspace of this eigenvalue, we form the matrix

$$A - (-1)I = A + I = \begin{bmatrix} 0 & 1 & 1 & 0 \\ 0 & 1 & 0 & -1 \\ 0 & 1 & 1 & 0 \\ 0 & 1 & 1 & 0 \end{bmatrix},$$

and solve the system $(A + I)\mathbf{x} = \mathbf{0}$. Doing so, we see that the eigenspace of $\lambda_1 = -1$ is the set of all vectors of the form

$$(s, t, -t, t) = s(1, 0, 0, 0) + t(0, 1, -1, 1).$$

A basis for this eigenspace is $\{(1, 0, 0, 0), (0, 1, -1, 1)\}$.

Computational note: As you can see from example 6.6, it is not very easy to determine the eigenvalues and eigenvectors of an $n \times n$ matrix by the procedure of theorem 1 even when n is small. Moreover this technique is not a very efficient one to use in computer applications and may lead to inaccurate answers because it is somewhat susceptible to the accumulation of round-off error. For an outline of a better way to use a computer to approximate eigenvalues, see the *computational note* of exercise 33 of section 6.2.

You may have noticed in example 6.5 that the eigenvalues of A turned out to be identical to the diagonal entries of this matrix. However, as you can see from examples 6.4 and 6.6, this is certainly not always true. The following theorem gives three special cases in which the diagonal entries of a matrix *are* its eigenvalues. Its proof follows from theorem 4 of section 3.1 and is left as an exercise.

Theorem 2 The eigenvalues of an upper triangular, a lower triangular, or a diagonal matrix are identical to its diagonal entries.

Example 6.7

Find the eigenvalues of the matrices

$$A = \begin{bmatrix} 8 & 0 & 0 \\ 0 & 0 & 0 \\ 0 & 0 & 7 \end{bmatrix} \quad \text{and} \quad B = \begin{bmatrix} 1 & 0 & 0 & 0 \\ 3 & -2 & 0 & 0 \\ 2 & -1 & 4 & 0 \\ 6 & 0 & 5 & -3 \end{bmatrix}.$$

The matrices A and B are diagonal and lower triangular, respectively. So applying theorem 2, the eigenvalues of A are 8, 0, and 7, while those of B are 1, -2, 4, and -3.

6.1 Exercises

In exercises 1 through 6 find the eigenvalues and eigenvectors of the given linear transformation.

1. The identity transformation, $I : \mathbf{R}^n \rightarrow \mathbf{R}^n$ defined by $I(\mathbf{x}) = \mathbf{x}$

2. The zero transformation, $0 : \mathbf{R}^n \rightarrow \mathbf{R}^n$ defined by $0(\mathbf{x}) = \mathbf{0}$

3. The transformation $T_{\pi/2} : \mathbf{R}^2 \rightarrow \mathbf{R}^2$, which rotates every vector $\mathbf{x}$ in $\mathbf{R}^2$ by $\pi/2$

4. The transformation $T_\pi : \mathbf{R}^2 \rightarrow \mathbf{R}^2$, which rotates every vector $\mathbf{x}$ in $\mathbf{R}^2$ by π

5. The transformation $T^{\pi/2} : \mathbf{R}^2 \rightarrow \mathbf{R}^2$, which reflects every vector $\mathbf{x}$ in $\mathbf{R}^2$ in the y-axis

6. The transformation $T^{\pi/4} : \mathbf{R}^2 \rightarrow \mathbf{R}^2$, which reflects every vector $\mathbf{x}$ in $\mathbf{R}^2$ in the line $y = x$

In exercises 7 through 18 find the characteristic equation and eigenvalues of the given matrix.

7. $\begin{bmatrix} 1 & 4 \\ 2 & -1 \end{bmatrix}$

8. $\begin{bmatrix} 1 & 2 \\ -2 & -1 \end{bmatrix}$

9. $\begin{bmatrix} 0 & 1 \\ -1 & 2 \end{bmatrix}$

10. $\begin{bmatrix} 0 & 0 \\ 0 & 0 \end{bmatrix}$

11. $\begin{bmatrix} 0 & -1 & 0 \\ -1 & 0 & -1 \\ 0 & -1 & 0 \end{bmatrix}$

12. $\begin{bmatrix} 2 & -1 & 0 \\ -1 & 2 & -1 \\ 0 & -1 & 2 \end{bmatrix}$

13. $\begin{bmatrix} 1 & 1 & 1 \\ 1 & 1 & 1 \\ 1 & 1 & 1 \end{bmatrix}$

14. $\begin{bmatrix} 1 & 0 & -1 \\ 1 & 1 & 2 \\ 2 & 1 & 1 \end{bmatrix}$

15. $\begin{bmatrix} 1 & 0 & 3 \\ 0 & -2 & 1 \\ 0 & 0 & 1 \end{bmatrix}$

16. $\begin{bmatrix} 1 & 0 & 0 & 0 \\ 0 & 2 & 0 & 0 \\ 3 & 1 & -1 & 0 \\ -2 & 0 & 0 & 2 \end{bmatrix}$

17. $\begin{bmatrix} 1 & 1 & 1 & 1 \\ 1 & 1 & 1 & 1 \\ 1 & 1 & 1 & 1 \\ 1 & 1 & 1 & 1 \end{bmatrix}$

18. $\begin{bmatrix} 1 & 0 & 0 & 0 \\ 0 & 0 & 1 & 0 \\ 0 & 0 & 0 & 1 \\ 0 & 1 & 0 & 0 \end{bmatrix}$

19–30. Find all eigenvectors and bases for the eigenspaces of the matrices in exercises 7 through 18.

In exercises 31 through 34 find the eigenvalues of the given matrix by inspection.

31. $\begin{bmatrix} 1 & 2 & 0 \\ 0 & 3 & -1 \\ 0 & 0 & 4 \end{bmatrix}$ **32.** $\begin{bmatrix} 1 & 0 & 0 \\ 0 & 5 & 0 \\ 0 & 0 & -6 \end{bmatrix}$

33. $\begin{bmatrix} 1 & 0 & 0 & 0 \\ 2 & 3 & 0 & 0 \\ 4 & 5 & 6 & 0 \\ 7 & 8 & 9 & 10 \end{bmatrix}$ **34.** $\begin{bmatrix} -1 & 0 & 1 & 0 \\ 0 & 0 & 0 & -2 \\ 0 & 0 & 2 & 0 \\ 0 & 0 & 0 & 0 \end{bmatrix}$

35. Prove theorem 2.

36. Prove that 0 is an eigenvalue of a square matrix A if and only if A is non-invertible. (*Hint:* use theorem 6 of section 2.5 or theorem 5 of section 3.2.)

37. Let $T_\theta : \mathbf{R}^2 \to \mathbf{R}^2$ be the rotation by θ (see example 5.5). Find the eigenvalues and eigenvectors of T_θ by (a) using the definition and (b) working with the matrix associated with T_θ.

38. Let $T^\theta : \mathbf{R}^2 \to \mathbf{R}^2$ be the reflection in the line through the origin that forms an angle θ with positive x-axis (see example 5.7). Follow the instructions of exercise 37 for T^θ.

39. Prove that A and A^T (the *transpose* of A) have the same eigenvalues.

40. Prove that λ is an eigenvalue of a nonsingular matrix A if and only if $1/\lambda$ is an eigenvalue of A^{-1}. What relationship holds between the eigenvectors of A and A^{-1}?

41. For the matrix

$$A = \begin{bmatrix} a & b \\ c & d \end{bmatrix},$$

show that A has (a) two eigenvalues if $(a - d)^2 + 4bc > 0$, (b) one eigenvalue if $(a - d)^2 + 4bc = 0$, and (c) no eigenvalues if $(a - d)^2 + 4bc < 0$.

42. Prove that the eigenvalues of a matrix are "invariant under similarity transformation." That is, if A is an $n \times n$ matrix, and P is an invertible matrix of order n, then the eigenvalues of A and $P^{-1}AP$ are identical.
(*Hint:* $\det[P^{-1}AP - \lambda I] = \det[P^{-1}(AP - \lambda PI)] = \det[P^{-1}(A - \lambda I)P]$.)

43–45. If $P(x) = a_n x^n + \cdots + a_1 x + a_0$ is a polynomial with coefficients a_i, $i = 0, 1, \ldots, n$, from the real numbers, and A is an order n square matrix, $P(A)$ is the matrix

$$P(A) = a_n A^n + \cdots + a_1 A + a_0 I.$$

The *Cayley-Hamilton theorem* states that if $P(x)$ is the characteristic polynomial of a square matrix A, then $P(A) = 0$. Verify the Cayley-Hamilton theorem for each matrix in exercises 9 through 11.

6.2   Diagonalization

As we know from section 5.4, the matrix of a linear transformation T from $\mathbf{R}^n$ into itself depends upon the choice of basis for $\mathbf{R}^n$. Moreover A and B are both matrices of a transformation T (corresponding to possibly different bases for $\mathbf{R}^n$) if and only if A and B are *similar*; that is, if and only if there is a nonsingular matrix P such that $B = P^{-1}AP$. Consequently the problem investigated in this section may be phrased in two equivalent ways:

1. If T is a linear transformation of $\mathbf{R}^n$ into itself, determine a basis for $\mathbf{R}^n$ for which the matrix associated with T is diagonal.
2. If A is a square matrix, determine a nonsingular matrix P such that $P^{-1}AP$ is a diagonal matrix.

In either formulation we refer to this problem as **diagonalization** of the transformation T or the matrix A. As in section 6.1 we will concentrate our efforts on the matrix rather than the transformation.

Definition   A square matrix A is **diagonalizable** (or can be **diagonalized**) if it is similar to a diagonal matrix.

In the first two theorems of this section we will show that the diagonalization problem can be solved completely in terms of the eigenvectors of the given matrix.

Theorem 1   Let A be a square matrix of order n, with n linearly independent eigenvectors, $\mathbf{x}_1, \mathbf{x}_2, \ldots, \mathbf{x}_n$. Let λ_i be the eigenvalue corresponding to $\mathbf{x}_i$, and let P be the $n \times n$ matrix whose ith column is $P^i = \mathbf{x}_i$. Then

i. P is invertible,
ii. $D = P^{-1}AP$ is a diagonal matrix,
iii. The ith diagonal entry of D is λ_i.

Before presenting the proof of this theorem, we will give an example of its use. This example shows that the λ_i of theorem 1 need not be distinct.

Example 6.8

Let

$$A = \begin{bmatrix} 1 & 0 & 0 \\ 0 & 0 & 1 \\ 0 & 1 & 0 \end{bmatrix}.$$

Show that A is diagonalizable, find a matrix P so that $P^{-1}AP$ is a diagonal matrix, and determine $P^{-1}AP$.

To apply theorem 1, we must be able to find three linearly independent eigenvectors of A. To do so, we use the procedure of theorem 1 of section 6.1. The characteristic polynomial for A is

$$\det(A - \lambda I) = (1 - \lambda)(\lambda^2 - 1) = -(\lambda - 1)^2(\lambda + 1)$$

Thus the eigenvalues of A are $\lambda_1 = 1$ and $\lambda_2 = -1$, and after some computation we see that bases for their respective eigenspaces are $\{(1, 0, 0), (0, 1, 1)\}$ and $\{(0, 1, -1)\}$. It is easy to check (for example, by theorem 3 of section 4.1) that the eigenvectors $(1, 0, 0)$, $(0, 1, 1)$, and $(0, 1, -1)$ form a linearly independent set. Consequently by theorem 1, A is diagonalizable. Moreover the matrix P that we are seeking may be taken to be the one whose columns are these three eigenvectors, so that

$$P = \begin{bmatrix} 1 & 0 & 0 \\ 0 & 1 & 1 \\ 0 & 1 & -1 \end{bmatrix}.$$

Finally, to find the diagonal matrix $P^{-1}AP$, we could perform the indicated inversion and multiplications, but it is much easier to apply part iii of theorem 1. Since the eigenvalues corresponding to the eigenvectors $(1, 0, 0)$, $(0, 1, 1)$, and $(0, 1, -1)$ are 1, 1, and -1, respectively, we have

$$P^{-1}AP = \begin{bmatrix} 1 & 0 & 0 \\ 0 & 1 & 0 \\ 0 & 0 & -1 \end{bmatrix}.$$

Note: Reordering the eigenvectors would produce a different diagonal matrix, so, if A is diagonalizable, it is usually similar to several diagonal matrices. Of course P changes accordingly.

Proof of theorem 1   By theorem 3 of section 4.1, P is invertible since the columns of P are linearly independent. The ith column of the product AP is given by $AP^i = A\mathbf{x}_i = \lambda_i \mathbf{x}_i = \lambda_i P^i$. Thus AP is a matrix whose ith column is $\lambda_i P^i$, and therefore the ith column of $D = P^{-1}AP$ is

$$D^i = P^{-1}(\lambda_i P^i) = \lambda_i P^{-1} P^i = \lambda_i \mathbf{e}_i,$$

where $\mathbf{e}_i$ is the ith elementary vector. Consequently the matrix D is diagonal with diagonal entries $\lambda_1, \lambda_2, \ldots, \lambda_n$, the eigenvalues corresponding to the eigenvectors $\mathbf{x}_1, \mathbf{x}_2, \ldots, \mathbf{x}_n$. ∎

The converse of theorem 1 is true as well, so the linear independence of the eigenvectors of a matrix gives us not only a sufficient condition but also a necessary one for its diagonalization.

Theorem 2 If an $n \times n$ matrix A is similar to a diagonal matrix D, then A has n linearly independent eigenvectors.

Proof Since A is similar to D, there exists an invertible matrix P with $P^{-1}AP = D$. Then $AP = PD$, and, letting the diagonal entries of D be $\lambda_1, \lambda_2, , \ldots, \lambda_n$, we have $AP^i = \lambda_i P^i$ for $i = 1, 2, \ldots, n$. But this implies that the columns of P are eigenvectors of A, and, since P is invertible, by theorem 3 of section 4.1, its columns are linearly independent. ∎

Example 6.9

Determine whether or not the matrix

$$A = \begin{bmatrix} 1 & 0 & 0 \\ 0 & 0 & -1 \\ 0 & 1 & 0 \end{bmatrix}$$

is diagonalizable.

The characteristic polynomial for A is $\det(A - \lambda I) = (1 - \lambda)(\lambda^2 + 1)$, so A has only one eigenvalue $\lambda = 1$. Moreover a basis for the eigenspace of $\lambda = 1$ is given by $\{(1, 0, 0)\}$ and is consequently one-dimensional. Thus A has no set of three independent eigenvectors and by theorem 2 is not diagonalizable.

The next example illustrates that even if *all* roots of the characteristic equation of a matrix are real, it still may not be diagonalizable.

Example 6.10

Determine whether or not the matrix

$$A = \begin{bmatrix} 1 & 2 & -1 \\ 0 & -1 & 3 \\ 0 & 0 & 1 \end{bmatrix}$$

is diagonalizable.

Since A is upper triangular, by theorem 2 of section 6.1 its eigenvalues are given by the diagonal entries. Therefore A has two distinct eigenvalues, $\lambda_1 = 1$ and $\lambda_2 = -1$. Computing bases for the corresponding eigenspaces, we see that $\{(1, 0, 0)\}$ is a basis for the eigenspace of λ_1 and $\{(-1, 1, 0)\}$ is a basis for that of λ_2. Thus all eigenvectors are nonzero scalar multiples of these two vectors, and A has no set of three independent eigenvectors. Consequently by theorem 2, A is not diagonalizable.

In example 6.10 there were only two distinct eigenvalues of the non-diagonalizable 3×3 matrix A. The following theorem tells us that, had there been *three* distinct eigenvalues, then this matrix would have been diagonalizable.

Theorem 3   Let A be an $n \times n$ matrix. If A has n distinct eigenvalues, then A is diagonalizable.

We will give the proof of this theorem after presenting an illustrative example.

Example 6.11

Determine whether or not the matrix

$$A = \begin{bmatrix} 1 & 1 \\ 1 & 1 \end{bmatrix}$$

is diagonalizable.

The characteristic polynomial $\det(A - \lambda I)$ is given by $\lambda^2 - 2\lambda$ which has zeros $\lambda_1 = 0$ and $\lambda_2 = 2$. Thus the 2×2 matrix A has two distinct eigenvalues, and theorem 3 tells us that A is diagonalizable.

Proof of theorem 3   Let $\lambda_1, \lambda_2, \ldots, \lambda_n$ be the eigenvalues of A, and let $\mathbf{x}_1, \mathbf{x}_2, \ldots, \mathbf{x}_n$ be corresponding eigenvectors. We will show that the set

$$\mathscr{S} = \{\mathbf{x}_1, \mathbf{x}_2, \ldots, \mathbf{x}_n\}$$

is linearly independent. Then by theorem 1, A is diagonalizable as desired.

Suppose $\mathscr{S}$ is linearly dependent. Then by theorem 5 of section 4.1, one of the vectors in $\mathscr{S}$, say $\mathbf{x}_n$, can be written as a linear combination of a linearly independent subset of $\mathscr{S}$, say, $\{\mathbf{x}_1, \mathbf{x}_2, \ldots, \mathbf{x}_m\}$, $m < n$:

$$\mathbf{x}_n = a_1\mathbf{x}_1 + a_2\mathbf{x}_2 + \cdots + a_m\mathbf{x}_m. \tag{1}$$

Then

$$Ax_n = A(a_1\mathbf{x}_1 + a_2\mathbf{x}_2 + \cdots + a_m\mathbf{x}_m)$$
$$= a_1 A\mathbf{x}_1 + a_2 A\mathbf{x}_2 + \cdots + a_m A\mathbf{x}_m.$$

Therefore

$$\lambda_n\mathbf{x}_n = a_1\lambda_1\mathbf{x}_1 + a_2\lambda_2\mathbf{x}_2 + \cdots + a_m\lambda_m\mathbf{x}_m. \tag{2}$$

Multiplying equation 1 by λ_n, we obtain

$$\lambda_n\mathbf{x}_n = a_1\lambda_n\mathbf{x}_1 + a_2\lambda_n\mathbf{x}_2 + \cdots + a_m\lambda_n\mathbf{x}_m. \tag{3}$$

Subtracting (2) from (3) results in the equation

$$0 = (\lambda_n - \lambda_1)a_1\mathbf{x}_1 + (\lambda_n - \lambda_2)a_2\mathbf{x}_2 + \cdots + (\lambda_n - \lambda_m)a_m\mathbf{x}_m.$$

But this last equation is a linear combination of $\mathbf{x}_1, \mathbf{x}_2, \ldots, \mathbf{x}_m$, and, since these vectors form a linearly independent set, by theorem 1 of section 4.1, we have

$$(\lambda_n - \lambda_1)a_1 = 0, \quad (\lambda_n - \lambda_2)a_2 = 0, \ldots, (\lambda_n - \lambda_m)a_m = 0.$$

Since the λ_i are distinct, we conclude that $a_1 = a_2 = \cdots = a_m = 0$.

Finally, substituting these values for the a_i into equation (1), we see that the assumption that $\mathscr{S}$ is linearly dependent leads to the statement $\mathbf{x}_n = \mathbf{0}$. But this contradicts the fact that $\mathbf{x}_n$ is an eigenvector of A, so the theorem is proved. ∎

Notice that theorem 3 just gives only a *sufficient* condition that a matrix be diagonalizable. Consequently, if an $n \times n$ matrix A has fewer than n eigenvalues, then A *may or may not* be diagonalizable. For example, the matrices of examples 6.8 and 6.10 have fewer than three eigenvalues, yet the first of these is diagonalizable while the second is not.

Diagonalization of Symmetric Matrices

We will now consider the class of *symmetric* matrices. These matrices can easily be identified by inspection. As we shall see, all symmetric matrices are diagonalizable.

Definition   A square matrix A is **symmetric** if it is equal to its transpose, A^T.

It is very easy to determine whether or not a given matrix is symmetric. Simply check for symmetry with respect to the main diagonal. For example, the matrices

$$\begin{bmatrix} 0 & 1 \\ 1 & 0 \end{bmatrix} \quad \text{and} \quad \begin{bmatrix} 1 & -1 & 3 \\ -1 & 2 & 0 \\ 3 & 0 & -5 \end{bmatrix}$$

are symmetric, but

$$\begin{bmatrix} 1 & 1 & -1 \\ 1 & 1 & -1 \\ -1 & 1 & 1 \end{bmatrix}$$

is not.

We will now proceed to investigate the diagonalization of symmetric matrices. First, recall from section 1.4 that two vectors $\mathbf{x}$ and $\mathbf{y}$ in $\mathbf{R}^n$ are *orthogonal* if and only if their dot product, $\mathbf{x} \cdot \mathbf{y}$, is equal to zero.

Theorem 4 Let λ_1 and λ_2 be distinct eigenvalues of a symmetric matrix A with corresponding eigenvectors $\mathbf{x}_1$ and $\mathbf{x}_2$. Then $\mathbf{x}_1$ and $\mathbf{x}_2$ are orthogonal.

Proof We must show that $\mathbf{x}_1 \cdot \mathbf{x}_2 = 0$. By definition $A\mathbf{x}_1 = \lambda_1\mathbf{x}_1$, so

$$\lambda_1(\mathbf{x}_1 \cdot \mathbf{x}_2) = (\lambda_1\mathbf{x}_1) \cdot \mathbf{x}_2 = (A\mathbf{x}_1) \cdot \mathbf{x}_2,$$

which by exercise 28 is equal to $\mathbf{x}_1 \cdot (A^T\mathbf{x}_2)$. Since A is symmetric, the last expression is equal to $\mathbf{x}_1 \cdot (A\mathbf{x}_2)$ which is the same as $\mathbf{x}_1 \cdot (\lambda_2\mathbf{x}_2)$ or $\lambda_2(\mathbf{x}_1 \cdot \mathbf{x}_2)$. Thus $\lambda_1(\mathbf{x}_1 \cdot \mathbf{x}_2) = \lambda_2(\mathbf{x}_1 \cdot \mathbf{x}_2)$. Now, if $\mathbf{x}_1 \cdot \mathbf{x}_2$ were not 0, then dividing by this quantity, we would get $\lambda_1 = \lambda_2$, a contradiction. Therefore $\mathbf{x}_1 \cdot \mathbf{x}_2$ must be 0, as desired. ■

Example 6.12

Verify theorem 4 for the matrix

$$A = \begin{bmatrix} 3 & 2 \\ 2 & 0 \end{bmatrix}.$$

Since $\det(A - \lambda I) = \lambda^2 - 3\lambda - 4$, the eigenvalues of A are $\lambda_1 = 4$ and $\lambda_2 = -1$. Theorem 4 tells us that, since A is symmetric and λ_1 and λ_2 are distinct, then any pair of eigenvectors corresponding to these eigenvalues should be orthogonal. The eigenspaces of λ_1 and λ_2 are easily computed to be $\{(2s, s)\}$ and $\{(-t, 2t)\}$, respectively. To confirm that theorem 4 holds in this case, we form the dot product

$$(2s, s) \cdot (-t, 2t) = -2st + 2st,$$

which is indeed equal to 0.

Since the matrix A of example 6.12 has distinct eigenvalues, theorem 3 implies that A is diagonalizable. In other words, there is a 2×2 invertible matrix P such that $P^{-1}AP = D$, a diagonal matrix. We can then apply theorem 1 to construct P by taking its columns to be linearly independent eigenvectors of A. By choosing these eigenvectors to have length 1 (that is, *normalizing* them), we obtain a matrix with particularly interesting properties. The matrix of normalized eigenvectors is

$$P = \begin{bmatrix} 2/\sqrt{5} & -1/\sqrt{5} \\ 1/\sqrt{5} & 2/\sqrt{5} \end{bmatrix}.$$

We not only have $P^1 \cdot P^2 = 0$ (reflecting the orthogonality of the eigenvectors) and $P^1 \cdot P^1 = P^2 \cdot P^2 = 1$ (reflecting the fact that we chose eigenvectors of norm 1), but, computing P^{-1}, we see that

$$P^{-1} \quad \begin{bmatrix} 2/\sqrt{5} & 1/\sqrt{5} \\ -1/\sqrt{5} & 2/\sqrt{5} \end{bmatrix},$$

which is just P^T. As we shall see, *every* symmetric matrix A (whether or not it has distinct eigenvalues) is diagonalizable by a matrix P having these properties.

Definition   A nonsingular matrix P is **orthogonal** if $P^{-1} = P^T$.

Example 6.13

Let

$$P = \begin{bmatrix} 1/3 & 2\sqrt{2}/3 & 0 \\ 2/3 & -\sqrt{2}/6 & \sqrt{2}/2 \\ -2/3 & \sqrt{2}/6 & \sqrt{2}/2 \end{bmatrix}.$$

Show that P is orthogonal.
  We could compute P^{-1} and then verify that $P^{-1} = P^T$, but there is an easier way. We simply compute the product $P^T P$. Since $P^T P = I$, by theorem 2 of section 2.4, $P^{-1} = P^T$, and thus P is orthogonal by definition.

Note: It is easy to see that the columns of the matrix P of example 6.13 have the following properties:

i. $P^i \cdot P^j = 0$ if $i \neq j$ $(i, j = 1, 2, 3)$,
ii. $P^i \cdot P^i = 1$ $(i = 1, 2, 3)$.

Recall that a set of vectors that possesses these two properties is called an *orthonormal set*.

Theorem 5   An $n \times n$ matrix P is orthogonal if and only if its columns

$$\{P^1, P^2, \ldots, P^n\}$$

form an orthonormal set.

Proof   Exercise 29.   ■

Definition   A square matrix A is **orthogonally diagonalizable** if there exists an orthogonal matrix P such that $P^{-1}AP (= P^T AP)$ is a diagonal matrix.

The next theorem states that *every* symmetric matrix is orthogonally diagonalizable. The proof is omitted.

Theorem 6 Let A be a symmetric matrix. Then there exists an orthogonal matrix P such that $D = P^{-1}AP$ is a diagonal matrix.

Note: If A is a symmetric matrix, then by theorem 6 it is diagonalizable, and for each i, $AP^i = PP^{-1}AP^i = PD^i = d_{ii}P^i$. Thus we have

i. the columns of P are linearly independent eigenvectors of A,
   $\mathbf{x}_i = P^i$, $i = 1, \ldots, n$,
ii. the ith diagonal entry $\lambda_i = d_{ii}$ of D is the eigenvalue of A corresponding to the eigenvector $\mathbf{x}_i$.

Moreover by theorem 5 the columns of P form an orthonormal set. Consequently $\{\mathbf{x}_1, \mathbf{x}_2, \ldots, \mathbf{x}_n\}$ is an orthonormal set of eigenvectors.

Procedure for orthogonally diagonalizing a symmetric matrix Let A be an $n \times n$ symmetric matrix. To find an orthogonal matrix P such that $D = P^{-1}AP$ is a diagonal matrix,

i. Find the distinct eigenvalues, $\lambda_1, \lambda_2, \ldots, \lambda_s$, of A.
ii. For each eigenvalue, λ_i, find a basis for its eigenspace, $\mathbf{S}_i$.
iii. With the aid of the Gram-Schmidt process (section 4.5) construct an orthonormal basis, $\mathscr{B}_i$, for each $\mathbf{S}_i$. The union of all $\mathscr{B}_i$ will be an orthonormal set of n vectors, $\{\mathbf{x}_1, \mathbf{x}_2, \ldots, \mathbf{x}_n\}$.
iv. Take the jth column of P to be $P^j = \mathbf{x}_j$.

To find the diagonal matrix D, set d_{kk} equal to the eigenvalue corresponding to the eigenvector $\mathbf{x}_k$, $k = 1, 2, \ldots, n$, and for all $i \neq j$ set $d_{ij} = 0$.

Example 6.14

Let

$$A = \begin{bmatrix} -1 & 0 & 0 \\ 0 & 0 & 2 \\ 0 & 2 & 3 \end{bmatrix}.$$

Is the matrix A orthogonally diagonalizable? If so, find an orthogonal matrix P and diagonal matrix D such that $D = P^{-1}AP$.

Since A is symmetric, by theorem 6, A is orthogonally diagonalizable. To find the matrices P and D, we apply the procedure just described.

We first find the eigenvalues of A by solving the characteristic equation $\det (A - \lambda I) = 0$, where

$$\det (A - \lambda I) = \det \begin{bmatrix} -1 - \lambda & 0 & 0 \\ 0 & -\lambda & 2 \\ 0 & 2 & 3 - \lambda \end{bmatrix} = -(\lambda + 1)^2(\lambda - 4).$$

Thus the eigenvalues of A are $\lambda_1 = -1$ and $\lambda_2 = 4$.

We now determine bases for the eigenspaces S_1 and S_2 of λ_1 and λ_2 by considering the linear homogeneous systems $(A + I)x = 0$ and $(A - 4I)x = 0$, respectively. Proceeding as in section 6.1, we see that a basis for S_1 is $\mathcal{T}_1 = \{(1, 0, 0), (0, -2, 1)\}$, while a basis for S_2 is $\mathcal{T}_2 = \{(0, 1, 2)\}$.

Applying the Gram-Schmidt process to $\mathcal{T}_1$, we obtain the orthonormal basis $\mathcal{B}_1 = \{(1, 0, 0), (0, -2/\sqrt{5}, 1/\sqrt{5})\}$ for the eigenspace of λ_1. (Since $\mathcal{T}_1$ was orthogonal, all we actually needed to do was to normalize each vector in this set.) Then, normalizing the vector $(0, 1, 2)$ of $\mathcal{T}_2$, we obtain an orthonormal basis $\mathcal{B}_2 = \{(0, 1/\sqrt{5}, 2/\sqrt{5})\}$ for S_2. Finally, taking the union of $\mathcal{B}_1$ and $\mathcal{B}_2$, we obtain an orthonormal set consisting of the three eigenvectors $x_1 = (1, 0, 0)$, $x_2 = (0, -2/\sqrt{5}, 1/\sqrt{5})$, and $x_3 = (0, 1/\sqrt{5}, 2/\sqrt{5})$.

Thus the matrix P is given by $P^j = x_j$ or

$$P = \begin{bmatrix} 1 & 0 & 0 \\ 0 & -2/\sqrt{5} & 1/\sqrt{5} \\ 0 & 1/\sqrt{5} & 2/\sqrt{5} \end{bmatrix}.$$

To construct D, we simply form the 3×3 diagonal matrix whose diagonal entries are the eigenvalues corresponding to the eigenvectors x_1, x_2, x_3. Therefore

$$D = \begin{bmatrix} -1 & 0 & 0 \\ 0 & -1 & 0 \\ 0 & 0 & 4 \end{bmatrix}.$$

We now know that every symmetric matrix is orthogonally diagonalizable. A natural question to ask at this point is whether any other types of matrices are orthogonally diagonalizable. The next theorem, which is the converse of theorem 6, provides a negative answer to this question.

Theorem 7 If a square matrix A is orthogonally diagonalizable, then it is symmetric.

Proof Exercise 30.   ∎

6.2 Exercises

In exercises 1 through 12 determine whether or not the given matrix A is diagonalizable. If it is, find a matrix P such that $P^{-1}AP$ is diagonal, and find this diagonal matrix.

1. $\begin{bmatrix} 1 & 4 \\ 2 & -1 \end{bmatrix}$

2. $\begin{bmatrix} 0 & 1 \\ 1 & 0 \end{bmatrix}$

3. $\begin{bmatrix} 1 & 2 \\ -2 & -1 \end{bmatrix}$

4. $\begin{bmatrix} 0 & 1 \\ -1 & 2 \end{bmatrix}$

5. $\begin{bmatrix} 0 & -1 & 0 \\ -1 & 0 & -1 \\ 0 & -1 & 0 \end{bmatrix}$

6. $\begin{bmatrix} 2 & -1 & 0 \\ -1 & 2 & -1 \\ 0 & -1 & 2 \end{bmatrix}$

7. $\begin{bmatrix} 1 & 1 & 1 \\ 1 & 1 & 1 \\ 1 & 1 & 1 \end{bmatrix}$

8. $\begin{bmatrix} 1 & 1 & -1 \\ -1 & -1 & 1 \\ 2 & 2 & -2 \end{bmatrix}$

9. $\begin{bmatrix} 1 & 0 & 3 \\ 0 & -2 & 1 \\ 0 & 0 & 1 \end{bmatrix}$

10. $\begin{bmatrix} 1 & 0 & -1 \\ 1 & 1 & 2 \\ 2 & 1 & 1 \end{bmatrix}$

11. $\begin{bmatrix} 1 & 1 & 1 & 1 \\ 1 & 1 & 1 & 1 \\ 1 & 1 & 1 & 1 \\ 1 & 1 & 1 & 1 \end{bmatrix}$

12. $\begin{bmatrix} 1 & 0 & 0 & 0 \\ 0 & 2 & 0 & 0 \\ 0 & 1 & -1 & 0 \\ -2 & 0 & 0 & 2 \end{bmatrix}$

In exercises 13 through 16 determine whether or not the given matrix is orthogonal.

13. $\begin{bmatrix} 1/3 & 2/3 & 2/3 \\ 2/3 & 1/3 & -2/3 \\ 2/3 & -2/3 & 1/3 \end{bmatrix}$

14. $\begin{bmatrix} 0 & 1 & 0 \\ 1 & 0 & 0 \\ 0 & 0 & 1 \end{bmatrix}$

15. $\begin{bmatrix} 1/\sqrt{5} & -2/\sqrt{5} & 1 \\ -2/\sqrt{5} & 1/\sqrt{5} & -1 \\ 0 & 0 & 1 \end{bmatrix}$

16. $\begin{bmatrix} 0 & 1 & 0 \\ 1 & 0 & 1 \\ 0 & 1 & 0 \end{bmatrix}$

In exercises 17 through 24 find an orthogonal matrix P such that $P^{-1}AP$ is diagonal, where A is the given symmetric matrix.

17. $\begin{bmatrix} 0 & 2 \\ 2 & 0 \end{bmatrix}$

18. $\begin{bmatrix} 1 & 1 \\ 1 & 1 \end{bmatrix}$

19. $\begin{bmatrix} 0 & -1 & 0 \\ -1 & 0 & -1 \\ 0 & -1 & 0 \end{bmatrix}$

20. $\begin{bmatrix} 2 & -1 & 0 \\ -1 & 2 & -1 \\ 0 & -1 & 2 \end{bmatrix}$

21. $\begin{bmatrix} 1 & 1 & 1 \\ 1 & 1 & 1 \\ 1 & 1 & 1 \end{bmatrix}$
22. $\begin{bmatrix} 0 & 1 & 0 \\ 1 & 0 & 0 \\ 0 & 0 & 1 \end{bmatrix}$

23. $\begin{bmatrix} 0 & 1 & 0 & 0 \\ 1 & 0 & 0 & 0 \\ 0 & 0 & 0 & 1 \\ 0 & 0 & 1 & 0 \end{bmatrix}$
24. $\begin{bmatrix} 1 & 1 & 1 & 1 \\ 1 & 1 & 1 & 1 \\ 1 & 1 & 1 & 1 \\ 1 & 1 & 1 & 1 \end{bmatrix}$

25. Show that the rotation matrix

$$A = \begin{bmatrix} \cos\theta & -\sin\theta \\ \sin\theta & \cos\theta \end{bmatrix}$$

is orthogonal.

26. Let A be an orthogonal matrix. Show that $\det(A) = \pm 1$. (*Hint:* $\det(A^TA) = \det(I) = 1$.)

27. For the matrix

$$A = \begin{bmatrix} a & b \\ c & d \end{bmatrix}$$

obtain conditions on a, b, c, and d such that (a) A is diagonalizable and (b) A is not diagonalizable. (*Hint:* see exercise 41 of section 6.1.)

28. Let A be an $n \times n$ matrix, and let x_1 and x_2 be in $\mathbf{R}^n$. Prove that $(Ax_1) \cdot x_2 = x_1 \cdot (A^Tx_2)$. (*Hint:* viewing $\mathbf{u}$ and $\mathbf{v}$ as column vectors, the dot product is given by $\mathbf{u} \cdot \mathbf{v} = \mathbf{v}^T\mathbf{u}$.)

29. Prove (theorem 5) that an $n \times n$ matrix P is orthogonal if and only if $\{P^1, P^2, \ldots, P^n\}$ is an orthonormal set.

30. Prove (theorem 7) that if a square matrix is orthogonally diagonalizable, then it is symmetric. (*Hint:* consider $(P^{-1}AP)^T = D^T$.)

31. Let P and Q be orthogonal matrices of the same order. Show that PQ and QP are also orthogonal. (*Hint:* show that $(PQ)(PQ)^T$.)

32. Let θ be a real number, and let s and t be integers with $1 \le s < t \le n$. Define an $n \times n$ matrix P by

$$P_{ss} = P_{tt} = \cos\theta, \quad P_{ts} = -P_{st} = \sin\theta,$$

and

$$P_{ii} = 1, \quad P_{ij} = 0 \quad (i \ne j), \quad \text{otherwise.}$$

Show that P is orthogonal. (The matrix P is called a *rotation matrix for the s, t-plane.*)

33. Let P be a third-order rotation matrix in the 1, 3-plane (see exercise 32), that is

$$P = \begin{bmatrix} \cos\theta & 0 & -\sin\theta \\ 0 & 1 & 0 \\ \sin\theta & 0 & \cos\theta \end{bmatrix}.$$

Let A be an arbitrary 3×3 symmetric matrix. Show that, if we choose θ so that $\tan 2\theta = 2a_{13}/(a_{11} - a_{33})$, then the matrix $B = P^T A P \ (= P^{-1} A P)$ satisfies $b_{13} = b_{31} = 0$. (*Note:* in general, if P is a nth rotation in the s, t-plane, and A is symmetric, then, by choosing θ so that $\tan 2\theta = 2a_{st}/(a_{ss} - a_{tt})$, both the st and the ts entries of $P^T A P$ are equal to 0.

Computational note: Jacobi's method approximates the eigenvalues of a symmetric matrix A by forming the sequence of products

$$(P_k^T P_{k-1}^T \ldots P_1^T) A (P_1 P_2 \ldots P_k), \quad k = 1, 2, 3, \ldots .$$

In this expression each P_i is a rotation matrix (see exercise 32) with θ chosen to "annihilate" (send to zero) the largest off-diagonal entry of

$$(P_{i-1}^T \ldots P_1^T) A (P_1 \ldots P_{i-1})$$

(see exercise 33). Unfortunately, as one off-diagonal entry is annihilated, a previously annihilated one may become nonzero. Nevertheless, the size of all off-diagonal entries does approach zero as the process is continued, and the diagonal entries will in turn get closer and closer to the eigenvalues of A. Jacobi's method is perhaps the most straightforward way of approximating the eigenvalues of a symmetric matrix, but there are more sophisticated techniques that are much more efficient.

6.3   Complex Eigenvalues and Eigenvectors

To this point in the text all scalars, including components of vectors and entries of matrices, have been real numbers. In this section we will allow scalars to be complex numbers. This generalization of the concept of scalar will alter very little of the material of the first five chapters and will make the subject of eigenvalues and eigenvectors more complete.

Complex m–Space

Recall that a **complex number** is one of the form $a + bi$, where a and b are real numbers and $i = \sqrt{-1}$. We will denote the set of all complex numbers by **C**.

The **absolute value (modulus)** of a complex number $z = a + bi$ is denoted $|z|$ and defined to be $|z| = \sqrt{a^2 + b^2}$. The **complex conjugate** of $z = a + bi$ is denoted $\bar{z}$ and given by $\bar{z} = a - bi$. We may represent any complex number $z = a + bi$ as the vector (a, b) in $\mathbf{R}^2$. Then the geometric interpretation of $|z|$ is simply $\|(a, b)\|$, the distance from the origin to the point (a, b), while that of $\bar{z}$ is the reflection of z in the x-axis (figure 6.1).

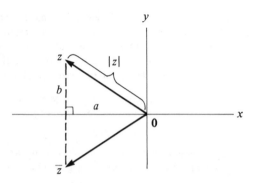

Figure 6.1

We do not wish to think of complex numbers as vectors in $\mathbf{R}^2$ but as the *scalars* for a new set of vectors. For this reason we will use scalar notation to symbolize complex numbers.

Of course the real numbers $\mathbf{R}$ form a subset of the set $\mathbf{C}$ of complex numbers, since any x in $\mathbf{R}$ may be written in the form $x + 0i$. Notice that the definition of absolute value for complex numbers is compatible with the one for real numbers, since $|x + 0i| = \sqrt{x^2 + 0^2} = \sqrt{x^2} = |x|$. Moreover a necessary and sufficient condition that a given complex number z be real is that $z = \bar{z}$ (since, if $z = a + bi$, $z = \bar{z}$ is equivalent to $a = a$ and $b = -b$, and the latter is true if and only if $b = 0$).

The set of all m-tuples of complex numbers, denoted $\mathbf{C}^m$, is called **complex m-space.** The definitions of addition and subtraction of vectors in $\mathbf{C}^m$ and multiplication of a vector in $\mathbf{C}^m$ by a scalar are identical to the $\mathbf{R}^m$ case, but the definitions of dot product and norm in $\mathbf{C}^m$ are not.

Definition Let $\mathbf{w} = (w_1, w_2, \ldots, w_m)$ and $\mathbf{z} = (z_1, z_2, \ldots, z_m)$ be vectors in $\mathbf{C}^m$. Then

a. $\mathbf{w} \cdot \mathbf{z} = w_1 \bar{z}_1 + w_2 \bar{z}_2 + \cdots + w_m \bar{z}_m$ (dot product),

b. $\|\mathbf{w}\| = \sqrt{|w_1|^2 + |w_2|^2 + \cdots + |w_m|^2}$ (*norm* or *length*).

Note: These definitions reduce to the corresponding ones for $\mathbf{R}^m$ in the case when $\mathbf{w}$ and $\mathbf{z}$ are in $\mathbf{R}^m$.

Notice that the dot product, $\mathbf{w} \cdot \mathbf{z}$, involves complex conjugates. It is defined in this manner to preserve the property that $\mathbf{w} \cdot \mathbf{w} = \|\mathbf{w}\|^2$. If we had defined the dot product as $\mathbf{w} \cdot \mathbf{z} = w_1 z_1 + w_2 z_2 + \cdots + w_m z_m$, then this would not always be the case (exercise 42). However, with the definition just stated we do have $\mathbf{w} \cdot \mathbf{w} = \|\mathbf{w}\|^2$ (exercise 41).

Example 6.15

Let $\mathbf{w} = (i, 1, 0)$ and $\mathbf{z} = (2i, 1 + i, 1 - i)$ be vectors in $\mathbf{C}^3$. Find $\mathbf{w} \cdot \mathbf{z}$, $\mathbf{z} \cdot \mathbf{w}$, and $\|\mathbf{w}\|$.

We have

  i. $\mathbf{w} \cdot \mathbf{z} = i(\overline{2i}) + 1(\overline{1 + i}) + 0(\overline{1 - i}) = i(-2i) + 1(1 - i)$
             $= -2i^2 + 1 - i = 3 - i,$

 ii. $\mathbf{z} \cdot \mathbf{w} = 2i(\bar{i}) + (1 + i)\bar{1} + (1 - i)\bar{0} = 2i(-i) + (1 + i)1$
             $= -2i^2 + 1 + i = 3 + i,$

iii. $\|\mathbf{w}\| = \sqrt{|i|^2 + 1^2 + 0^2} = \sqrt{1^2 + 1^2} = \sqrt{2}.$

Notice that for the $\mathbf{w}$ and $\mathbf{z}$ of example 6.15 we have $\mathbf{w} \cdot \mathbf{z} = \overline{\mathbf{z} \cdot \mathbf{w}}$. In fact this equation holds for every $\mathbf{w}$ and $\mathbf{z}$ in $\mathbf{C}^m$. The next theorem presents some properties of dot product in $\mathbf{C}^m$, including this one.

Theorem 1   Let $\mathbf{w}$ and $\mathbf{z}$ be vectors in $\mathbf{C}^m$, and let c be a scalar (a complex number). Then

a. $\mathbf{w} \cdot \mathbf{w} \geq 0$, $\mathbf{w} \cdot \mathbf{w} = 0$ if and only if $\mathbf{w} = \mathbf{0}$,
b. $\mathbf{w} \cdot \mathbf{z} = \overline{\mathbf{z} \cdot \mathbf{w}}$,
c. $(c\mathbf{w}) \cdot \mathbf{z} = c(\mathbf{w} \cdot \mathbf{z})$, $\mathbf{w} \cdot (c\mathbf{z}) = \bar{c}(\mathbf{w} \cdot \mathbf{z})$,
d. $\mathbf{w} \cdot \mathbf{w} = \|\mathbf{w}\|^2.$

Proof of b   The proofs of a, c, and d are left as an exercise.
    Let $\mathbf{w} = (w_1, w_2, \ldots, w_m)$, $\mathbf{z} = (z_1, z_2, \ldots, z_m)$. Then

$$
\begin{aligned}
\overline{\mathbf{z} \cdot \mathbf{w}} &= \overline{z_1\bar{w}_1 + z_2\bar{w}_2 + \cdots + z_m\bar{w}_m} \\
&= \overline{z_1\bar{w}_1} + \overline{z_2\bar{w}_2} + \cdots + \overline{z_m\bar{w}_m} \quad \text{(exercise 43a)} \\
&= \bar{z}_1\bar{\bar{w}}_1 + \bar{z}_2\bar{\bar{w}}_2 + \cdots + \bar{z}_m\bar{\bar{w}}_m \quad \text{(exercise 43b)} \\
&= \bar{z}_1 w_1 + \bar{z}_2 w_2 + \cdots + \bar{z}_m w_m \quad \text{(exercise 43c)} \\
&= w_1\bar{z}_1 + w_2\bar{z}_2 + \cdots + w_m\bar{z}_m = \mathbf{w} \cdot \mathbf{z}.
\end{aligned}
$$
                                                           ■

Matrices with Entries in C

In this section we shall allow the entries of matrices to be complex numbers. All definitions and theorems concerning matrices and determinants given in chapters 2 and 3 remain completely unchanged with one minor exception. In section 2.3 we defined the product of two matrices in terms of dot product. The product of an $m \times n$ matrix A and an $n \times p$ matrix

B, whether the entries are complex or real, is defined to be the $m \times p$ matrix C in which

$$c_{ij} = a_{i1}b_{1j} + a_{i2}b_{2j} + \cdots + a_{in}b_{nj}.$$

However, because the complex dot product involves complex conjugates, this expression can no longer be given as simply $c_{ij} = A_i \cdot B^j$.

Example 6.16

Let

$$A = \begin{bmatrix} i & 1-i & 2 \\ 0 & -2i & 1+i \\ 1 & 0 & 3-i \end{bmatrix}, \quad B = \begin{bmatrix} 1-i & i \\ -i & 0 \\ 0 & 2+i \end{bmatrix}.$$

Find $AB - 2B$.

We have

$$AB - 2B = \begin{bmatrix} i & 1-i & 2 \\ 0 & -2i & 1+i \\ 1 & 0 & 3-i \end{bmatrix}\begin{bmatrix} 1-i & i \\ -i & 0 \\ 0 & 2+i \end{bmatrix} - 2\begin{bmatrix} 1-i & i \\ -i & 0 \\ 0 & 2+i \end{bmatrix}$$

$$= \begin{bmatrix} 0 & 3+2i \\ -2 & 1+3i \\ 1-i & 7+2i \end{bmatrix} - \begin{bmatrix} 2-2i & 2i \\ -2i & 0 \\ 0 & 4+2i \end{bmatrix}$$

$$= \begin{bmatrix} -2+2i & 3 \\ -2+2i & 1+3i \\ 1-i & 3 \end{bmatrix}.$$

Before proceeding to consider complex eigenvalues and eigenvectors, we should point out that all definitions and theorems of chapters 4 and 5 (linear independence, basis, linear transformations, and so on) are exactly the same for the $\mathbf{C}^m$ case as they were in $\mathbf{R}^m$. This will be elaborated upon to some extent in chapter 7.

Complex Eigenvalues, Eigenvectors in C^n

Although much of the material dealing with complex eigenvalues and eigenvectors (section 6.1) is completely analogous to the real case, we shall also see some interesting differences.

Definition Let A be an $n \times n$ matrix with complex entries. A scalar λ is an **eigenvalue** of A if there is a nonzero vector z in $\mathbf{C}^n$ such that $Az = \lambda z$. The vector z is said to be an **eigenvector** corresponding to λ. The **eigenspace** of λ is the set of all solutions of the homogeneous linear system $(A - \lambda I)z = 0$.

The computation of complex eigenvalues and their corresponding eigenvectors proceeds in a manner similar to that of the real case (see theorem 1 of section 6.1 and its accompanying procedure). However, since the characteristic equation, $\det(A - \lambda I) = 0$, is a polynomial equation, it must have at least one complex root by the *fundamental theorem of algebra*. Consequently we now have the following theorem.

Theorem 2 Every $n \times n$ matrix has at least one (complex) eigenvalue.

Example 6.17

Find the eigenvalues and associated eigenspaces of the matrix

$$A = \begin{bmatrix} i & 0 & 0 \\ 0 & 0 & 1 \\ 0 & -1 & 0 \end{bmatrix}.$$

To find the eigenvalues of A, we compute the characteristic polynomial

$$\det(A - \lambda I) = \det \begin{bmatrix} i - \lambda & 0 & 0 \\ 0 & -\lambda & 1 \\ 0 & -1 & -\lambda \end{bmatrix} = (i - \lambda)(\lambda^2 + 1).$$

Thus the eigenvalues of A, the roots of the characteristic equation, are $\lambda_1 = i$ and $\lambda_2 = -i$ (since $\lambda^2 + 1 = 0$ has the solution $\lambda = \pm i$).

To compute the eigenspaces of λ_1 and λ_2, we solve the homogeneous linear systems $(A - iI)z = 0$ and $(A + iI)z = 0$, respectively. Doing so, we see that the eigenspace of $\lambda_1 = i$ is the set of all vectors of the form $(r, -is, s)$, while that of $\lambda_2 = -i$ consists of all vectors of the form $(0, it, t)$, where r, s, and t are scalars (complex numbers). (*Note:* Bases for the eigenspaces of λ_1 and λ_2 are, respectively, $\{(1, 0, 0), (0, -i, 1)\}$ and $\{(0, i, 1)\}$.)

Diagonalization of Complex Matrices

The basic theory concerning diagonalization of matrices carries over to the case of matrices with complex entries. Specifically theorems 1, 2, and

3 of section 6.2 continue to hold, and the proofs are essentially the same. We restate these theorems here in a more concise form.

Theorem 3   Let A be an $n \times n$ matrix with complex entries. Then

    a. A is diagonalizable if and only if A has n linearly independent eigenvectors in $\mathbf{C}^n$,

    b. if A has n distinct eigenvalues, then A is diagonalizable.

The next example illustrates the procedure of theorem 1 of section 6.2 in the case of complex matrices.

Example 6.18

Determine whether or not the matrix

$$A = \begin{bmatrix} i & 0 & 0 \\ 0 & 0 & 1 \\ 0 & -1 & 0 \end{bmatrix}$$

is diagonalizable. If it is, find an invertible matrix P such that $P^{-1}AP$ is diagonal. Also find this diagonal matrix.

Notice that the given matrix is the same as the one of example 6.17. Since it has only two distinct eigenvalues, $\lambda_1 = i$ and $\lambda_2 = -i$, we cannot apply theorem 3b. However, the union of the bases for the eigenspaces of λ_1 and λ_2 (computed in example 6.17) is $\{(1, 0, 0), (0, -i, 1), (0, i, 1)\}$, a set of three linearly independent eigenvectors. Therefore by theorem 3a, A is diagonalizable.

Moreover, following theorem 1 of section 6.2, we may take P to be the matrix whose columns are these eigenvectors:

$$P = \begin{bmatrix} 1 & 0 & 0 \\ 0 & -i & i \\ 0 & 1 & 1 \end{bmatrix}.$$

Finally, the diagonal matrix $P^{-1}AP$ is the one with diagonal entries that are the eigenvalues of A. Thus

$$P^{-1}AP = \begin{bmatrix} i & 0 & 0 \\ 0 & i & 0 \\ 0 & 0 & -i \end{bmatrix}.$$

We next consider the complex analog of the orthogonal diagonalization of real symmetric matrices.

Definition The **conjugate transpose** of a matrix A, denoted A^H, is the transpose of the matrix obtained from A by replacing each entry by its complex conjugate.
We say that a square matrix A is **hermitian** if $A = A^H$.

For example the matrix

$$A = \begin{bmatrix} 1 & 1+i & 0 \\ 1-i & -3 & -2i \\ 0 & 2i & 0 \end{bmatrix}$$

is hermitian, since

$$A^H = \left(\begin{bmatrix} 1 & 1-i & 0 \\ 1+i & -3 & 2i \\ 0 & -2i & 0 \end{bmatrix} \right)^T = A.$$

Note: The *diagonal* entries of a hermitian matrix must be real, since they are equal to their complex conjugates. Moreover, if *all* entries of a square matrix A are real, then $A^H = A^T$. Thus *hermitian* is the complex generalization of *symmetric*.

The reason why we consider *conjugate transpose* instead of just *transpose* in the complex case is to preserve the validity of theorems such as theorems 4, 6 and 7 of section 6.2 when we replace *symmetric* by *hermitian*. For example, if a square matrix A has imaginary entries, it is *not* true that $A = A^T$ implies that A is diagonalizable (see exercise 44). The underlying reason for the need for *conjugate* transpose in these situations is that complex dot products involve conjugates. Notice how this factor enters into the proofs of the following two theorems. (Theorem 5 is a complex analog of theorem 4 of section 6.2.)

Theorem 4 Let A be a hermitian matrix. Then all eigenvalues of A are real numbers.

Proof Let λ be an eigenvalue of A with corresponding eigenvector $\mathbf{z}$. Then $A\mathbf{z} = \lambda\mathbf{z}$ and

$$\lambda(\mathbf{z} \cdot \mathbf{z}) = (\lambda\mathbf{z}) \cdot \mathbf{z} = (A\mathbf{z}) \cdot \mathbf{z} = \mathbf{z} \cdot (A^H\mathbf{z})$$

by exercise 45. Moreover A is hermitian, so

$$\mathbf{z} \cdot (A^H\mathbf{z}) = \mathbf{z} \cdot (A\mathbf{z}) = \mathbf{z} \cdot (\lambda\mathbf{z}) = \bar{\lambda}(\mathbf{z} \cdot \mathbf{z})$$

by theorem 1c. Hence $\lambda(\mathbf{z} \cdot \mathbf{z}) = \bar{\lambda}(\mathbf{z} \cdot \mathbf{z})$. But $\mathbf{z}$ is an eigenvector, and therefore $\mathbf{z} \neq \mathbf{0}$ which in turn implies that $\mathbf{z} \cdot \mathbf{z} \neq 0$ (theorem 1a). Consequently $\lambda = \bar{\lambda}$, and we conclude that λ is real. ∎

Theorem 5 Let λ_1 and λ_2 be distinct eigenvalues of a hermitian matrix A with corresponding eigenvectors $\mathbf{z}_1$ and $\mathbf{z}_2$, respectively. Then $\mathbf{z}_1$ and $\mathbf{z}_2$ are *orthogonal*.

Proof The proof is virtually identical to that of theorem 4 of section 6.2. We have

$$\lambda_1(\mathbf{z}_1 \cdot \mathbf{z}_2) = (\lambda_1 \mathbf{z}_1) \cdot \mathbf{z}_2 = (A\mathbf{z}_1) \cdot \mathbf{z}_2 = \mathbf{z}_1 \cdot (A^H \mathbf{z}_2)$$

by exercise 45. But A is hermitian, so

$$\mathbf{z}_1 \cdot (A^H \mathbf{z}_2) = \mathbf{z}_1 \cdot (A\mathbf{z}_2) = \mathbf{z}_1 \cdot (\lambda_2 \mathbf{z}_2) = \bar{\lambda}_2(\mathbf{z}_1 \cdot \mathbf{z}_2)$$

by theorem 1c. Thus $\lambda_1(\mathbf{z}_1 \cdot \mathbf{z}_2) = \bar{\lambda}_2(\mathbf{z}_1 \cdot \mathbf{z}_2) = \lambda_2(\mathbf{z}_1 \cdot \mathbf{z}_2)$ by theorem 4. Now if $\mathbf{z}_1 \cdot \mathbf{z}_2$ were not equal to zero, we would have $\lambda_1 = \lambda_2$, a contradiction. Thus $\mathbf{z}_1 \cdot \mathbf{z}_2 = 0$, as desired. ∎

In order to state the main theorem concerning the diagonalization of hermitian matrices, we must first introduce the complex analog of an orthogonal matrix.

Definition A nonsingular matrix P with complex entries is **unitary** if $P^{-1} = P^H$. For example, the matrix

$$P = \begin{bmatrix} i/\sqrt{2} & i/\sqrt{2} \\ i/\sqrt{2} & -i/\sqrt{2} \end{bmatrix}$$

is unitary. It is an easy matter to check that $P^H P = I$, which implies that $P^{-1} = P^H$.

The next theorem tells us that the columns of unitary matrices behave in exactly the same manner with respect to the complex dot product as do those of orthogonal matrices relative to the real dot product.

Theorem 6 An $n \times n$ matrix P with complex entries is unitary if and only if its columns $\{P^1, P^2, \ldots, P^n\}$ form an orthonormal set, that is, if and only if $P^i \cdot P^j = 0$ if $i \neq j$ while $P^i \cdot P^i = 1$.

Proof Exercise 46. ∎

Our last theorem for this section gives a generalization of theorems 6 and 7 of section 6.2 to the complex case. We omit the proof.

Theorem 7 Let A be a square matrix with complex entries. Then there exists a unitary matrix P such that $P^{-1}AP (= P^H AP)$ is diagonal with real entries if and only if A is hermitian.

Example 6.19

Let

$$A = \begin{bmatrix} 0 & -i & 0 \\ i & 0 & 0 \\ 0 & 0 & 2 \end{bmatrix}.$$

Find a unitary matrix P such that $P^{-1}AP$ is diagonal.

To find the matrix P, we follow the procedure given in section 6.2 for orthogonally diagonalizing a real symmetric matrix. Since the characteristic polynomial of A is $\det(A - \lambda I) = (2 - \lambda)(\lambda^2 - 1)$, the eigenvalues of A are $\lambda_1 = -1$, $\lambda_2 = 1$, and $\lambda_3 = 2$. Bases for the eigenspaces corresponding to these eigenvalues are then computed to be

$$\{(i, 1, 0)\} \quad \text{for} \quad \lambda_1 = -1,$$
$$\{(-i, 1, 0)\} \quad \text{for} \quad \lambda_2 = 1,$$
and
$$\{(0, 0, 1)\} \quad \text{for} \quad \lambda_3 = 2.$$

Finally, to obtain the columns of P, we normalize these three eigenvectors to get $(i/\sqrt{2}, 1/\sqrt{2}, 0)$, $(-i/\sqrt{2}, 1/\sqrt{2}, 0)$, and $(0, 0, 1)$. Hence

$$P = \begin{bmatrix} i/\sqrt{2} & -i/\sqrt{2} & 0 \\ 1/\sqrt{2} & 1/\sqrt{2} & 0 \\ 0 & 0 & 1 \end{bmatrix}.$$

Note: In general, to obtain a unitary matrix that diagonalizes a hermitian one, it may be necessary to employ the Gram-Schmidt process in the same manner as in section 6.2. We will give an extension of the Gram-Schmidt process of section 4.5 in section 7.4.

6.3 Exercises

In exercises 1 through 4 letting $\mathbf{w} = (i, 0, 1 - i, 1)$ and $\mathbf{z} = (-i, 2 + i, 0, -1)$, evaluate the expressions.

1. $\mathbf{w} \cdot \mathbf{z}$　　　　**2.** $\mathbf{z} \cdot \mathbf{w}$　　　　**3.** $\|\mathbf{w}\|$　　　　**4.** $\mathbf{w} \cdot \mathbf{w}$

In exercises 5 through 13 let

$$A = \begin{bmatrix} 1 & 0 & -i \\ 0 & 1+i & 1 \\ i & -1 & 2+i \end{bmatrix}, \quad B = \begin{bmatrix} i & -i & 0 \\ i & i & 1 \\ 0 & 1 & i \end{bmatrix}.$$

5. Solve $A\mathbf{z} = \mathbf{b}$, where $\mathbf{b} = (i, 0, -1)$

6. Solve $B\mathbf{z} = \mathbf{c}$, where $\mathbf{c} = (1 + i, 1 - i, 1)$

7. Find AB **8.** Find BA

9. Find A^{-1} **10.** Find B^{-1}

11. Find $\det (A)$ **12.** Find $\det (B)$

13. Is either A or B hermitian?

14. Show that the dimension of $\mathbf{C}^m$ is equal to m.

In exercises 15 through 18 determine whether or not the given set is a basis for $\mathbf{C}^3$.

15. $\{(i, 0, -1), (1, 1, 1), (0, -i, i)\}$

16. $\{(i, -i, 1), (1, 1+i, 1-i), (0, 2+i, 1)\}$

17. $\{(i, 1, 0), (0, 0, 1)\}$

18. $\{(0, i, 1), (i, 0, 0), (1+i, -1, 0), (1-i, 0, 1)\}$

In exercises 19 through 22 find bases for the given subspace of $\mathbf{C}^m$.

19. The subspace of $\mathbf{C}^3$ of all vectors of the form (z_1, z_2, z_3), where
$z_1 + z_2 + z_3 = 0$

20. The subspace of $\mathbf{C}^4$ of all vectors whose first and last components are zero

21. The kernel of the transformation $T : \mathbf{C}^2 \to \mathbf{C}^2$ given by $T(z_1, z_2) = (z_1 - z_2, 0)$

22. The kernel of the transformation $S : \mathbf{C}^3 \to \mathbf{C}^2$ given by $S(\mathbf{z}) = A\mathbf{z}$, where

$$A = \begin{bmatrix} i & 0 & 1 \\ 1 & -i & 0 \end{bmatrix}$$

In exercises 23 through 30 find the eigenvalues and associated eigenspaces for each matrix. Also find a basis for each eigenspace.

23. $\begin{bmatrix} 0 & 3+4i \\ 3-4i & 0 \end{bmatrix}$ **24.** $\begin{bmatrix} 0 & -2 \\ 2 & 0 \end{bmatrix}$

25. $\begin{bmatrix} 2 & 0 & -4 \\ 0 & 1 & 0 \\ 2 & 0 & -2 \end{bmatrix}$ **26.** $\begin{bmatrix} 1 & i & 0 \\ -i & 1 & -i \\ 0 & i & 1 \end{bmatrix}$

27. $\begin{bmatrix} i & 0 & 1 \\ 0 & i & 0 \\ 0 & 0 & i \end{bmatrix}$ **28.** $\begin{bmatrix} 1+i & 0 & 0 \\ -1 & 1-i & 0 \\ 0 & -1 & 0 \end{bmatrix}$

29. $\begin{bmatrix} 1 & 0 & 0 \\ 0 & 0 & -i \\ 0 & i & 0 \end{bmatrix}$ **30.** $\begin{bmatrix} 0 & 0 & -1 \\ 0 & 1 & 0 \\ 1 & 0 & 0 \end{bmatrix}$

31–38. Determine whether or not each matrix in exercises 23 through 30 is diagonalizable. If it is diagonalizable, find an invertible matrix P that performs the diagonalization as well as the resultant diagonal matrix. If the given matrix is hermitian, find a unitary matrix P that performs the diagonalization as well as the resultant diagonal matrix.

39. Which, if any, of the matrices in exercises 23 through 30 are unitary?

40. Let A be a unitary matrix. Show that $\det(A) = \pm 1$. (See exercise 26 of section 6.2.)

41. Prove parts a, c, and d of theorem 1.

42. Let $\mathbf{w}$ and $\mathbf{z}$ be in $\mathbf{C}^m$. Define a scalar-valued product, $\mathbf{w} \circ \mathbf{z}$, by
$$\mathbf{w} \circ \mathbf{z} = w_1 z_1 + w_2 z_2 + \cdots + w_m z_m.$$
Show that $\mathbf{w} \circ \mathbf{w}$ is not necessarily equal to $\|\mathbf{w}\|^2$.

43. Let z_1 and z_2 be complex numbers. Show that
   (a) $\overline{z_1 + z_2} = \bar{z}_1 + \bar{z}_2$ (b) $\overline{z_1 z_2} = \bar{z}_1 \bar{z}_2$
   (c) $\bar{\bar{z}}_1 = z_1$ (d) $z_1 \bar{z}_1 = |z_1|^2$

44. Let
$$A = \begin{bmatrix} 1 & 0 & 0 \\ 0 & 1 & i \\ 0 & i & 3 \end{bmatrix}.$$
Show that A is not diagonalizable. (Notice that $A = A^T$.)

45. Let A be an $n \times n$ matrix with complex entries, and let $\mathbf{w}$ and $\mathbf{z}$ be in $\mathbf{C}^n$. Show that $(A\mathbf{w}) \cdot \mathbf{z} = \mathbf{w} \cdot (A^H \mathbf{z})$. (*Hint:* see exercise 28 of section 6.2.)

46. Prove (theorem 6) that an $n \times n$ matrix P with complex entries is unitary if and only if its columns form an orthonormal set (under the complex dot product).

Vector Spaces

7

In the preceding chapters we developed rules of arithmetic for three different entities—vectors in $\mathbf{R}^m$, matrices, and linear transformations. For all three the operations of addition and multiplication by scalars obey many of the same properties. A *vector space* is a mathematical structure that collects these entities, as well as others, under one name and allows us to study their common properties all at once. We will only consider vector spaces *over the real numbers*, for which the set of scalars consists of the real numbers.

7.1 Vector Spaces and Subspaces

A *vector space* is intuitively a set that, together with operations of addition and multiplication by scalars, behaves in essentially the same manner as $\mathbf{R}^m$. The following definition makes this precise.

Definition   A set **V** together with operations of *addition* and *multiplication by scalars* (or *scalar multiplication*) is a **vector space** if the following properties are satisfied for every choice of elements **u**, **v**, **w** in **V** and scalars c and d. The elements of **V** are called **vectors**.

a. $\mathbf{u} + \mathbf{v}$ is in **V** (**V** is closed under addition).
b. $c\mathbf{u}$ is in **V** (**V** is closed under scalar multiplication).
c. $\mathbf{u} + \mathbf{v} = \mathbf{v} + \mathbf{u}$ (addition is commutative).
d. $(\mathbf{u} + \mathbf{v}) + \mathbf{w} = \mathbf{u} + (\mathbf{v} + \mathbf{w})$ (addition is associative).
e. There is an element, denoted by **0**, in **V** such that $\mathbf{u} + \mathbf{0} = \mathbf{u}$ for all **u** in **V** (**0** is the **zero** of **V**).
f. For each **u** in **V** there is an element, denoted by $-\mathbf{u}$, such that $\mathbf{u} + (-\mathbf{u}) = \mathbf{0}$ ($-\mathbf{u}$ is the **negative** of **u**).
g. $(cd)\mathbf{u} = c(d\mathbf{u})$.
h. $(c + d)\mathbf{u} = c\mathbf{u} + d\mathbf{u}$ $\left.\right\}$ (distributive laws).
i. $c(\mathbf{u} + \mathbf{v}) = c\mathbf{u} + c\mathbf{v}$
j. $1\mathbf{u} = \mathbf{u}$.

Note: Since we have restricted the scalars to be the set of real numbers, this is called a vector space *over the real numbers*. If the scalars were allowed to be complex numbers, the same definition would apply, and the resulting structure would be called a vector space *over the complex numbers*. Throughout the rest of the text we will usually simply say *vector space* rather than *vector space over the real numbers*.

Example 7.1

The set $\mathbf{R}^m$ of all m-vectors, with the usual operations of addition and multiplication by scalars, is a vector space.

This statement is a consequence of theorem 2 of section 1.4.

Example 7.2

The set $\mathbf{M}^{m,\,n}$ of all $m \times n$ matrices, with the usual operations of addition and scalar multiplication, is a vector space.

This statement is a consequence of theorem 1 of section 2.3.

Example 7.3

The set $\mathbf{L}(\mathbf{R}^n, \mathbf{R}^m)$ of all linear transformations from $\mathbf{R}^n$ to $\mathbf{R}^m$, with the usual operations of addition and scalar multiplication, is a vector space.

This statement is a consequence of theorem 7 of section 5.2.

Example 7.4

The set $\mathbf{C}$ of all complex numbers, with the usual operations of addition of complex numbers and multiplication by real numbers, is a vector space (over the real numbers).

Recall that $\mathbf{C}$ is the set of all elements of the form $z = a + bi$, where a and b are real numbers. Two complex numbers $a + bi$ and $c + di$ are equal if and only if $a = c$ and $b = d$. Addition is defined by

$$(a + bi) + (c + di) = (a + c) + (b + d)i$$

and scalar multiplication by $k(a + bi) = (ka) + (kb)i$. Since each of these is an element of $\mathbf{C}$, properties a and b of the definition of vector space are satisfied.

c. For $z_1 = a_1 + b_1 i$ and $z_2 = a_2 + b_2 i$,

$$\begin{aligned} z_1 + z_2 &= (a_1 + b_1 i) + (a_2 + b_2 i) = (a_1 + a_2) + (b_1 + b_2)i \\ &= (a_2 + a_1) + (b_2 + b_1)i = (a_2 + b_2 i) + (a_1 + b_1 i) \\ &= z_2 + z_1. \end{aligned}$$

d. For $z_1 = a_1 + b_1 i$, $z_2 = a_2 + b_2 i$, and $z_3 = a_3 + b_3 i$,

$$(z_1 + z_2) + z_3 = [(a_1 + b_1 i) + (a_2 + b_2 i)] + (a_3 + b_3 i)$$

$$= [(a_1 + a_2) + (b_1 + b_2)i] + (a_3 + b_3 i)$$
$$= [(a_1 + a_2) + a_3] + [(b_1 + b_2) + b_3]i$$
$$= [a_1 + (a_2 + a_3)] + [b_1 + (b_2 + b_3)]i$$
$$= (a_1 + b_1 i) + [(a_2 + a_3) + (b_2 + b_3)i]$$
$$= (a_1 + b_1 i) + [(a_2 + b_2 i) + (a_3 + b_3 i)]$$
$$= z_1 + (z_2 + z_3).$$

e. $0 + 0i$ serves as the zero element, since

$$(a + bi) + (0 + 0i) = (a + 0) + (b + 0)i = a + bi.$$

f. For $z = a + bi$,

$$-z = -a - bi,$$

since $(a + bi) + (-a - bi) = 0 + 0i$.

g. For $z = a + bi$ and real numbers c and d,

$$(cd)(a + bi) = (cd)a + (cd)bi = c(da) + c(db)i$$
$$= c(da + dbi) = c(d(a + bi)).$$

h. For $z = a + bi$ and real numbers c and d,

$$(c + d)(a + bi) = (c + d)a + (c + d)bi = (ca + da) + (cb + db)i$$
$$= (ca + cbi) + (da + dbi) = c(a + bi) + d(a + bi).$$

i. For $z_1 = a_1 + b_1 i, z_2 = a_2 + b_2 i$, and real number c,

$$c(z_1 + z_2) = c[(a_1 + b_1 i) + (a_2 + b_2 i)]$$
$$= c[(a_1 + a_2) + (b_1 + b_2)i]$$
$$= c(a_1 + a_2) + c(b_1 + b_2)i$$
$$= (ca_1 + ca_2) + (cb_1 + cb_2)i$$
$$= (ca_1 + cb_1 i) + (ca_2 + cb_2 i)$$
$$= c(a_1 + b_1 i) + c(a_2 + b_2 i)$$
$$= cz_1 + cz_2.$$

j. For $z = a + bi$,

$$1z = 1(a + bi) = (1a) + (1b)i = a + bi = z.$$

Thus **C** is a vector space.

Note: No mention was made in example 7.4 of the fact that two complex numbers can be *multiplied*. This added structure is ignored when **C** is being viewed as a vector space. You may notice that special characteristics of m-vectors, matrices, and linear transformations have been ignored as well. In the study of vector spaces only properties a through j and their consequences are examined.

Example 7.5

Let **P** be the set of all polynomials in x with the usual addition and scalar multiplication. Then **P** is a vector space.

Recall that a **polynomial p** is an expression of the form

$$p(x) = a_n x^n + a_{n-1} x^{n-1} + \cdots + a_1 x + a_0,$$

where each a_i is a real number called the **coefficient** of x^i. The **degree** of **p** is n, provided $a_n \neq 0$. Two polynomials are **equal** if all their corresponding coefficients are equal. The **sum** of two polynomials is the polynomial obtained by adding their corresponding coefficients, that is, adding "like" terms. The product of a polynomial by a scalar is the polynomial obtained by multiplying each of its coefficients by the scalar.

Clearly, properties a and b of the definition of vector space are satisfied by these operations. The **0** for **P** is the zero polynomial, the one whose graph is the x-axis. The negative of **p**, $-$**p**, is the polynomial with each coefficient replaced by its negative. (The graph of $-$**p** is the reflection of the graph of **p** in the x-axis.) The other vector space properties follow easily, the arguments are very similar to those of example 7.4.

Example 7.6 (from calculus)

Let $\mathbf{C}(-\infty, \infty)$ denote the set of all continuous real-valued functions of a real variable x. Addition of **f** and **g** in $\mathbf{C}(-\infty, \infty)$ and multiplication by a scalar c are defined by

i. $(\mathbf{f} + \mathbf{g})(x) = \mathbf{f}(x) + \mathbf{g}(x)$, for all real numbers x,

ii. $(c\mathbf{f})(x) = c(\mathbf{f}(x))$, for all real numbers x.

Then $\mathbf{C}(-\infty, \infty)$ is a vector space over **R**.

Vector space properties a and b follow from the fact that the sum of two continuous functions is continuous and a scalar multiple of a continuous function is continuous. Further properties rest on the fact that two functions are equal if they agree at every point. That is, $\mathbf{f} = \mathbf{g}$ if and only if $\mathbf{f}(x) = \mathbf{g}(x)$ for each real number x. For example, to check property c, let **f** and **g** be in $\mathbf{C}(-\infty, \infty)$. Then for each real number x

$$(\mathbf{f} + \mathbf{g})(x) = \mathbf{f}(x) + \mathbf{g}(x) = \mathbf{g}(x) + \mathbf{f}(x) = (\mathbf{g} + \mathbf{f})(x).$$

Thus $\mathbf{f} + \mathbf{g} = \mathbf{g} + \mathbf{f}$. The other properties are also easily checked. It should be noted that **0** is the zero function of example 7.5 and for **f** in $\mathbf{C}(-\infty, \infty)$ the negative of **f** is the function $-\mathbf{f}$, defined by $(-\mathbf{f})(x) = -(\mathbf{f}(x))$ for each x. As in example 7.5 the graph of $-\mathbf{f}$ is the reflection of the graph of **f** in the x-axis.

Example 7.7

Let **V** be the set $\mathbf{R}^2$ of all ordered pairs of real numbers. Let addition be defined in the usual way but define scalar multiplication · by

$$c \cdot \mathbf{v} = |c|\mathbf{v}$$

for any scalar c and ordered pair $\mathbf{v}$. That is, $c \cdot \mathbf{v}$ is the usual scalar product of the *absolute value* of c times $\mathbf{v}$. Show that $\mathbf{V}$ is *not* a vector space.

All we need show is that *one* of properties a through j fails to hold for *some* choice of elements of $\mathbf{V}$ and *some* choice of scalars. We show that property h fails when $c = 1$, $d = -1$, and $\mathbf{u} = (2, 3)$. Since

$$(c + d) \cdot \mathbf{u} = (1 + -1) \cdot (2, 3) = 0 \cdot (2, 3) = 0(2, 3) = (0, 0),$$

but

$$c \cdot \mathbf{u} + d \cdot \mathbf{u} = 1 \cdot \mathbf{u} + (-1) \cdot \mathbf{u} = 1\mathbf{u} + 1\mathbf{u} = \mathbf{u} + \mathbf{u} = (4, 6),$$

we have that $(c + d) \cdot \mathbf{u} \neq c \cdot \mathbf{u} + d \cdot \mathbf{u}$. Thus $\mathbf{V}$ is not a vector space under these operations. (You can check that the rest of the properties *do* hold.)

The next theorem gives three additional properties common to all vector spaces. Although all of them may look "obvious," these properties must be proved from our existing knowledge of vector spaces.

Theorem 1   Let $\mathbf{V}$ be a vector space, let $\mathbf{v}$ be an element of $\mathbf{V}$, and let c be a scalar. Then

a. $0\mathbf{v} = \mathbf{0}$,
b. $(-1)\mathbf{v} = -\mathbf{v}$,
c. $c\mathbf{0} = \mathbf{0}$.

Proof of a   Since $0\mathbf{v} = (0 + 0)\mathbf{v} = 0\mathbf{v} + 0\mathbf{v}$ by property h, and $0\mathbf{v} = 0\mathbf{v} + \mathbf{0}$ by property e, we have

$$\mathbf{0} + 0\mathbf{v} = 0\mathbf{v} + 0\mathbf{v},$$

because addition is commutative. Every element has a negative (property f), so there is an element $-(0\mathbf{v})$ that we add to both sides,

$$(\mathbf{0} + 0\mathbf{v}) + -(0\mathbf{v}) = (0\mathbf{v} + 0\mathbf{v}) + -(0\mathbf{v}).$$

Applying property d, and then f and e, we have

$$\mathbf{0} + (0\mathbf{v} + -(0\mathbf{v})) = 0\mathbf{v} + (0\mathbf{v} + -(0\mathbf{v}))$$
$$\mathbf{0} + \mathbf{0} = 0\mathbf{v} + \mathbf{0}$$
$$\mathbf{0} = 0\mathbf{v}.$$

Proof of b   We first show that $(-1)\mathbf{v}$ "acts like" $-\mathbf{v}$, that is, $(-1)\mathbf{v} + \mathbf{v} = 0$. Using properties h and j, we compute

$$(-1)\mathbf{v} + \mathbf{v} = (-1)\mathbf{v} + 1\mathbf{v} = (-1 + 1)\mathbf{v} = 0\mathbf{v} = \mathbf{0},$$

with the last equality being part a of this theorem. But also $-\mathbf{v} + \mathbf{v} = 0$, and thus we have

$$(-1)\mathbf{v} + \mathbf{v} = -\mathbf{v} + \mathbf{v}.$$

Adding $-\mathbf{v}$ to both sides of the equation, we obtain

$$(-1)\mathbf{v} = -\mathbf{v}.$$

Proof of c   Since $c(\mathbf{0}) = c(\mathbf{0} + \mathbf{0}) = c(\mathbf{0}) + c(\mathbf{0})$, we have

$$\mathbf{0} + c(\mathbf{0}) = c(\mathbf{0}) + c(\mathbf{0}).$$

Now adding $-c(\mathbf{0})$ to each side, we have the result.   ∎

Since polynomials are examples of continuous functions, the set **P** of example 7.5 can be viewed as a subset of $\mathbf{C}(-\infty, \infty)$ of example 7.6. Furthermore the operations are "compatible," which means that addition of polynomials is the same as addition of functions and multiplication by scalars is also the same in both vector spaces. Under the following definition, **P** is a *subspace* of $\mathbf{C}(-\infty, \infty)$.

Definition   Let **V** be a vector space, and let **W** be a subset of **V**. Then **W** is a **subspace** of **V** if **W** is a vector space using the operations of **V**.

Note: By virtue of the next theorem this definition of subspace agrees with that of section 4.2 in the case of subspaces of $\mathbf{R}^m$. Thus all of the subspaces of $\mathbf{R}^m$ considered in earlier chapters are subspaces under this new definition.

Theorem 2   Let **V** be a vector space, and let **W** be a nonempty subset of **V**. Then **W** is a subspace of **V** if and only if vector space properties a and b are satisfied.

Proof   If **W** is a subspace, *all* of properties a through j must be satisfied, so in particular a and b must hold. Conversely, suppose that a and b are satisfied. Properties c, d, and g through j are "inherited" from **V**. In other words, since they hold for all vectors in **V**, and **W** is a subset of **V**, they hold for all vectors in **W** as well.

Since **V** is a vector space, there is an element **0** in **V** by property e, and for each **w** in **W**, $-\mathbf{w}$ is in **V** by property f. We need to know that each of these is in **W** to complete the argument that **W** is a subspace. By property a, for any **w** in **W** it follows that $0\mathbf{w}$ and $(-1)\mathbf{w}$ are in **W**. But from theorem 1 these are seen to be simply **0** and $-\mathbf{w}$. Hence **W** is a subspace.   ∎

In chapter 4 we saw many examples of subspaces of the vector space $\mathbf{R}^m$. Furthermore, as we have already indicated, $\mathbf{P}$ is a subspace of the vector space $\mathbf{C}(-\infty, \infty)$. We will close this section with three additional examples—the first giving a subspace of the vector space $\mathbf{M}^{m,n}$, the second a subspace of $\mathbf{P}$, and the third a *subset* of $\mathbf{P}$ that is *not* a *subspace* of $\mathbf{P}$.

Example 7.8

Let $\mathbf{D}^{2,2}$ be the set of all 2×2 *diagonal* matrices. Then $\mathbf{D}^{2,2}$ is a subspace of $\mathbf{M}^{2,2}$, the set of *all* 2×2 matrices.

By theorem 2 all we need show is that $\mathbf{D}^{2,2}$ satisfies properties a and b. (In other words, all we need show is that $\mathbf{D}^{2,2}$ is *closed* under the addition and scalar multiplication of $\mathbf{M}^{2,2}$.) To do this, let A and B be diagonal matrices, and let c be a scalar. Then A and B have the form

$$A = \begin{bmatrix} a_1 & 0 \\ 0 & a_2 \end{bmatrix}, \quad B = \begin{bmatrix} b_1 & 0 \\ 0 & b_2 \end{bmatrix},$$

and we have

$$A + B = \begin{bmatrix} a_1 + b_1 & 0 \\ 0 & a_2 + b_2 \end{bmatrix}$$

and

$$cA = \begin{bmatrix} ca_1 & 0 \\ 0 & ca_2 \end{bmatrix}.$$

Since $A + B$ and cA are diagonal matrices—that is, elements of $\mathbf{D}^{2,2}$—the $\mathbf{D}^{2,2}$ is a subspace of $\mathbf{M}^{2,2}$.

Note: By virtue of theorem 2 often the easiest way to show that a set with two operations is a vector space is to show that it is a subspace of a known vector space. The following is an example of this technique.

Example 7.9

Let $\mathbf{P}_n$ be the set of all polynomials of degree less than or equal to n together with the usual polynomial arithmetic. Then $\mathbf{P}_n$ is a vector space for each positive integer n.

We show that $\mathbf{P}_n$ is a subspace of $\mathbf{P}$ for each positive integer n. Since adding two polynomials does not introduce new terms of degree higher than those of the polynomials, property a is satisfied. Likewise, multiplication by a scalar leaves the degree unchanged unless the scalar is zero, which *lowers* the degree. Thus property b is satisfied as well, and $\mathbf{P}_n$ is a subspace of $\mathbf{P}$. Since $\mathbf{P}_n$ is a subspace of $\mathbf{P}$, it is a vector space in its own right.

Example 7.10

Let $\mathbf{P}_3'$ be the set of all polynomials of degree *exactly equal* to three together with $\mathbf{0}$. Then $\mathbf{P}_3'$ is *not* a subspace of $\mathbf{P}_3$, the set of all polynomials of degree *less than or equal* to three.

To show this we need only show that either property a or b *fails* to hold for a specific choice of polynomials and/or scalars. For example, let $\mathbf{p}(x) = x^3 + 2x - 1$, and let $\mathbf{q}(x) = -x^3 + x^2 - 3$. Then

$$(\mathbf{p} + \mathbf{q})(x) = x^2 + 2x - 4,$$

a polynomial of degree two. Thus $\mathbf{p}$ and $\mathbf{q}$ are two elements in $\mathbf{P}_3'$ whose sum is not in $\mathbf{P}_3'$, so $\mathbf{P}_3'$ is not a subspace of $\mathbf{P}_3$.

7.1 Exercises

In exercises 1 through 10 determine whether or not the given set is a vector space under the given operations. If it is not a vector space, list all properties that fail to hold.

1. The set of all 2×3 matrices whose second column consists of zeros: the usual matrix operations

2. The set of all 2×3 matrices the sum of whose entries is 1: the usual matrix operations

3. The set of all polynomials of degree greater than or equal to 3: the usual polynomial operations

4. The set of all polynomials of degree less than or equal to 2 the sum of whose coefficients is 0: the usual operations

5. The set of all linear transformations from $\mathbf{R}^2$ to $\mathbf{R}^2$ that leave the x-axis *as a set* fixed: the usual operations

6. The set of all vectors in $\mathbf{R}^3$: the usual scalar multiplication but addition of $\mathbf{u}$ and $\mathbf{v}$ defined by taking the cross product of $\mathbf{u}$ and $\mathbf{v}$

7. The set $\{\mathbf{x}, \mathbf{y}\}$, where $\mathbf{x} \neq \mathbf{y}$ with addition: $\mathbf{x} + \mathbf{x} = \mathbf{x}$, $\mathbf{x} + \mathbf{y} = \mathbf{y} + \mathbf{x} = \mathbf{y}$, $\mathbf{y} + \mathbf{y} = \mathbf{x}$ and scalar multiplication: $c\mathbf{x} = \mathbf{x}$, $c\mathbf{y} = \mathbf{x}$ for all scalars c

8. Same as exercise 7 except scalar multiplication: $c\mathbf{x} = \mathbf{x}$, $c\mathbf{y} = \mathbf{y}$ for all scalars c

9. The set $\mathbf{C}[a, b]$ of all continuous functions on the interval $[a, b]$, where a and b are real numbers: the same operations as in example 7.6

10. The set consisting solely of a single element, $\mathbf{0}$ with addition: $\mathbf{0} + \mathbf{0} = \mathbf{0}$ and scalar multiplication: $c\mathbf{0} = \mathbf{0}$ for all scalars c

In exercises 11 through 14 determine whether or not the given set is a subspace of $\mathbf{M}^{2,2}$ (see example 7.2).

11. The set of all 2×2 matrices the sum of whose entries is zero

12. The set of all 2×2 matrices whose determinant is zero

13. The set of all 2×2 matrices of the form

$$\begin{bmatrix} a & b \\ c & 1 \end{bmatrix},$$

where a, b, and c are real numbers

14. The set of all 2×2 matrices of the form

$$\begin{bmatrix} a & b \\ b & a + b \end{bmatrix},$$

where a and b are real numbers

In exercises 15 and 16 determine whether or not the given set is a subspace of

$$C(-\infty, \infty)$$

(see example 7.6).

15. The set of all functions of the form $a \sin x + b \cos x$, where a and b are real numbers

16. The set, **D**, of all differentiable functions

17. If $\mathscr{S} = \{v_1, v_2, \cdots, v_n\}$ is a set of vectors in a vector space **V**, show that

$$c_1 v_1 + c_2 v_2 + \cdots + c_n v_n$$

is also in **V**, where the c_i are arbitrary scalars.

7.2 Linear Independence, Basis, and Dimension

The concepts of *linear independence*, *span*, *basis*, and *dimension* were studied in chapter 4 in the context of Euclidean m-space $\mathbf{R}^m$. In this section we will generalize these ideas to arbitrary vector spaces.

Linear Independence

We will follow theorem 1 of section 4.1 for the extended definition of linear independence of a set of vectors. Consequently the definition given here is compatible with the one for the special case of vectors in $\mathbf{R}^m$.

Definition   Let $\mathscr{S} = \{v_1, v_2, \ldots, v_n\}$ be a set of vectors in a vector space **V**. The set $\mathscr{S}$ is **linearly independent** if the equation

$$c_1 v_1 + c_2 v_2 + \cdots + c_n v_n = 0$$

has *only* the trivial solution, $c_i = 0$, for all i. If $\mathscr{S}$ is not linearly independent, we say that it is **linearly dependent.**

Note: If $\mathscr{S} = \{\mathbf{v}_1, \mathbf{v}_2, \ldots, \mathbf{v}_n\}$ is a set of vectors in a vector space **V**, a **linear combination** of the $\mathbf{v}_i$ is any expression **v** of the form

$$\mathbf{v} = c_1\mathbf{v}_1 + c_2\mathbf{v}_2 + \cdots + c_n\mathbf{v}_n,$$

where the c_i are scalars. By exercise 17 of section 7.1 all such expressions are in **V**. Thus $\mathscr{S}$ is linearly independent if and only if the only linear combination of the $\mathbf{v}_i$ equal to zero is the one in which every scalar c_i is zero.

Example 7.11

Determine whether or not the set

$$\mathscr{S} = \left\{ \begin{bmatrix} 1 & 0 \\ 0 & -1 \end{bmatrix}, \begin{bmatrix} 0 & 1 \\ 0 & -1 \end{bmatrix}, \begin{bmatrix} 0 & 0 \\ 1 & -1 \end{bmatrix} \right\}$$

is linearly independent in $\mathbf{M}^{2,2}$ (see example 7.2).
We consider the matrix equation

$$c_1 \begin{bmatrix} 1 & 0 \\ 0 & -1 \end{bmatrix} + c_2 \begin{bmatrix} 0 & 1 \\ 0 & -1 \end{bmatrix} + c_3 \begin{bmatrix} 0 & 0 \\ 1 & -1 \end{bmatrix} = \begin{bmatrix} 0 & 0 \\ 0 & 0 \end{bmatrix}.$$

If the only solution is the trivial one, $c_1 = c_2 = c_3 = 0$, then $\mathscr{S}$ is linearly independent. To determine whether or not this is the case, we perform the scalar multiplications and additions on the left side to obtain

$$\begin{bmatrix} c_1 & c_2 \\ c_3 & -c_1 - c_2 - c_3 \end{bmatrix} = \begin{bmatrix} 0 & 0 \\ 0 & 0 \end{bmatrix}$$

and then equate corresponding entries. This yields the homogeneous system of four linear equations

$$\begin{aligned} c_1 \qquad\qquad &= 0 \\ c_2 \qquad &= 0 \\ c_3 &= 0 \\ -c_1 - c_2 - c_3 &= 0. \end{aligned}$$

From the first three equations it is clear that the only solution of this system is the trivial one. Thus $\mathscr{S}$ is linearly independent.

Example 7.12

Let $\mathscr{S} = \{S, T, U, V, W\}$ be the set of linear transformations in

$$\mathbf{L}(\mathbf{R}^2, \mathbf{R}^2)$$

(see example 7.3) given by

$$S(x, y) = (2x - y, y)$$
$$T(x, y) = (x + y, x - y)$$
$$U(x, y) = (2x + 3y, 2x - 3y)$$
$$V(x, y) = (x + y, 0)$$
$$W(x, y) = (y, x).$$

Determine whether or not $\mathscr{S}$ is linearly independent.

To do this, we form a linear combination of the members of $\mathscr{S}$ and set it equal to 0 (the zero transformation):

$$c_1 S + c_2 T + c_3 U + c_4 V + c_5 W = 0.$$

By theorem 6 of section 5.2 this equation is equivalent to the one obtained by replacing each transformation by the matrix associated with it:

$$c_1 \begin{bmatrix} 2 & -1 \\ 0 & 1 \end{bmatrix} + c_2 \begin{bmatrix} 1 & 1 \\ 1 & -1 \end{bmatrix} + c_3 \begin{bmatrix} 2 & 3 \\ 2 & -3 \end{bmatrix}$$
$$+ c_4 \begin{bmatrix} 1 & 1 \\ 0 & 0 \end{bmatrix} + c_5 \begin{bmatrix} 0 & 1 \\ 1 & 0 \end{bmatrix} = \begin{bmatrix} 0 & 0 \\ 0 & 0 \end{bmatrix}.$$

The problem is now reduced to one similar to that of example 7.11. Proceeding as we did there, we obtain a homogeneous system of four linear equations in the five unknowns c_1, c_2, c_3, c_4, and c_5. By corollary 5 of section 2.5 this system must have a *nontrivial* solution. Thus $\mathscr{S}$ is linearly dependent.

Example 7.13   (from calculus)

Determine whether or not the set $\mathscr{S} = \{1, x, \sin x\}$ is linearly independent in the vector space, $C(-\infty, \infty)$, of continuous functions (see example 7.6).

Here we consider the equation

$$c_1(1) + c_2 x + c_3 \sin x = 0,$$

where 0 is the zero function. Since this equation must hold for *all* x, it holds for $x = 0$, $\pi/2$, and π. Successively substituting these values for x into the previous equation yields

$$c_1 \qquad\qquad = 0$$
$$c_1 + (\pi/2)c_2 + c_3 = 0$$
$$c_1 + (\pi)c_2 \qquad = 0.$$

It is easily checked that this homogeneous linear system has only the trivial solution. Hence $\mathscr{S}$ is linearly independent.

Basis for a Vector Space

We will now investigate the notions of *span* and *basis* of an arbitrary vector space. The definitions are virtually identical to the ones given in the context of $\mathbf{R}^m$ in sections 4.2 and 4.3.

Definition   Let $\mathbf{V}$ be a vector space. A set $\mathscr{S}$ of vectors in $\mathbf{V}$ **spans** (or **generates**) $\mathbf{V}$ if every vector in $\mathbf{V}$ can be expressed as a linear combination of those in $\mathscr{S}$.

Example 7.14

Show that the set $\mathscr{S} = \{1, x - 1, x^2 - 1\}$ spans $\mathbf{P}_2$, the subspace of $\mathbf{P}$ consisting of all polynomials of degree less than or equal to 2 (see example 7.9).

Let $\mathbf{p}(x) = a_0 + a_1 x + a_2 x^2$ be an arbitrary vector in $\mathbf{P}_2$. We must show that there exist scalars c_1, c_2, and c_3 such that

$$c_1(1) + c_2(x - 1) + c_3(x^2 - 1) = a_0 + a_1 x + a_2 x^2.$$

Collecting like terms on the left side, we obtain

$$(c_1 - c_2 - c_3) + c_2 x + c_3 x^2 = a_0 + a_1 x + a_2 x^2.$$

Since two polynomials are equal if and only if their corresponding coefficients are equal, this last equation implies that

$$
\begin{aligned}
c_1 - c_2 - c_3 &= a_0 \\
c_2 &= a_1 \\
c_3 &= a_2.
\end{aligned}
$$

Now this linear system has a (unique) solution (since, for example, the determinant of the coefficient matrix is nonzero) for c_1, c_2, and c_3 no matter what a_0, a_1, and a_2 are. Hence $\mathscr{S}$ spans $\mathbf{P}_2$.

Definition   Let $\mathbf{V}$ be a vector space. A set $\mathscr{S}$ of vectors in $\mathbf{V}$ is a **basis** for $\mathbf{V}$ if

i. $\mathscr{S}$ is linearly independent,
ii. $\mathscr{S}$ spans $\mathbf{V}$.

Just as the set $\{\mathbf{e}_1, \mathbf{e}_2, \ldots, \mathbf{e}_m\}$ provides a *standard basis* for $\mathbf{R}^m$, there are standard bases for $\mathbf{P}_n$, $\mathbf{M}^{m, n}$, and $\mathbf{L}(\mathbf{R}^n, \mathbf{R}^m)$. The next theorem describes these especially simple bases.

Theorem 1   Each of the following are bases for the respective vector spaces:

a. the set $\mathscr{S} = \{1, x, x^2, \ldots, x^n\}$ for $\mathbf{P}_n$, the vector space of all polynomials of degree less than or equal to n;
b. the set $\{E^{ij} | i = 1, \ldots, m; j = 1, \ldots, n\}$, where E^{ij} is the $m \times n$
matrix with one in the i, j position and zeros elsewhere, for $\mathbf{M}^{m, n}$,
the vector space of all $m \times n$ matrices;
c. the set $\{T^{ij} | i = 1, \ldots, m; j = 1, \ldots, n\}$, where $T^{ij}: \mathbf{R}^n \rightarrow \mathbf{R}^m$ is the
linear transformation such that $T^{ij}(x_1, \ldots, x_j, \ldots, x_n)$ is the vector
in $\mathbf{R}^m$ with x_j in the ith position and zeros elsewhere, for $\mathbf{L}(\mathbf{R}^n, \mathbf{R}^m)$,
the vector space of all linear transformations from $\mathbf{R}^n$ to $\mathbf{R}^m$.

Proof   We shall prove only part a, leaving the proofs of b and c as exercises.

To prove part a, we must show that $\mathscr{S}$ is linearly independent and
spans $\mathbf{P}_n$. First consider the equation

$$c_1(1) + c_2(x) + \cdots + c_{n+1}(x^n) = \mathbf{0} = 0(1) + 0(x) + \cdots + 0(x^n).$$

Since two polynomials are equal if and only if their corresponding coefficients are equal, we must have $c_i = 0$ for $i = 1, \ldots, n + 1$. Hence
$\mathscr{S}$ is linearly independent.

Now suppose $\mathbf{p}(x) = a_0 + a_1 x + \cdots + a_n x^n$ is an arbitrary member
of $\mathbf{P}_n$. Then $\mathbf{p}$ is *already* expressed as a linear combination of $\mathscr{S}$ with the
a_i as the scalars (since $a_0 = a_0 \cdot 1$). Thus $\mathscr{S}$ spans $\mathbf{P}_n$ as well.   ∎

Example 7.15

Find a basis for (a) $\mathbf{P}_3$, (b) $\mathbf{M}^{2, 3}$, and (c) $\mathbf{L}(\mathbf{R}^2, \mathbf{R}^2)$.

We apply theorem 1 to these special cases:

a. $\mathscr{S} = \{1, x, x^2, x^3\}$ is a basis for $\mathbf{P}_3$.

b. $\mathscr{S} = \left\{ \begin{bmatrix} 1 & 0 & 0 \\ 0 & 0 & 0 \end{bmatrix}, \begin{bmatrix} 0 & 1 & 0 \\ 0 & 0 & 0 \end{bmatrix}, \begin{bmatrix} 0 & 0 & 1 \\ 0 & 0 & 0 \end{bmatrix}, \begin{bmatrix} 0 & 0 & 0 \\ 1 & 0 & 0 \end{bmatrix}, \right.$
$\left. \begin{bmatrix} 0 & 0 & 0 \\ 0 & 1 & 0 \end{bmatrix}, \begin{bmatrix} 0 & 0 & 0 \\ 0 & 0 & 1 \end{bmatrix} \right\}$

is a basis for $\mathbf{M}^{2, 3}$.

c. $\mathscr{S} = \{T^{11}, T^{12}, T^{21}, T^{22}\}$, where $T^{11}(x, y) = (x, 0)$, $T^{12}(x, y)$
$= (y, 0)$, $T^{21}(x, y) = (0, x)$, and $T^{22}(x, y) = (0, y)$ is a basis for
$\mathbf{L}(\mathbf{R}^2, \mathbf{R}^2)$.

Example 7.16

Show that the set

$$\mathscr{S} = \left\{ \begin{bmatrix} 1 & 0 \\ 0 & -1 \end{bmatrix}, \begin{bmatrix} 0 & 1 \\ 0 & -1 \end{bmatrix}, \begin{bmatrix} 0 & 0 \\ 1 & -1 \end{bmatrix} \right\}$$

is a basis for the subspace **S** of $\mathbf{M}^{2,2}$ consisting of all 2×2 matrices, the sum of whose entries is zero.

We know that **S** is a subspace of $\mathbf{M}^{2,2}$ by exercise 11 of section 7.1 and that $\mathscr{S}$ is certainly a subset of **S**. Therefore all we need do is verify that $\mathscr{S}$ is linearly independent and spans **S**. We have already shown the first part in example 7.11. To show that $\mathscr{S}$ spans **S**, let

$$A = \begin{bmatrix} a_1 & a_2 \\ a_3 & a_4 \end{bmatrix}, \; a_1 + a_2 + a_3 + a_4 = 0,$$

be an arbitrary vector in **S**. We must find scalars c_1, c_2, and c_3 such that

$$c_1 \begin{bmatrix} 1 & 0 \\ 0 & -1 \end{bmatrix} + c_2 \begin{bmatrix} 0 & 1 \\ 0 & -1 \end{bmatrix} + c_3 \begin{bmatrix} 0 & 0 \\ 1 & -1 \end{bmatrix} = \begin{bmatrix} a_1 & a_2 \\ a_3 & a_4 \end{bmatrix}.$$

Performing the scalar multiplications and additions on the left side and equating corresponding entries yields the linear system

$$
\begin{aligned}
c_1 & & & = a_1 \\
& c_2 & & = a_2 \\
& & c_3 & = a_3 \\
-c_1 & - c_2 & - c_3 & = a_4.
\end{aligned}
$$

From the first three equations we see that $c_1 = a_1$, $c_2 = a_2$, and $c_3 = a_3$. Substituting these values for the c_i into the fourth equation yields $-a_1 - a_2 - a_3 = a_4$, or equivalently $a_1 + a_2 + a_3 + a_4 = 0$. But the latter is true by hypothesis, so A can be expressed as a linear combination of the vectors in $\mathscr{S}$ for all choices of the a_i, that is, the set $\mathscr{S}$ spans **S**. Thus $\mathscr{S}$ is a basis for **S**.

The following theorems are the counterparts of theorems in chapter 4. For example, the next theorem is a generalization of theorem 2 of section 4.2. That theorem was stated in terms of a *subspace* of $\mathbf{R}^m$, whereas the next theorem refers to a *vector space*. Since any subspace of a vector space is also a vector space, the following result does encompass the former one. The proof is exactly the same as was given before and hence omitted.

Theorem 2 Let **V** be a vector space spanned by the set $\mathscr{T}$. If one of the vectors in $\mathscr{T}$ is a linear combination of the others, then the set obtained by deleting that vector still spans **V**.

If $\mathscr{T}$ in theorem 2 is a finite set, repeated application of theorem 2 results in a set that both spans and is linearly independent—a basis. The result is true in general, and we state this as the next theorem.

Theorem 3   Let **V** be a vector space spanned by a set $\mathcal{T}$. Then there is a subset of $\mathcal{T}$ that is a basis for **V**. In other words, a generating set may be reduced to a basis.

Example 7.17

The set $\mathcal{T} = \{1, x + 1, x - 1\}$ generates $\mathbf{P}_1$, the vector space of polynomials of degree less than or equal to 1 (exercise 13). Theorem 3 tells us that there is a subset of $\mathcal{T}$ that is a basis for $\mathbf{P}_1$. In fact there are three of them. Each of $\mathcal{T}_1 = \{1, x + 1\}$, $\mathcal{T}_2 = \{1, x - 1\}$, and $\mathcal{T}_3 = \{x + 1, x - 1\}$ is a basis for $\mathbf{P}_1$ (exercise 38).

Theorem 4   Let **V** be a vector space spanned by a set $\mathcal{S}$, and let $\mathcal{T}_1$ be a linearly independent subset of **V**. Then there is a subset $\mathcal{T}_2$ of $\mathcal{S}$ such that $\mathcal{T} = \mathcal{T}_1 \cup \mathcal{T}_2$, the union of $\mathcal{T}_1$ and $\mathcal{T}_2$, is a basis for **V**. In other words, a linearly independent set may be extended to a basis.

Proof   We only prove the case where $\mathcal{T}_1$ and $\mathcal{S}$ are finite sets. Consider the set $\mathcal{T}_1 \cup \mathcal{S}$. Fixing the order of the elements, we take first the elements of $\mathcal{T}_1$ and then those of $\mathcal{S}$, and we apply theorem 2. Since $\mathcal{S}$ alone spans **V**, the set $\mathcal{T}_1 \cup \mathcal{S}$ does as well. Since no element of $\mathcal{T}_1$ is a linear combination of the others ($\mathcal{T}_1$ is linearly independent), if $\mathcal{T}_1 \cup \mathcal{S}$ is not linearly independent, some element of $\mathcal{S}$ is a linearly dependent combination of the others and can be deleted. Repeated application of this process yields a set that spans **V**, is linearly independent, and contains $\mathcal{T}_1$. The remaining elements constitute $\mathcal{T}_2$.   ∎

Example 7.18

From example 7.11 we know that

$$\mathcal{T}_1 = \left\{ \begin{bmatrix} 1 & 0 \\ 0 & -1 \end{bmatrix}, \begin{bmatrix} 0 & 1 \\ 0 & -1 \end{bmatrix}, \begin{bmatrix} 0 & 0 \\ 1 & -1 \end{bmatrix} \right\}$$

is a linearly independent subset of $\mathbf{M}^{2,2}$ and from theorem 1b that the set

$$\mathcal{S} = \left\{ \begin{bmatrix} 1 & 0 \\ 0 & 1 \end{bmatrix}, \begin{bmatrix} 0 & 1 \\ 0 & 0 \end{bmatrix}, \begin{bmatrix} 0 & 0 \\ 1 & 0 \end{bmatrix}, \begin{bmatrix} 0 & 0 \\ 0 & 1 \end{bmatrix} \right\}$$

is a generating set (in fact a basis) for $\mathbf{M}^{2,2}$. Taking

$$\mathcal{T}_2 = \left\{ \begin{bmatrix} 0 & 0 \\ 0 & 1 \end{bmatrix} \right\}$$

to be the subset of $\mathscr{S}$, it can be shown that the union of $\mathscr{T}_1$ and $\mathscr{T}_2$ is a basis for $\mathbf{M}^{2,2}$ (exercises 5 and 15).

Dimension of a Vector Space

Theorem 5   Let $\mathbf{V}$ be a vector space with bases $\mathscr{B}$ and $\mathscr{B}'$. Then $\mathscr{B}$ and $\mathscr{B}'$ have the same number of elements.

Proof   We only consider the case where $\mathscr{B}$ and $\mathscr{B}'$ are finite sets. Let

$$\mathscr{B} = \{\mathbf{v}_1, \ldots, \mathbf{v}_n\} \quad \text{and} \quad \mathscr{B}' = \{\mathbf{w}_1, \ldots, \mathbf{w}_m\}.$$

We must show that $m = n$. It suffices to assume that $n \leq m$ and show that $n = m$.

Since $\mathscr{B}$ spans $\mathbf{V}$, and $\mathbf{w}_1$ is in $\mathbf{V}$, there are scalars $c_1, \ldots, c_n$ such that

$$\mathbf{w}_1 = c_1\mathbf{v}_1 + \cdots + c_n\mathbf{v}_n.$$

If every $c_i = 0$, $\mathbf{w}_1$ would be the zero vector, and thus $\mathbf{0}$ would be in $\mathscr{B}'$. It is easy to show (exercise 40) that if $\mathbf{0}$ is in a set, the set is not linearly independent but $\mathscr{B}'$ *is* linearly independent. Thus at least one $c_i \neq 0$. By renumbering the elements of $\mathscr{B}$, if necessary, we may assume that $c_1 \neq 0$. Now we solve for $\mathbf{v}_1$,

$$\mathbf{v}_1 = (1/c_1)\mathbf{w}_1 - (c_2/c_1)\mathbf{v}_2 - \cdots - (c_n/c_1)\mathbf{v}_n.$$

Thus $\mathbf{v}_1$ is a linear combination of the elements of the set

$$\mathscr{B}_1 = \{\mathbf{w}_1, \mathbf{v}_2, \ldots, \mathbf{v}_n\}.$$

Letting $\mathscr{B}^*$ be the set $\mathscr{B}$ together with $\mathbf{w}_1$, we have a spanning set (since $\mathscr{B}$ alone is a spanning set) with the element $\mathbf{v}_1$ as a linear combination of the others. By theorem 2 the set $\mathscr{B}_1$, obtained by deleting $\mathbf{v}_1$, still spans $\mathbf{V}$.

Since $\mathscr{B}_1$ spans $\mathbf{V}$, and $\mathbf{w}_2$ is in $\mathbf{V}$, there are scalars $d_1, \ldots, d_n$ such that

$$\mathbf{w}_2 = d_1\mathbf{w}_1 + d_2\mathbf{v}_2 + \cdots + d_n\mathbf{v}_n.$$

If $d_2 = d_3 = \cdots = d_n = 0$, we would have $\mathbf{w}_2 = d_1\mathbf{w}_1$, from which it is easy to show that the set $\mathscr{B}'$ is not linearly independent. Thus one of the $d_i \neq 0$ for $i = 2, \ldots, n$. By renumbering the remaining elements of $\mathscr{B}$, if necessary, we may assume that $d_2 \neq 0$. Now we solve for $\mathbf{v}_2$,

$$\mathbf{v}_2 = -(d_1/d_2)\mathbf{w}_1 + (1/d_2)\mathbf{w}_2 - (d_3/d_2)\mathbf{v}_3 - \cdots - (d_n/d_2)\mathbf{v}_n.$$

Thus $\mathbf{v}_2$ is a linear combination of the elements of the set

$$\mathscr{B}_2 = \{\mathbf{w}_1, \mathbf{w}_2, \mathbf{v}_3, \ldots, \mathbf{v}_n\}.$$

Letting $\mathscr{B}^*$ be the set $\mathscr{B}_1$ together with $\mathbf{w}_2$, we have a spanning set with the element $\mathbf{v}_2$ as a linear combination of the others. By theorem 2 the set $\mathscr{B}_2$ still spans $\mathbf{V}$.

Continuing this same idea, we can use each $\mathbf{w}_i$ to replace some $\mathbf{v}_i$ and obtain

$$\mathscr{B}_k = \{\mathbf{w}_1, \ldots, \mathbf{w}_k, \mathbf{v}_{k+1}, \ldots, \mathbf{v}_n\}$$

as a spanning set for $\mathbf{V}$ until $k = n$. In other words, we continue until we run out of elements of $\mathscr{B}$ and have

$$\mathscr{B}_n = \{\mathbf{w}_1, \ldots, \mathbf{w}_n\}$$

as a spanning set for $\mathbf{V}$. If $n < m$, we would have $\mathbf{w}_{n+1}$ in the span of $\mathscr{B}_n$. It follows easily that $\mathscr{B}'$ could not be linearly independent, a contradiction. Therefore $n = m$. ∎

Definition   The **dimension** of a vector space $\mathbf{V}$ is equal to the number of vectors in a basis for $\mathbf{V}$.

Note: For the zero vector space $\{\mathbf{0}\}$ it is customary to take the empty set (no elements) as a basis. Thus the dimension of the zero vector space is 0. If a basis for a vector space $\mathbf{V}$ is not finite, we will not consider its dimension, although it is possible to do so using the notion of infinite cardinal numbers. If a vector space $\mathbf{V}$ has a finite basis, or more generally, a finite generating set, it can be shown by an argument similar to the proof of theorem 5 that all bases are finite and hence have the same number of elements. Such a vector space is said to be **finite dimensional.**

As a direct consequence of theorem 1 and the definition of dimension, we have the following theorem.

Theorem 6   The dimension of

a. $\mathbf{P}_n$ is $n + 1$,
b. $\mathbf{M}^{m,\,n}$ is mn,
c. $\mathbf{L}(\mathbf{R}^n, \mathbf{R}^m)$ is mn.

Example 7.19

Find the dimension of the subspace $\mathbf{S}$ of $\mathbf{M}^{3,\,3}$ consisting of all 3×3 diagonal matrices.

It can be easily shown that a basis for $\mathbf{S}$ is given by

$$\left\{ \begin{bmatrix} 1 & 0 & 0 \\ 0 & 0 & 0 \\ 0 & 0 & 0 \end{bmatrix}, \begin{bmatrix} 0 & 0 & 0 \\ 0 & 1 & 0 \\ 0 & 0 & 0 \end{bmatrix}, \begin{bmatrix} 0 & 0 & 0 \\ 0 & 0 & 0 \\ 0 & 0 & 1 \end{bmatrix} \right\}$$

Therefore $\mathbf{S}$ has dimension equal to three.

We will close this section with a theorem that is sometimes useful in quickly answering questions concerning a given subset of a vector space: "Is it linearly independent?" "Does it span?" "Is it a basis?" We leave the proof as an exercise.

Theorem 7   Let $\mathscr{S}$ be a subset of a vector space **V** of dimension n.

   i. If $\mathscr{S}$ contains fewer than n vectors, it does not span **V**.

   ii. If $\mathscr{S}$ contains more than n vectors, it is not linearly independent (it is linearly dependent).

   iii. If $\mathscr{S}$ contains exactly n vectors and is either linearly independent or spans **V**, then it is a basis for **V**.

Example 7.20

Demonstrate each of the following:

a. The set $\{x, 1 + x, 1 - x\}$ is linearly dependent in $\mathbf{P}_1$.

b. The set

$$\left\{ \begin{bmatrix} 1 & 1 & 1 \\ 0 & 0 & 0 \end{bmatrix}, \begin{bmatrix} 0 & 0 & 0 \\ 1 & 1 & 1 \end{bmatrix} \right\}$$

does not span $\mathbf{M}^{2,\,3}$.

c. The set $\{S, T\}$, where $S : \mathbf{R}^1 \to \mathbf{R}^2$ and $T : \mathbf{R}^1 \to \mathbf{R}^2$ are linear transformations given by $S(x) = (x, 2x)$, $T(x) = (0, -x)$, is a basis for $\mathbf{L}(\mathbf{R}^1, \mathbf{R}^2)$.

Parts a and b follow immediately from parts iii and i of theorem 7, conditions ii and i of theorem 7, respectively. In part a the set consists of three elements and is a subset of the two-dimensional vector space $\mathbf{P}_1$. Therefore it is linearly dependent. In part b the given set contains two vectors in the six-dimensional $\mathbf{M}^{2,\,3}$, so it cannot span.

For part c it can be shown that $\{S, T\}$ is linearly independent (exercise 9). Since this set contains two vectors in the two-dimensional space $\mathbf{L}(\mathbf{R}^1, \mathbf{R}^2)$, it is a basis by part iii of theorem 7.

7.2 Exercises

In exercises 1 through 10 determine whether or not the given set is linearly independent. If the set is linearly dependent, write one of its vectors as a linear combination of the others.

1. $\{1, 1 + x, 1 + x + x^2\}$ in $\mathbf{P}_2$
2. $\{1 + x^2, 1 + x + 2x^2, x + x^2\}$ in $\mathbf{P}_2$
3. $\{1, x - 1, x + 1\}$ in $\mathbf{P}_1$

4. $\{x - 1, x + 1\}$ in P_1

5. $\left\{\begin{bmatrix} 1 & 0 \\ 0 & -1 \end{bmatrix}, \begin{bmatrix} 0 & 1 \\ 0 & -1 \end{bmatrix}, \begin{bmatrix} 0 & 0 \\ 1 & -1 \end{bmatrix}, \begin{bmatrix} 0 & 0 \\ 0 & 1 \end{bmatrix}\right\}$ in $M^{2,2}$

6. $\left\{\begin{bmatrix} 1 & 2 \\ -1 & 0 \end{bmatrix}, \begin{bmatrix} 0 & -1 \\ 1 & 1 \end{bmatrix}, \begin{bmatrix} 1 & 0 \\ 1 & 2 \end{bmatrix}\right\}$ in $M^{2,2}$

7. $\{1, \sin^2 x, \cos^2 x\}$ in $C(-\infty, \infty)$

8. $\{1, e^x, e^{-x}\}$ in $C(-\infty, \infty)$

9. $\{S, T\}$, where $S : \mathbf{R}^1 \to \mathbf{R}^2$ and $T : \mathbf{R}^1 \to \mathbf{R}^2$ are the linear transformations defined by $S(x) = (x, 2x)$ and $T(x) = (0, -x)$, in $L(\mathbf{R}^1, \mathbf{R}^2)$

10. $\{S, T, U, V\}$, where S, T, U, and V are linear transformations from $\mathbf{R}^2$ to $\mathbf{R}^2$ given by $S(x, y) = (x, 0)$, $T(x, y) = (y, 0)$, $U(x, y) = (0, x)$, and $V(x, y) = (0, y)$

11-20. Determine whether or not each set in exercises 1 through 10 generates (that is, spans) the indicated vector space. (Theorem 7 may come in handy here.)

21-30. Determine whether or not each set in exercises 1 through 10 is a basis for the indicated vector space.

In exercises 31 through 37 find the dimension of the given vector space.

31. $\mathbf{M}^{3,4}$ 32. $\mathbf{P}_5$ 33. $L(\mathbf{R}^1, \mathbf{R}^4)$

34. The subspace of $\mathbf{M}^{2,2}$ consisting of all diagonal 2×2 matrices

35. The subspace of $\mathbf{M}^{2,2}$ consisting of all 2×2 matrices whose diagonal entries are zero

36. The subspace of $\mathbf{P}_1$ consisting of all polynomials $\mathbf{p}(x) = a_0 + a_1 x$ such that $a_0 + a_1 = 0$

37. The subspace of $\mathbf{P}_3$ consisting of all polynomials

$$\mathbf{p}(x) = a_0 + a_1 x + a_2 x^2 + a_3 x^3$$

with $a_2 = 0$

38. Show that $\mathcal{T}_1 = \{1, x + 1\}$, $\mathcal{T}_2 = \{1, x - 1\}$, and $\mathcal{T}_3 = \{x + 1, x - 1\}$ are all bases for $\mathbf{P}_1$.

39. Prove parts b and c of theorem 1

40. If $\mathcal{S}$ is a subset of a vector space V, and 0 is in $\mathcal{S}$, then $\mathcal{S}$ is linearly dependent.

41. Let $\mathcal{S} = \{v_1, v_2, \cdots, v_n\}$ be a linearly independent set in a vector space V.
(a) Show that the set $\{v_1, v_2, \cdots, v_m\}$ is linearly independent if $m < n$.
(b) More generally show that any subset of $\mathcal{S}$ is linearly independent.

42. Let $\mathcal{S} = \{v_1, v_2, \cdots, v_n\}$ be a linearly dependent set in a vector space V.
(a) If w is a vector in V, show that the set $\{w, v_1, v_2, \cdots, v_n\}$ is linearly dependent.
(b) More generally, if $\mathcal{T}$ is any set in V that contains $\mathcal{S}$, show $\mathcal{T}$ is linearly dependent.

43. If $\mathcal{B}$ is a basis for a vector space V, and v is in V, show that the representation of v as a linear combination of elements of $\mathcal{B}$ is unique.

7.3   Coordinates and Linear Transformations

In this section we show how the theory of $\mathbf{R}^m$ developed in chapters 1 through 6 can sometimes be usefully applied to more general finite dimensional vector spaces.

Coordinate Vectors

If $\mathscr{B}$ is a basis for a vector space $\mathbf{V}$, it follows from exercise 43 of section 7.2 that each vector $\mathbf{x}$ in $\mathbf{V}$ is expressible as a linear combination of the elements of $\mathscr{B}$ in one and only one way.

Definition   Let $\mathscr{B} = \{\mathbf{v}_1, \ldots, \mathbf{v}_n\}$ be a basis for a vector space $\mathbf{V}$. The unique vector $\mathbf{y} = (y_1, \ldots, y_n)$ such that $\mathbf{x} = y_1\mathbf{v}_1 + \cdots + y_n\mathbf{v}_n$ is called the **coordinate vector** of $\mathbf{x}$ with respect to the basis $\mathscr{B}$.

Example 7.21

Find the coordinate vector of the polynomial $\mathbf{p}(x) = 1 + 2x - 3x^2$ with respect to the basis $\{1, x, x^2, x^3\}$ for $\mathbf{P}_3$, the vector space of polynomials of degree less than or equal to 3.

Since $\mathbf{p}(x) = 1(1) + 2(x) + (-3)(x^2) + 0(x^3)$, the coordinate vector is $(1, 2, -3, 0)$.

Note: If $\mathbf{V}$ is a vector space with basis $\mathscr{B} = \{\mathbf{v}_1, \ldots, \mathbf{v}_n\}$, then the coordinate vector of each $\mathbf{v}_i$ is $\mathbf{e}_i$ the ith standard basis vector for $\mathbf{R}^n$.

Theorem 1   Let $\mathbf{V}$ be a vector space with basis $\mathscr{B} = \{\mathbf{v}_1, \ldots, \mathbf{v}_n\}$, let $\mathbf{x}, \mathbf{x}_1, \ldots, \mathbf{x}_m$ be elements of $\mathbf{V}$, and let $\mathbf{y}, \mathbf{y}_1, \ldots, \mathbf{y}_m$ be their respective coordinate vectors in $\mathbf{R}^n$. Then for scalars $d_1, \ldots, d_m$,

$$\mathbf{x} = d_1\mathbf{x}_1 + \cdots + d_m\mathbf{x}_m$$

if and only if

$$\mathbf{y} = d_1\mathbf{y}_1 + \cdots + d_m\mathbf{y}_m.$$

Proof   For each $i = 1, \ldots, m$ we have $\mathbf{y}_i = (y_{i1}, \ldots, y_{in})$, where

$$\mathbf{x}_i = y_{i1}\mathbf{v}_1 + \cdots + y_{in}\mathbf{v}_n$$

and

$$\mathbf{x} = y_1\mathbf{v}_1 + \cdots + y_n\mathbf{v}_n.$$

If $\mathbf{x} = d_1\mathbf{x}_1 + \cdots + d_m\mathbf{x}_m$, then

$$\mathbf{x} = d_1(y_{11}\mathbf{v}_1 + \cdots + y_{1n}\mathbf{v}_n) + \cdots + d_m(y_{m1}\mathbf{v}_1 + \cdots + y_{mn}\mathbf{v}_n)$$
$$= (d_1 y_{11} + \cdots + d_m y_{m1})\mathbf{v}_1 + \cdots + (d_1 y_{1n} + \cdots + d_m y_{mn})\mathbf{v}_n.$$

Since $\mathbf{y} = (y_1, \ldots, y_n)$ is the coordinate vector for $\mathbf{x}$, we have for each i

$$y_i = d_1 y_{1i} + \cdots + d_m y_{mi}.$$

Taken collectively,

$$\mathbf{y} = (y_1, \ldots, y_n) = (d_1 y_{11} + \cdots + d_m y_{m1}, \ldots, d_1 y_{1n} + \cdots + d_m y_{mn})$$
$$= d_1(y_{11}, \ldots, y_{1n}) + \cdots + d_m(y_{m1}, \ldots, y_{mn})$$
$$= d_1 y_1 + \cdots + d_m y_m.$$

The statements in this proof are reversible and thus the result is proved. ∎

Note: By letting $\mathbf{x} = \mathbf{0}$ in theorem 1, we have that $\{\mathbf{x}_1, \ldots, \mathbf{x}_m\}$ is a linearly independent set if and only if $\{\mathbf{y}_1, \ldots, \mathbf{y}_m\}$ is linearly independent.

Example 7.22

Determine whether or not the set

$$\mathscr{S} = \left\{ \begin{bmatrix} 2 & 1 \\ 1 & 1 \end{bmatrix}, \begin{bmatrix} 1 & 0 \\ 1 & 0 \end{bmatrix}, \begin{bmatrix} 0 & 1 \\ -1 & 1 \end{bmatrix}, \begin{bmatrix} 5 & 1 \\ 4 & 3 \end{bmatrix} \right\}$$

is linearly independent in $\mathbf{M}^{2,2}$, the set of all 2×2 matrices.

The coordinate vectors of the elements of $\mathscr{S}$ with respect to the basis $\{E^{11}, E^{12}, E^{21}, E^{22}\}$ (section 7.2, theorem 1b) are

$$\begin{bmatrix} 2 & 1 \\ 1 & 1 \end{bmatrix} \leftrightarrow (2, 1, 1, 1),$$

$$\begin{bmatrix} 1 & 0 \\ 1 & 0 \end{bmatrix} \leftrightarrow (1, 0, 1, 0),$$

$$\begin{bmatrix} 0 & 1 \\ -1 & 1 \end{bmatrix} \leftrightarrow (0, 1, -1, 1),$$

$$\begin{bmatrix} 5 & 1 \\ 4 & 3 \end{bmatrix} \leftrightarrow (5, 1, 4, 3).$$

As a consequence of theorem 1 (with $\mathbf{x} = \mathbf{0}$) $\mathscr{S}$ is linearly independent if and only if $\mathscr{S}' = \{(2, 1, 1, 1), (1, 0, 1, 0), (0, 1, -1, 1), (5, 1, 4, 3)\}$ is linearly independent. But these are vectors in $\mathbf{R}^4$, so we may use the methods of chapter 4. Since

$$\det \begin{bmatrix} 2 & 1 & 0 & 5 \\ 1 & 0 & 1 & 1 \\ 1 & 1 & -1 & 4 \\ 1 & 0 & 1 & 3 \end{bmatrix} = 0,$$

$\mathscr{S}'$ is linearly dependent. Hence $\mathscr{S}$ is as well.

Note: Since the *order* in which basis elements are listed affects the coordinate vectors with respect to that basis, an order must be chosen and fixed throughout the computation. The order for $\{E^{ij}\}$ and $\{T^{ij}\}$ of theorem 1 of section 7.2 will be taken *along rows*. That is, let $i = 1$, and take $j = 1, \ldots , n$ in natural order, and then proceed with $i = 2$, and so on.

Example 7.23

Show that $\mathscr{B} = \{1, x - 1, (x - 1)^2\}$ is a basis for $\mathbf{P}_2$, the set of all polynomials of degree less than or equal to 2, and express

$$\mathbf{p}(x) = 2x^2 + x - 1$$

as a linear combination of the elements of $\mathscr{B}$.

We know from theorem 1a of section 7.2 that $\mathscr{B}' = \{1, x, x^2\}$ is a basis for $\mathbf{P}_2$. The coordinate vectors of each element of $\mathscr{B}$ with respect to the basis $\mathscr{B}'$ are

$$1 \leftrightarrow (1, 0, 0),$$
$$x - 1 = -1 + x \leftrightarrow (-1, 1, 0),$$
$$(x - 1)^2 = 1 - 2x + x^2 \leftrightarrow (1, -2, 1).$$

Applying theorem 1 to the coordinate vectors, we have that $\mathscr{B}$ is linearly independent, since

$$\det \begin{bmatrix} 1 & -1 & 1 \\ 0 & 1 & -2 \\ 0 & 0 & 1 \end{bmatrix} \neq 0.$$

Since $\mathscr{B}$ is linearly independent, and $\mathbf{P}_2$ is a three dimensional vector space, $\mathscr{B}$ is a basis.

For the second part we need the coordinate vector of $\mathbf{p}$,

$$2x^2 + x - 1 = -1 + x + 2x^2 \leftrightarrow (-1, 1, 2).$$

According to theorem 1 the coefficients needed to express $(-1, 1, 2)$ as a linear combination of $\{(1, 0, 0), (-1, 1, 0), (1, -2, 1)\}$ are exactly the coefficients needed to express $\mathbf{p}$ as a linear combination of elements of $\mathscr{B}$. But this is also a problem from chapter 4. We perform the row reduction

$$\begin{bmatrix} 1 & -1 & 1 & -1 \\ 0 & 1 & -2 & 1 \\ 0 & 0 & 1 & 2 \end{bmatrix} \rightarrow \begin{bmatrix} 1 & 0 & 0 & 2 \\ 0 & 1 & 0 & 5 \\ 0 & 0 & 1 & 2 \end{bmatrix},$$

and as a consequence of theorem 5 of section 4.1 we obtain from the last column

$$\mathbf{p}(x) = 2(1) + 5(x - 1) + 2(x - 1)^2.$$

Example 7.24

Show that $\mathscr{S} = \{x^2 - 1, x^2 + 1\}$ is a linearly independent subset of $\mathbf{P}_3$, and extend $\mathscr{S}$ to a basis.

Letting $\mathscr{B} = \{1, x, x^2, x^3\}$, we have the coordinate vectors

$$x^2 - 1 = -1 + 0x + x^2 + 0x^3 \leftrightarrow (-1, 0, 1, 0),$$
$$x^2 + 1 = 1 + 0x + x^2 + 0x^3 \leftrightarrow (1, 0, 1, 0).$$

Forming the matrix of columns and row reducing, we have

$$\begin{bmatrix} -1 & 1 \\ 0 & 0 \\ 1 & 1 \\ 0 & 0 \end{bmatrix} \rightarrow \begin{bmatrix} 1 & 0 \\ 0 & 1 \\ 0 & 0 \\ 0 & 0 \end{bmatrix}.$$

Since the columns of the row-reduced matrix are distinct elementary columns, by theorem 4 of section 4.1, the coordinate vectors are linearly independent. Hence the original polynomials are as well.

To extend $\mathscr{S}$ to a basis, we augment the matrix of coordinate vector columns by a basis for $\mathbf{R}^4$, specifically the standard basis, and row reduce:

$$\begin{bmatrix} -1 & 1 & 1 & 0 & 0 & 0 \\ 0 & 0 & 0 & 1 & 0 & 0 \\ 1 & 1 & 0 & 0 & 1 & 0 \\ 0 & 0 & 0 & 0 & 0 & 1 \end{bmatrix} \rightarrow \begin{bmatrix} 1 & 0 & -1/2 & 0 & 1/2 & 0 \\ 0 & 1 & 1/2 & 0 & 1/2 & 0 \\ 0 & 0 & 0 & 1 & 0 & 0 \\ 0 & 0 & 0 & 0 & 0 & 1 \end{bmatrix}.$$

From the row-reduced matrix we see that $\mathbf{e}_2$ and $\mathbf{e}_4$ are needed to form a basis for $\mathbf{R}^4$. The corresponding polynomials are x and x^3. Thus $\{x^2 - 1, x^2 + 1, x, x^3\}$ is a basis for $\mathbf{P}_3$.

Linear Transformations

A *linear transformation* between two general vector spaces is defined in the same manner as in chapter 5. Again we will see that matrices are an important tool in the study of this topic.

Definition   Let **V** and **W** be vector spaces. A **linear transformation** from **V** to **W** is a function $T : \mathbf{V} \to \mathbf{W}$ that satisfies the following two conditions. For each **u** and **v** in **V** and a in **R**,

   i. $T(a\mathbf{u}) = aT(\mathbf{u})$ (scalars factor out),
   ii. $T(\mathbf{u} + \mathbf{v}) = T(\mathbf{u}) + T(\mathbf{v})$ (T is additive).

Of course the examples of linear transformations in chapter 5 provide many linear transformations from $\mathbf{R}^n$ to $\mathbf{R}^m$. The next two examples provide linear transformation between other spaces.

Example 7.25

Define $T : \mathbf{M}^{2,3} \to \mathbf{P}_2$ by

$$T(A) = T\begin{bmatrix} a_{11} & a_{12} & a_{13} \\ a_{21} & a_{22} & a_{23} \end{bmatrix} = (a_{11} + a_{13})x^2 + (a_{21} - a_{22})x + a_{23}.$$

To verify condition i of the definition, let A be in $\mathbf{M}^{2,3}$, and let c be a scalar. Applying T to cA, we have

$$T(cA) = T\begin{bmatrix} ca_{11} & ca_{12} & ca_{13} \\ ca_{21} & ca_{22} & ca_{23} \end{bmatrix} = (ca_{11} + ca_{13})x^2$$
$$+ (ca_{21} - ca_{22})x + ca_{23}$$
$$= c[(a_{11} + a_{13})x^2 + (a_{21} - a_{22})x + a_{23}] = cT(A).$$

To verify condition ii, let A and B be in $\mathbf{M}^{2,3}$. Applying T to $A + B$, we have

$$T(A + B) = T\begin{bmatrix} a_{11} + b_{11} & a_{12} + b_{12} & a_{13} + b_{13} \\ a_{21} + b_{21} & a_{22} + b_{22} & a_{23} + b_{23} \end{bmatrix}$$
$$= [(a_{11} + b_{11}) + (a_{13} + b_{13})]x^2 + [(a_{21} + b_{21})$$
$$- (a_{22} + b_{22})]x + [a_{23} + b_{23}]$$
$$= (a_{11} + a_{13})x^2 + (a_{21} - a_{22})x + a_{23}$$
$$+ (b_{11} + b_{13})x^2 + (b_{21} - b_{22})x + b_{23}$$
$$= T(A) + T(B).$$

Thus T is a linear transformation.

The next example is one that is very important in a more advanced study of linear algebra.

Example 7.26

Let A be a square matrix of order n. Define $T : \mathbf{P} \to \mathbf{M}^{n,n}$ as follows. For each polynomial

$$\mathbf{p}(x) = a_m x^m + a_{m-1} x^{m-1} + \cdots + a_1 x + a_0$$

define $T(\mathbf{p})$ to be the matrix

$$T(\mathbf{p}) = a_m A^m + a_{m-1} A^{m-1} + \cdots + a_1 A + a_0 I,$$

where I is the identity matrix of order n. Then T is a linear transformation.

To show condition i, let $\mathbf{p}$ be the polynomial

$$\mathbf{p}(x) = a_m x^m + \cdots + a_1 x + a_0$$

and c a scalar. Then $c\mathbf{p}$ is the polynomial

$$(c\mathbf{p})(x) = (ca_m)x^m + \cdots + (ca_1)x + ca_0.$$

Applying T to $c\mathbf{p}$, we have

$$T(c\mathbf{p}) = (ca_m)A^m + \cdots + (ca_1)A + (ca_0)I$$
$$= c(a_m A^m + \cdots + a_1 A + a_0 I) = cT(\mathbf{p}).$$

To show condition ii, let $\mathbf{p}$ and $\mathbf{q}$ be the polynomials $\mathbf{p}(x) = a_m x^m + \cdots + a_1 x + a_0$ and $\mathbf{q}(x) = b_m x^m + \cdots + b_1 x + b_0$. (By using zero coefficients, if necessary, $\mathbf{p}$ and $\mathbf{q}$ can be made to appear as if they have the same degree.) Then $\mathbf{p} + \mathbf{q}$ is the polynomial

$$(\mathbf{p} + \mathbf{q})(x) = (a_m + b_m)x^m + \cdots + (a_1 + b_1)x + (a_0 + b_0).$$

Applying T to $\mathbf{p} + \mathbf{q}$, we have

$$T(\mathbf{p} + \mathbf{q}) = (a_m + b_m)A^m + \cdots + (a_1 + b_1)A + (a_0 + b_0)I$$
$$= (a_m A^m + \cdots + a_1 A + a_0 I) + (b_m A^m + \cdots$$
$$+ b_1 A + b_0 I)$$
$$= T(\mathbf{p}) + T(\mathbf{q}).$$

Thus T is a linear transformation.

Most of the theorems in chapter 5 are valid for general vector spaces. The concepts of *kernel* and *image* generalize easily and play the same role as before. We redefine them here and collect their elementary properties as the next theorem. The proofs of the results are routine modifications of those given earlier (with the exception of part h) and will be omitted.

Definition   Let $T : \mathbf{V} \to \mathbf{W}$ be a linear transformation. The set of all vectors $\mathbf{v}$ in $\mathbf{V}$ that satisfy $T(\mathbf{v}) = \mathbf{0}$ is called the **kernel** (or **nullspace**) of T and is denoted by $\mathrm{Ker}(T)$. The set of all vectors $\mathbf{w}$ such that $T(\mathbf{v}) = \mathbf{w}$, for some $\mathbf{v}$, is called the **image** of T and is denoted by $\mathrm{Im}(T)$.

Theorem 2   Let $T : V \to W$ be a linear transformation. Then

a. $T(0) = 0$

b. $T(-u) = -T(u)$ (section 5.1, theorem 1),

c. $T(u - v) = T(u) - T(v)$

d. $\text{Ker}(T)$ is a subspace of V (section 5.3, theorem 1),

e. T is one-to-one if and only if $\text{Ker}(T) = \{0\}$ (section 5.3, theorem 3),

f. $\text{Im}(T)$ is a subspace of W (section 5.3, theorem 4),

g. T is onto if and only if $\dim(\text{Im}(T)) = \dim W$ (section 5.3, theorem 6),

h. If $n = \dim(V)$, then $\dim(\text{Ker}(T)) + \dim(\text{Im}(T)) = n$ (section 5.3, theorem 7),

i. If $n = \dim(V) = \dim(W)$, then T is one-to-one if and only if it is onto (section 5.3, theorem 8).

Note: If $\mathscr{B} = \{v_1, \ldots, v_n\}$ is a basis for a vector space V, then the function $T : V \to R^n$ defined by $T(x) = y$, the coordinate vector of x with respect to $\mathscr{B}$, is a linear transformation that is one-to-one and onto: that T is a linear transformation follows from theorem 1 (with $m = 2$); that it is one-to-one follows from theorem 2e; and that it is onto follows from theorem 2i.

The last topic we will consider in this section is the use of matrices to represent linear transformations between general vector spaces. The key is again coordinate vectors with respect to some basis; in fact we need bases for *both* spaces, V and W. In case $V = W$, we can take the same basis for each, but in general this is not possible. The following theorem describes the situation. It is the analog of theorem 1 of section 5.2.

Theorem 3   Let $T : V \to W$ be a linear transformation, and let $\mathscr{B} = \{v_1, \ldots, v_n\}$ and $\mathscr{B}' = \{w_1, \ldots, w_m\}$ be bases for V and W, respectively. Let A be the $m \times n$ matrix with the ith column A^i given by the coordinate vector of $T(v_i)$ with respect to the basis $\mathscr{B}'$. Then for each x in V, $T(x)$ has as its coordinate vector with respect to $\mathscr{B}'$ the vector Ay, where y is the coordinate vector of x with respect to $\mathscr{B}$. (We say A **represents** T with respect to the bases $\mathscr{B}$ and $\mathscr{B}'$.)

Proof   By exercise 15 it is enough to show that the theorem holds for each element in the basis $\mathscr{B}$. In other words, we must show that the coordinate vector for each $T(v_i)$ with respect to $\mathscr{B}'$ is Ay_i, where y_i is the coordinate vector of v_i with respect to $\mathscr{B}$. But the coordinate vector of v_i with respect to $\mathscr{B}$ is e_i, the ith elementary vector of R^n. Since $Ae_i = A^i$ for each i, and A^i was taken to be the coordinate vector of $T(v_i)$ with respect to $\mathscr{B}'$, the result is proved.   ∎

Example 7.27

Let $T : \mathbf{M}^{2,3} \to \mathbf{P}_2$ be as in example 7.25,

$$T \begin{bmatrix} a_{11} & a_{12} & a_{13} \\ a_{21} & a_{22} & a_{23} \end{bmatrix} = (a_{11} + a_{13})x^2 + (a_{21} - a_{22})x + a_{23}.$$

Find the matrix A in $\mathbf{M}^{6,3}$ that represents T with respect to the bases $\mathcal{B} = \{E^{ij} | i = 1, 2; j = 1, 2, 3\}$ and $\mathcal{B}' = \{1, x, x^2\}$.

Following theorem 3 we compute $T(E^{ij})$ for each i, j and represent the result by means of the coordinate vector with respect to $\mathcal{B}'$,

$$T(E^{11}) = (1 + 0)x^2 + (0 - 0)x + 0 = 0(1) + 0(x) + 1(x^2)$$
$$\leftrightarrow (0, 0, 1),$$
$$T(E^{12}) = (0 + 0)x^2 + (0 - 0)x + 0 = 0(1) + 0(x) + 0(x^2)$$
$$\leftrightarrow (0, 0, 0),$$
$$T(E^{13}) = (0 + 1)x^2 + (0 - 0)x + 0 = 0(1) + 0(x) + 1(x^2)$$
$$\leftrightarrow (0, 0, 1),$$
$$T(E^{21}) = (0 + 0)x^2 + (1 - 0)x + 0 = 0(1) + 1(x) + 0(x^2)$$
$$\leftrightarrow (0, 1, 0),$$
$$T(E^{22}) = (0 + 0)x^2 + (0 - 1)x + 0 = 0(1) - 1(x) + 0(x^2)$$
$$\leftrightarrow (0, -1, 0),$$
$$T(E^{23}) = (0 + 0)x^2 + (0 - 0)x + 1 = 1(1) + 0(x) + 0(x^2)$$
$$\leftrightarrow (1, 0, 0).$$

The resulting matrix is then

$$A = \begin{bmatrix} 0 & 0 & 0 & 0 & 0 & 1 \\ 0 & 0 & 0 & 1 & -1 & 0 \\ 1 & 0 & 1 & 0 & 0 & 0 \end{bmatrix}.$$

Example 7.28

Let $D : \mathbf{P}_3 \to \mathbf{P}_2$ be defined as follows. For each

$$\mathbf{p}(x) = a_3x^3 + a_2x^2 + a_1x + a_0$$

define

$$D(\mathbf{p}) = 3a_3x^2 + 2a_2x + a_1.$$

Find the matrix X that represents D with respect to the bases $\mathcal{B} = \{1, x, x^2, x^3\}$ and $\mathcal{B}' = \{1, x, x^2\}$.

We have

$$D(1) = 0 = 0(1) + 0(x) + 0(x^2) = (0, 0, 0),$$
$$D(x) = 1 = 1(1) + 0(x) + 0(x^2) = (1, 0, 0),$$

$$D(x^2) = 2x = 0(1) + 2(x) + 0(x^2) = (0, 2, 0),$$
$$D(x^3) = 3x^2 = 0(1) + 0(x) + 3(x^2) = (0, 0, 3).$$

The resulting matrix is then

$$A = \begin{bmatrix} 0 & 1 & 0 & 0 \\ 0 & 0 & 2 & 0 \\ 0 & 0 & 0 & 3 \end{bmatrix}.$$

Note: (From calculus) the linear transformation D of example 7.28 simply gives the *derivative* of $\mathbf{p}(x)$. Thus we can give the derivative of any polynomial of degree less than or equal to 3 by multiplication by A. That is,

$$\frac{d}{dx}(a_3x^3 + a_2x^2 + a_1x + a_0) \leftrightarrow \begin{bmatrix} 0 & 1 & 0 & 0 \\ 0 & 0 & 2 & 0 \\ 0 & 0 & 0 & 3 \end{bmatrix} \begin{bmatrix} a_0 \\ a_1 \\ a_2 \\ a_3 \end{bmatrix}.$$

For example, if $\mathbf{p}(x) = 1 - 3x^2 + x^3$, then $\dfrac{d}{dx}(\mathbf{p}(x))$ can be obtained by computing

$$\begin{bmatrix} 0 & 1 & 0 & 0 \\ 0 & 0 & 2 & 0 \\ 0 & 0 & 0 & 3 \end{bmatrix} \begin{bmatrix} 1 \\ 0 \\ -3 \\ 1 \end{bmatrix} = \begin{bmatrix} 0 \\ -6 \\ 3 \end{bmatrix}.$$

Now $(0, -6, 3)$ is the coordinate vector of $\dfrac{d}{dx}(\mathbf{p}(x)) = -6x + 3x^2$.

7.3 Exercises

In exercises 1 through 5 find the coordinate vector of $\mathbf{x}$ with respect to the given basis $\mathscr{B}$ for the vector space $\mathbf{V}$.

1. $\mathbf{x} = 2 - x + 3x^3$, $\mathscr{B} = \{1, x, x^2, x^3\}$, $\mathbf{V} = \mathbf{P}_3$

2. $\mathbf{x} = \begin{bmatrix} 1 & 2 & 1 \\ -1 & 1 & 2 \end{bmatrix}$, $\mathscr{B} = \{E^{ij} | i = 1, 2; j = 1, 2, 3\}$, $\mathbf{V} = \mathbf{M}^{2, 3}$

3. $\mathbf{x} = 2 - 5x$, $\mathscr{B} = \{x + 1, x - 1\}$, $\mathbf{V} = \mathbf{P}_1$

4. $\mathbf{x} = \begin{bmatrix} -2 & 0 \\ 0 & 3 \end{bmatrix}$, $\mathscr{B} = \left\{ \begin{bmatrix} 1 & 0 \\ 0 & 0 \end{bmatrix}, \begin{bmatrix} 0 & 0 \\ 0 & 1 \end{bmatrix} \right\}$,

   $\mathbf{V}$ is the vector space of all 2×2 diagonal matrices

5. $\mathbf{x} = \begin{bmatrix} -2 & 0 \\ 0 & 3 \end{bmatrix}$, $\mathcal{B} = \left\{ \begin{bmatrix} 3 & 1 \\ 0 & 2 \end{bmatrix}, \begin{bmatrix} 1 & -1 \\ 0 & 1 \end{bmatrix}, \begin{bmatrix} 2 & 2 \\ 0 & 0 \end{bmatrix} \right\}$,

V is the vector space of all 2×2 matrices A such that $a_{21} = 0$

In exercises 6 through 10 use coordinate vectors to decide whether or not the given set is linearly independent. If it is linearly dependent, express one of the vectors as a linear combination of the others.

6. $\mathcal{S}_1 = \{x^2 + x - 1, x^2 - 2x + 3, x^2 + 4x - 5\}$ in $\mathbf{P}_2$

7. $\mathcal{S}_2 = \{x^2 + x - 1, x^2 - 2x + 3, x^2 + 4x - 3\}$ in $\mathbf{P}_3$

8. $\mathcal{S}_3 = \left\{ \begin{bmatrix} 2 & -1 \\ 1 & 0 \end{bmatrix}, \begin{bmatrix} 1 & 0 \\ 2 & 1 \end{bmatrix}, \begin{bmatrix} 3 & 1 \\ -1 & 1 \end{bmatrix}, \begin{bmatrix} -1 & 4 \\ 2 & 0 \end{bmatrix} \right\}$ in $\mathbf{M}^{2,2}$

9. $\mathcal{S}_4 = \left\{ \begin{bmatrix} 1 & 2 \\ -1 & 0 \end{bmatrix}, \begin{bmatrix} 0 & -1 \\ 1 & 1 \end{bmatrix}, \begin{bmatrix} 1 & 0 \\ 1 & 2 \end{bmatrix} \right\}$ in $\mathbf{M}^{2,2}$

10. $\mathcal{S}_5 = \left\{ \begin{bmatrix} 2 & 1 \\ 1 & 3 \end{bmatrix}, \begin{bmatrix} 3 & -1 \\ -1 & 2 \end{bmatrix}, \begin{bmatrix} 1 & 0 \\ 0 & 1 \end{bmatrix}, \begin{bmatrix} 2 & 2 \\ 2 & 4 \end{bmatrix} \right\}$

in the subspace of $\mathbf{M}^{2,2}$ of all matrices of the form $\begin{bmatrix} a & b \\ b & a+b \end{bmatrix}$

In exercises 11 through 14 find the matrix A that represents the linear transformation T with respect to the bases $\mathcal{B}$ and $\mathcal{B}'$.

11. $T : \mathbf{R}^3 \rightarrow \mathbf{M}^{2,2}$ given by

$$T(x, y, z) = \begin{bmatrix} y & z \\ -x & 0 \end{bmatrix},$$

where $\mathcal{B} = \{\mathbf{e}_1, \mathbf{e}_2, \mathbf{e}_3\}$ and $\mathcal{B}' = \{E^{ij} | i = 1, 2; j = 1, 2\}$

12. $T : \mathbf{P}_3 \rightarrow \mathbf{P}_3$ given by

$$T(a_0 + a_1 x + a_2 x^2 + a_3 x^3) = (a_0 + a_2) - (a_1 + 2a_3)x^2,$$

where $\mathcal{B} = \mathcal{B}' = \{x^i | i = 0, 1, 2, 3\}$

13. $T : \mathbf{P}_2 \rightarrow \mathbf{M}^{2,2}$ given by

$$T(a_0 + a_1 x + a_2 x^2) = \begin{bmatrix} a_0 & -a_2 \\ -a_2 & a_0 - a_1 \end{bmatrix},$$

where $\mathcal{B} = \{x^i | i = 0, 1, 2\}$ and $\mathcal{B}' = \{E^{ij} | i = 1, 2; j = 1, 2\}$

14. $T : \mathbf{P}_2 \rightarrow \mathbf{P}_2$ given by

$$T(a_0 + a_1 x + a_2 x^2) = a_1 - (a_0 + a_2)x + (a_1 + a_2)x^2,$$

where $\mathcal{B} = \{x^i | i = 0, 1, 2\}$ and $\mathcal{B}' = \{x + 1, x - 1, x^2 + 1\}$

15. Prove that a linear transformation $T : \mathbf{V} \rightarrow \mathbf{W}$ is determined by its *action* on a basis. In other words, if $\mathcal{B} = \{\mathbf{v}_1, \cdots, \mathbf{v}_n\}$ is a basis for $\mathbf{V}$, and S and T are linear transformations from V to W such that $S(\mathbf{v}_i) = T(\mathbf{v}_i)$ for $i = 1, \cdots,$ n, then $S = T$(that is $S(\mathbf{v}) = T(\mathbf{v})$ for every $\mathbf{v}$ in V).

16. Let $T: \mathbf{V} \rightarrow \mathbf{W}$ be a linear transformation, let $\mathscr{B}$ and $\mathscr{B}'$ be bases for V and W, respectively, and let A be the matrix that represents T with respect to $\mathscr{B}$ and $\mathscr{B}'$. For any $\mathbf{x}$ in V, let $\mathbf{y}$ be the coordinate vector of $\mathbf{x}$ with respect to $\mathscr{B}$. Prove that $\mathbf{x}$ is in Ker(T) if and only if $A\mathbf{y} = \mathbf{0}$.

17–20. Use exercise 16 to compute Ker(T) for each linear transformation in exercises 11 through 14. Which ones are one-to-one?

21. Find the 4×5 matrix associated with the derivative transformation

$$D : \mathbf{P}_4 \rightarrow \mathbf{P}_3$$

with respect to $\mathscr{B} = \{1, x, x^2, x^3, x^4\}$ and $\mathscr{B}' = \{1, x, x^2, x^3\}$ (see example 7.28).

22. Use the matrix of exercise 21 to find the derivative of

$$\mathbf{p}(x) = 1 + x^2 - 3x^3 + 2x^4.$$

23. Let $S : \mathbf{P}_2 \rightarrow \mathbf{P}_3$ be defined as follows. For each $\mathbf{p}(x) = a_2 x^2 + a_1 x + a_0$, define $S(\mathbf{p}) = a_2 x^3/3 + a_1 x^2/2 + a_0 x$. Find the matrix A that represents S with respect to the bases $\mathscr{B} = \{1, x, x^2\}$ and $\mathscr{B}' = \{1, x, x^2, x^3\}$. (The linear transformation S gives the *integral* of $\mathbf{p}(x)$, with the constant term equal to zero.)

24. Use the matrix of exercise 23 to find the integral of $\mathbf{p}(x) = 1 - x + 2x^2$.

7.4   Inner-Product Spaces

In section 1.4 we introduced the notion of the *dot product* (also called *inner product* or *scalar product*) of two vectors in $\mathbf{R}^m$. In this section we will generalize this concept to vectors in an arbitrary vector space. This will in turn lead to a natural definition of *orthogonality* of such vectors and a generalization of the *Gram-Schmidt process* for constructing orthonormal bases.

In theorem 3 of section 1.4 we gave several important properties of the dot product in $\mathbf{R}^m$. We shall now use these properties to define our extended notion of dot product. Following standard convention, we will use the alternate terminology *inner product* as well as change the previous notation in our generalized definition.

Definition   Let V be a vector space over the real numbers. An **inner product** for V is a function that associates with every pair of vectors, **u** and **v**, in V, a real number, (**u**, **v**), satisfying the following properties. For all **u**, **v**, and **w** in **V** and scalars c:

a. (**u**, **v**) = (**v**, **u**),
b. (**u**, **v** + **w**) = (**u**, **v**) + (**u**, **w**),
c. (c**u**, **v**) = c(**u**, **v**),
d. (**u**, **u**) $\geq$ 0 and (**u**, **u**) = 0 if and only if **u** = **0**.

A vector space **V** together with an inner product is called an **inner-product space.**

Example 7.29

Let **u** = $(u_1, u_2, \cdots, u_m)$ and **v** = $(v_1, v_2, \cdots, v_m)$ be arbitrary vectors in the vector space $\mathbf{R}^m$. Define

$$(\mathbf{u}, \mathbf{v}) = u_1 v_1 + u_2 v_2 + \cdots + u_m v_m,$$

the dot product of **u** and **v**. Then (**u**, **v**) defines an inner product for $\mathbf{R}^m$.

By theorem 3 of section 1.4, this function satisfies the four properties of the definition and is therefore an inner product.

Example 7.29 verifies that the dot product for $\mathbf{R}^m$ defined in section 1.4 is an inner product for this vector space. The next example demonstrates that one can define other inner products for $\mathbf{R}^m$—there may be more than one inner product for a given vector space.

Example 7.30

Let **u** = $(u_1, u_2, \cdots, u_m)$ and **v** = $(v_1, v_2, \cdots, v_m)$ be arbitrary vectors in $\mathbf{R}^m$. Define

$$(\mathbf{u}, \mathbf{v}) = u_1 v_1 + 2u_2 v_2 + 3u_3 v_3 + \cdots + mu_m v_m.$$

Then (**u**, **v**) is an inner product for $\mathbf{R}^m$.

We must verify properties a through d of the definition:

a.

$$
\begin{aligned}
(\mathbf{u}, \mathbf{v}) &= u_1 v_1 + 2u_2 v_2 + \cdots + mu_m v_m \\
&= v_1 u_1 + 2v_2 u_2 + \cdots + mv_m u_m \\
&= (\mathbf{v}, \mathbf{u}).
\end{aligned}
$$

b.

$$
\begin{aligned}
(\mathbf{u}, \mathbf{v} + \mathbf{w}) &= u_1(v_1 + w_1) + 2u_2(v_2 + w_2) + \cdots + mu_m(v_m + w_m) \\
&= (u_1 v_1 + u_1 w_1) + (2u_2 v_2 + 2u_2 w_2) + \cdots \\
&\quad + (mu_m v_m + mu_m w_m)
\end{aligned}
$$

$$= (u_1 v_1 + 2u_2 v_2 + \cdots + mu_m v_m) + (u_1 w_1 + 2u_2 w_2$$
$$+ \cdots + mu_m w_m)$$
$$= (\mathbf{u}, \mathbf{v}) + (\mathbf{u}, \mathbf{w}).$$

c.

$$(c\mathbf{u}, \mathbf{v}) = cu_1 v_1 + 2cu_2 v_2 + \cdots + mcu_m v_m$$
$$= c(u_1 v_1 + 2u_2 v_2 + \cdots + mu_m v_m)$$
$$= c(\mathbf{u}, \mathbf{v}).$$

d.

$$(\mathbf{u}, \mathbf{u}) = u_1^2 + 2u_2^2 + \cdots + mu_m^2,$$

which is ≥ 0, and if

$$u_1^2 + 2u_2^2 + \cdots + mu_m^2 = 0,$$

then

$$u_1 = u_2 = \cdots = u_m = 0.$$

Thus $\mathbf{u} = \mathbf{0}$, as desired.

Example 7.31

Let $\mathbf{p}$ and $\mathbf{q}$ be arbitrary vectors in $\mathbf{P}_1$, the vector space of all polynomials in x of degree less than or equal to 1. Define

$$(\mathbf{p}, \mathbf{q}) = \mathbf{p}(0)\mathbf{q}(0) + \mathbf{p}(1)\mathbf{q}(1).$$

Then $(\mathbf{p}, \mathbf{q})$ is an inner product for $\mathbf{P}_1$.

Properties a, b, and c are easily verified. For property d we have $(\mathbf{p}, \mathbf{p}) = [\mathbf{p}(0)]^2 + [\mathbf{p}(1)]^2$ which is ≥ 0. Moreover $(\mathbf{p}, \mathbf{p}) = 0$ if and only if $\mathbf{p}(0) = 0$ and $\mathbf{p}(1) = 0$. But the latter implies that $\mathbf{p}(x)$ is the zero polynomial, since the graph of $\mathbf{p}(x)$, a line, goes through the points $(0, 0)$ and $(1, 0)$ and is therefore the graph of the zero polynomial.

Example 7.32 (from calculus)

Let $\mathbf{f}$ and $\mathbf{g}$ be vectors in $\mathbf{C}[-1, 1]$, the vector space of all real-valued functions continuous on the interval $[-1, 1]$. (See example 7.6 and exercise 9 of section 7.1.) Define

$$(\mathbf{f}, \mathbf{g}) = \int_{-1}^{1} \mathbf{f}(x)\mathbf{g}(x)dx.$$

Then $(\mathbf{f}, \mathbf{g})$ is an inner product for $\mathbf{C}[-1, 1]$.

The verification that $(\mathbf{f}, \mathbf{g})$ is an inner product rests on some properties of the definite integral learned in calculus. Specifically properties b, c, and d hold because, respectively,

$$\int_{-1}^{1} \mathbf{f}(x)(\mathbf{g}(x) + \mathbf{h}(x))dx = \int_{-1}^{1} \mathbf{f}(x)\mathbf{g}(x)dx + \int_{-1}^{1} \mathbf{f}(x)\mathbf{h}(x)dx,$$

$$\int_{-1}^{1} c\mathbf{f}(x)\mathbf{g}(x)dx = c\int_{-1}^{1} \mathbf{f}(x)\mathbf{g}(x)dx,$$

and

$$\int_{-1}^{1} (\mathbf{f}(x))^2 dx \geq 0 \quad \text{with} \quad \int_{-1}^{1} (\mathbf{f}(x))^2 dx = 0$$

if and only if $\mathbf{f}(x)$ is the zero function on $[-1, 1]$. (Property a holds simply because $\mathbf{f}(x)\mathbf{g}(x) = \mathbf{g}(x)\mathbf{f}(x)$ for all x in $[-1, 1]$.)

From the defining properties of the inner product, we can derive several other properties satisfied by all inner products. Some of these are presented in the next two theorems.

Theorem 1   Let $\mathbf{V}$ be an inner product space. Then for all vectors $\mathbf{u}_1$, $\mathbf{u}_2$ and $\mathbf{v}$ in $\mathbf{V}$ and scalars c_1 and c_2

a. $(c_1\mathbf{u}_1 + c_2\mathbf{u}_2, \mathbf{v}) = (\mathbf{v}, c_1\mathbf{u}_1 + c_2\mathbf{u}_2) = c_1(\mathbf{u}_1, \mathbf{v}) + c_2(\mathbf{u}_2, \mathbf{v}),$
b. $(\mathbf{0}, \mathbf{v}) = (\mathbf{v}, \mathbf{0}) = 0.$

Proof

a. $\begin{aligned}(c_1\mathbf{u}_1 + c_2\mathbf{u}_2, \mathbf{v}) &= (\mathbf{v}, c_1\mathbf{u}_1 + c_2\mathbf{u}_2) && \text{(property a),}\\ &= (\mathbf{v}, c_1\mathbf{u}_1) + (\mathbf{v}, c_2\mathbf{u}_2) && \text{(property b),}\\ &= (c_1\mathbf{u}_1, \mathbf{v}) + (c_2\mathbf{u}_2, \mathbf{v}) && \text{(property a),}\\ &= c_1(\mathbf{u}_1, \mathbf{v}) + c_2(\mathbf{u}_2, \mathbf{v}) && \text{(property c).}\end{aligned}$

b. Exercise 17.   ∎

In section 1.4 we saw that for any $\mathbf{u}$ in $\mathbf{R}^m$ the *norm* of $\mathbf{u}$, $\|\mathbf{u}\|$, is equal to $\sqrt{\mathbf{u} \cdot \mathbf{u}}$. In a general inner-product space we use this for a definition. If $\mathbf{u}$ is a vector in an inner-product space, then the **norm** of $\mathbf{u}$, denoted $\|\mathbf{u}\|$, is given by $\|\mathbf{u}\| = \sqrt{(\mathbf{u}, \mathbf{u})}$. The next theorem, known as the **Cauchy-Schwarz inequality,** relates the inner product of vectors $\mathbf{u}$ and $\mathbf{v}$ to the norms of $\mathbf{u}$ and $\mathbf{v}$.

Theorem 2   Let $\mathbf{u}$ and $\mathbf{v}$ be vectors in an inner-product space. Then

$$|(\mathbf{u}, \mathbf{v})| \leq \|\mathbf{u}\| \, \|\mathbf{v}\|,$$

where $|(\mathbf{u}, \mathbf{v})|$ denotes the *absolute value* of $(\mathbf{u}, \mathbf{v})$.

Proof   First notice that, if $\mathbf{u} = \mathbf{0}$, then $(\mathbf{u}, \mathbf{v}) = 0$ by theorem 1b, and

$$\|\mathbf{u}\| = \sqrt{(\mathbf{u}, \mathbf{u})} = 0,$$

as well. In this case the inequality becomes $0 \le 0$, which implies that the theorem holds if $\mathbf{u} = \mathbf{0}$.

Now assume that $\mathbf{u} \ne \mathbf{0}$. If x is an arbitrary real number, we have

$$
\begin{aligned}
0 &\le (x\mathbf{u} + \mathbf{v}, x\mathbf{u} + \mathbf{v}) && \text{(property d)}, \\
&= x^2(\mathbf{u}, \mathbf{u}) + x(\mathbf{u}, \mathbf{v}) + x(\mathbf{v}, \mathbf{u}) + (\mathbf{v}, \mathbf{v}) && \text{(theorem 1a)}, \\
&= x^2(\mathbf{u}, \mathbf{u}) + 2x(\mathbf{u}, \mathbf{v}) + (\mathbf{v}, \mathbf{v}) && \text{(property a)}.
\end{aligned}
$$

Now, setting $(\mathbf{u}, \mathbf{u}) = a$, $2(\mathbf{u}, \mathbf{v}) = b$, and $(\mathbf{v}, \mathbf{v}) = c$, this inequality becomes $ax^2 + bx + c \ge 0$, which is quadratic in x. In geometric terms this implies that the parabola $y = ax^2 + bx + c$ must be contained in the region on and above the x-axis (figures 7.1 and 7.2). But this in turn implies that the polynomial $ax^2 + bx + c$ cannot have two distinct real zeros, since, if it did, the curve would dip below the x-axis (figure 7.3).

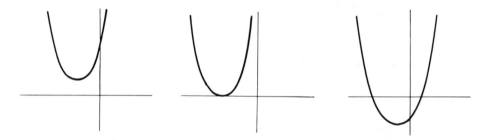

Figure 7.1 No real roots   **Figure 7.2** One real root   **Figure 7.3** Two real roots

Now the equation $ax^2 + bx + c = 0$ fails to have two real roots (it has either no roots or one root) if and only if $b^2 - 4ac \le 0$. But this implies that $b^2 \le 4ac$, or

$$4(\mathbf{u}, \mathbf{v})^2 \le 4(\mathbf{u}, \mathbf{u})(\mathbf{v}, \mathbf{v}),$$

or

$$(\mathbf{u}, \mathbf{v})^2 \le (\mathbf{u}, \mathbf{u})(\mathbf{v}, \mathbf{v}).$$

Finally, taking square roots of both sides, we obtain

$$\sqrt{(\mathbf{u}, \mathbf{v})^2} \le \sqrt{(\mathbf{u}, \mathbf{u})}\, \sqrt{(\mathbf{v}, \mathbf{v})}$$

or

$$|(\mathbf{u}, \mathbf{v})| \le \|\mathbf{u}\|\, \|\mathbf{v}\|. \qquad \blacksquare$$

Example 7.33

With the inner products of examples 7.29 and 7.32 the Cauchy-Schwarz inequality becomes

$$|u_1 v_1 + u_2 v_2 + \cdots + u_m v_m|$$
$$\leq \sqrt{u_1^2 + u_2^2 + \cdots + u_m^2} \sqrt{v_1^2 + v_2^2 + \cdots + v_m^2}$$

and

$$\left| \int_{-1}^{1} f(x)g(x)dx \right| \leq \sqrt{\int_{-1}^{1} (f(x))^2 dx} \sqrt{\int_{-1}^{1} (g(x))^2 dx},$$

respectively.

Recall that two vectors in $\mathbf{R}^m$ are *orthogonal* if their dot product is equal to zero. The next definition generalizes the concept of orthogonality to inner-product spaces.

Definition   Two vectors, $\mathbf{u}$ and $\mathbf{v}$, in an inner product space $\mathbf{V}$ are **orthogonal** if $(\mathbf{u}, \mathbf{v}) = 0$. A set of vectors, $\mathscr{S}$, is an **orthogonal set** if every pair of distinct vectors in $\mathscr{S}$ is orthogonal. A set $\mathscr{T}$ is an **orthonormal set** if it is orthogonal and every vector in $\mathscr{T}$ has norm equal to one.

Example 7.34

Define an inner product on the vector space $\mathbf{M}^{2,2}$, the set of all 2×2 matrices, by

$$(A, B) = a_{11}b_{11} + a_{12}b_{12} + a_{21}b_{21} + a_{22}b_{22}.$$

In this inner-product space show that the matrices

$$C_1 = \frac{\sqrt{3}}{3}\begin{bmatrix} 1 & -1 \\ 1 & 0 \end{bmatrix}, \quad C_2 = \frac{\sqrt{6}}{6}\begin{bmatrix} -1 & -1 \\ 0 & 2 \end{bmatrix}, \quad C_3 = \frac{1}{3}\begin{bmatrix} 0 & 2 \\ 2 & 1 \end{bmatrix}$$

form an orthonormal set. (It is exercise 4 to show that (A, B) is an inner product.)

We first show that $(C_1, C_2) = (C_1, C_3) = (C_2, C_3) = 0$ by computing each of these inner products. We have

$$(C_1, C_3) = (\sqrt{3}/3)(0) + (-\sqrt{3}/3)(2/3) + (\sqrt{3}/3)(2/3) + (0)(1/3) = 0,$$

as desired. Similarly $(C_1, C_2) = 0$ and $(C_2, C_3) = 0$. Thus $\{C_1, C_2, C_3\}$ is orthogonal. We then show that each matrix has norm 1. We have

$$\|C_1\| = \sqrt{(C_1, C_1)}$$
$$= \sqrt{(\sqrt{3}/3)(\sqrt{3}/3) + (-\sqrt{3}/3)(-\sqrt{3}/3) + (\sqrt{3}/3)(\sqrt{3}/3) + (0)(0)}$$
$$= \sqrt{3/9 + 3/9 + 3/9} = 1,$$

as desired. Similarly $\|C_2\| = 1$ and $\|C_3\| = 1$. Therefore $\{C_1, C_2, C_3\}$ is an orthonormal set.

We are now in a position to extend the *Gram-Schmidt process* of section 4.5 (for producing orthonormal bases in $\mathbf{R}^m$) to arbitrary inner-product spaces. By an **orthogonal** (or **orthonormal**) **basis** for an inner-product space $\mathbf{V}$, we of course mean an orthogonal (or orthonormal) set that is a basis for $\mathbf{V}$.

Theorem 3   Let $\mathcal{T} = \{\mathbf{u}_1, \mathbf{u}_2, \ldots, \mathbf{u}_n\}$ be a basis for an inner-product space $\mathbf{V}$. Let $\mathcal{T}' = \{\mathbf{v}_1, \mathbf{v}_2, \ldots, \mathbf{v}_n\}$ be defined as follows:

$$\mathbf{v}_1 = \mathbf{u}_1,$$

$$\mathbf{v}_2 = \mathbf{u}_2 - \frac{(\mathbf{u}_2, \mathbf{v}_1)}{(\mathbf{v}_1, \mathbf{v}_1)}\mathbf{v}_1,$$

$$\mathbf{v}_3 = \mathbf{u}_3 - \frac{(\mathbf{u}_3, \mathbf{v}_1)}{(\mathbf{v}_1, \mathbf{v}_1)}\mathbf{v}_1 - \frac{(\mathbf{u}_3, \mathbf{v}_2)}{(\mathbf{v}_2, \mathbf{v}_2)}\mathbf{v}_2,$$

$$\vdots$$

$$\mathbf{v}_n = \mathbf{u}_n - \frac{(\mathbf{u}_n, \mathbf{v}_1)}{(\mathbf{v}_1, \mathbf{v}_1)}\mathbf{v}_1 - \frac{(\mathbf{u}_n, \mathbf{v}_2)}{(\mathbf{v}_2, \mathbf{v}_2)}\mathbf{v}_2 - \cdots - \frac{(\mathbf{u}_n, \mathbf{v}_{n-1})}{(\mathbf{v}_{n-1}, \mathbf{v}_{n-1})}\mathbf{v}_{n-1}.$$

Then the set $\mathcal{T}'$ is an orthogonal basis for $\mathbf{V}$. An orthonormal basis for $\mathbf{V}$ is given by $\mathcal{T}'' = \{\mathbf{w}_1, \mathbf{w}_2, \ldots, \mathbf{w}_n\}$, where $\mathbf{w}_i = (1/\|\mathbf{v}_i\|)\mathbf{v}_i$ for $i = 1, 2, \ldots, n$.

We shall prove theorem 3 after presenting an example of its use.

Example 7.35

Find an orthonormal basis for $\mathbf{P}_2$, the vector space of all polynomials of degree ≤ 2, under the inner product $(\mathbf{p}, \mathbf{q}) = \int_{-1}^{1} \mathbf{p}(x)\mathbf{q}(x)dx$.

We know that a basis for $\mathbf{P}_2$ (the *standard basis* for $\mathbf{P}_2$) is given by $\mathcal{T} = \{\mathbf{u}_1, \mathbf{u}_2, \mathbf{u}_3\}$, where $\mathbf{u}_1 = 1$, $\mathbf{u}_2 = x$, and $\mathbf{u}_3 = x^2$ (theorem 1 of section 7.2). Applying theorem 3, $\mathcal{T}' = \{\mathbf{v}_1, \mathbf{v}_2, \mathbf{v}_3\}$ will be an *orthogonal* basis for $\mathbf{P}_2$ if

$$\mathbf{v}_1 = 1,$$

$$\mathbf{v}_2 = x - \left(\frac{\int_{-1}^{1}(x)(1)dx}{\int_{-1}^{1}(1)(1)dx}\right)(1) = x - \left(\frac{0}{2}\right)1 = x,$$

$$\mathbf{v}_3 = x^2 - \left(\frac{\int_{-1}^{1}(x^2)(1)dx}{\int_{-1}^{1}(1)(1)dx}\right)(1) - \left(\frac{\int_{-1}^{1}(x^2)(x)dx}{\int_{-1}^{1}(x)(x)dx}\right)(x)$$

$$= x^2 - \left(\frac{2/3}{2}\right)(1) - \left(\frac{0}{2/3}\right)(x) = x^2 - \frac{1}{3}.$$

Finally, $\mathscr{T}'' = \{\mathbf{w}_1, \mathbf{w}_2, \mathbf{w}_3\}$ is an ortho*normal* basis for P_2 if

$$\mathbf{w}_1 = \left(\frac{1}{\sqrt{\int_{-1}^{1} (1)(1)dx}}\right)(1) = \frac{1}{\sqrt{2}},$$

$$\mathbf{w}_2 = \left(\frac{1}{\sqrt{\int_{-1}^{1} (x)(x)dx}}\right)(x) = \frac{1}{\sqrt{2/3}}x = \sqrt{\frac{3}{2}}x,$$

$$\mathbf{w}_3 = \left(\frac{1}{\sqrt{\int_{-1}^{1} (x^2 - 1/3)(x^2 - 1/3)dx}}\right)(x^2 - 1/3)$$

$$= \frac{1}{(2/3)\sqrt{2/5}}(x^2 - 1/3) = \frac{1}{2}\sqrt{\frac{5}{2}}(3x^2 - 1).$$

The following two lemmas will be used in the proof of theorem 3.

Lemma 1 Let $\mathscr{T}$ and $\mathscr{T}'$ be as in theorem 3, and let

$$a_{ij} = \frac{(\mathbf{u}_i, \mathbf{v}_j)}{(\mathbf{v}_j, \mathbf{v}_j)}.$$

Then $(a_{ij}\mathbf{v}_j, \mathbf{v}_j) = (\mathbf{u}_i, \mathbf{v}_j)$, and, if $\mathbf{v}_i$ and $\mathbf{v}_j$ are orthogonal, $(a_{ij}\mathbf{v}_j, \mathbf{v}_i) = 0$.

Proof We have

$$(a_{ij}\mathbf{v}_j, \mathbf{v}_j) = \frac{(\mathbf{u}_i, \mathbf{v}_j)}{(\mathbf{v}_j, \mathbf{v}_j)}(\mathbf{v}_j, \mathbf{v}_j) = (\mathbf{u}_i, \mathbf{v}_j).$$

If $\mathbf{v}_i$ and $\mathbf{v}_j$ are orthogonal, we have $(a_{ij}\mathbf{v}_j, \mathbf{v}_i) = a_{ij}(\mathbf{v}_j, \mathbf{v}_i) = 0$, as desired. ∎

Lemma 2 Let $\mathscr{T}$ and $\mathscr{T}'$ be as in theorem 3. Then every vector in $\mathscr{T}'$ is nonzero.

Proof Suppose a vector in $\mathscr{T}'$, say, $\mathbf{v}_k$, is zero. Then, using the notation of lemma 1,

$$\mathbf{u}_k = a_{k1}\mathbf{v}_1 + a_{k2}\mathbf{v}_2 + \cdots + a_{k,k-1}\mathbf{v}_{k-1}.$$

Now each $\mathbf{v}_i$ is a linear combination of $\mathbf{u}_i$ and the $\mathbf{v}_j$ that precede it. So by successive substitutions into this equation, we can write $\mathbf{u}_k$ as a linear combination of $\mathbf{u}_1, \mathbf{u}_2, \ldots, \mathbf{u}_{k-1}$. Hence $\{\mathbf{u}_1, \mathbf{u}_2, \ldots, \mathbf{u}_k\}$ is linearly dependent, and by exercise 42 of section 7.2 so is $\mathscr{T}$ because $\mathscr{T}$ contains this set. But this is a contradiction, since $\mathscr{T}$ is a basis for $\mathbf{V}$, so all members of $\mathscr{T}'$ must be nonzero. ∎

Lemma 3 Let $\mathbf{V}$ be an inner-product space, and let $\mathscr{S}$ be either an orthonormal subset of **V** or an orthogonal set of nonzero vectors in **V**. Then $\mathscr{S}$ is linearly independent.

The proof is virtually identical to that of theorem 1 of section 4.5, so we will omit it.

Proof of theorem 3   We will first show that $\mathscr{T}'$ is an orthogonal set and then that it is a basis for **V**. The fact that $\mathscr{T}''$ is an orthonormal basis for **V** then follows easily, and we will leave it as exercise 18.

Using the notation and results of lemma 1, we have

$$(\mathbf{v}_2, \mathbf{v}_1) = (\mathbf{u}_2 - a_{21}\mathbf{v}_1, \mathbf{v}_1) = (\mathbf{u}_2, \mathbf{v}_1) - (a_{21}\mathbf{v}_1, \mathbf{v}_1)$$
$$= (\mathbf{u}_2, \mathbf{v}_1) - (\mathbf{u}_2, \mathbf{v}_1) = 0.$$

Therefore $\mathbf{v}_2$ and $\mathbf{v}_1$ are orthogonal. Furthermore

$$(\mathbf{v}_3, \mathbf{v}_1) = (\mathbf{u}_3 - a_{31}\mathbf{v}_1 - a_{32}\mathbf{v}_2, \mathbf{v}_1)$$
$$= (\mathbf{u}_3, \mathbf{v}_1) - (a_{31}\mathbf{v}_1, \mathbf{v}_1) - (a_{32}\mathbf{v}_2, \mathbf{v}_1)$$
$$= (\mathbf{u}_3, \mathbf{v}_1) - (\mathbf{u}_3, \mathbf{v}_1) - 0 = 0.$$

Therefore $\mathbf{v}_3$ and $\mathbf{v}_1$ are orthogonal. In a similar manner we can show that $\mathbf{v}_3$ and $\mathbf{v}_2$ are orthogonal and that $\mathbf{v}_4$ is orthogonal, successively, to $\mathbf{v}_1$, $\mathbf{v}_2$, and $\mathbf{v}_3$. For example,

$$(\mathbf{v}_4, \mathbf{v}_2) = (\mathbf{u}_4 - a_{41}\mathbf{v}_1 - a_{42}\mathbf{v}_2 - a_{43}\mathbf{v}_3, \mathbf{v}_2)$$
$$= (\mathbf{u}_4, \mathbf{v}_2) - (a_{41}\mathbf{v}_1, \mathbf{v}_2) - (a_{42}\mathbf{v}_2, \mathbf{v}_2) - (a_{43}\mathbf{v}_3, \mathbf{v}_3)$$
$$= (\mathbf{u}_4, \mathbf{v}_2) - 0 - (\mathbf{u}_4, \mathbf{v}_2) - 0 = 0$$

(by once again applying lemma 1). We continue this process until we finally show that $\mathbf{v}_n$ is orthogonal, successively, to $\mathbf{v}_1$, $\mathbf{v}_2$, . . . , $\mathbf{v}_{n-1}$. Thus $\mathscr{T}'$ is an orthogonal set.

To show that $\mathscr{T}'$ is a basis for **V**, notice that it contains n vectors as does the known basis, $\mathscr{T}$. Thus by part iii of theorem 7 of section 7.2 all we need show is that $\mathscr{T}'$ is linearly independent. But since $\mathscr{T}'$ is an orthogonal set of nonzero vectors (lemma 2), $\mathscr{T}'$ is linearly independent and therefore a basis—an orthogonal basis—for **V** by lemma 3.   ■

7.4   Exercises

In exercises 1 through 8 determine whether or not each of the following is an inner product for the indicated vector space. If it is not, list all properties violated.

1. $(\mathbf{p}, \mathbf{q}) = a_0 b_0 + a_1 b_1$; $\mathbf{p}(x) = a_0 + a_1 x$, $\mathbf{q}(x) = b_0 + b_1 x$ in $\mathbf{P}_1$

2. $(\mathbf{p}, \mathbf{q}) = \mathbf{p}(0)\mathbf{q}(0)$; $\mathbf{p}, \mathbf{q}$ in $\mathbf{P}_1$

3. $(A, B) = \det(A)\det(B)$; A, B in $\mathbf{M}^{2, 2}$

4. $(A, B) = a_{11}b_{11} + a_{12}b_{12} + a_{21}b_{21} + a_{22}b_{22}$; A, B in $\mathbf{M}^{2,2}$

5. $(S, T) = S(\mathbf{e}_1) \cdot T(\mathbf{e}_1)$; S, T in $\mathbf{L}(\mathbf{R}^3, \mathbf{R}^2)$ ($\mathbf{x} \cdot \mathbf{y}$ is dot product in $\mathbf{R}^2$)

6. $(S, T) = S(\mathbf{e}_1)$; S, T in $\mathbf{L}(\mathbf{R}^3, \mathbf{R}^1)$

7. $(\mathbf{f}, \mathbf{g}) = \mathbf{f}'(0)\mathbf{g}'(0)$; $\mathbf{f}$, $\mathbf{g}$ in $\mathbf{C}^1(-1, 1)$, the vector space of all differentiable functions on the interval $(-1, 1)$

8. $(\mathbf{f}, \mathbf{g}) = \left(\max_{[-1, 1]} |\mathbf{f}(x)| \right) \left(\max_{[-1, 1]} |\mathbf{g}(x)| \right)$; $\mathbf{f}$, $\mathbf{g}$ in $\mathbf{C}[-1, 1]$

In exercises 9 through 12 determine whether or not the given set of vectors is (a) orthogonal or (b) orthonormal for the indicated vector space and inner product.

9. $\{1, x, x^2\}$ in $\mathbf{P}_2$; $(\mathbf{p}, \mathbf{q}) = \int_0^1 \mathbf{p}(x)\mathbf{q}(x)dx$

10. $\{x/2, x^2 - 1\}$ in $\mathbf{P}_2$; $(\mathbf{p}, \mathbf{q}) = \mathbf{p}(0)\mathbf{q}(0) + \mathbf{p}(1)\mathbf{q}(1) + \mathbf{p}(-1)\mathbf{q}(-1)$

11. $\left\{ \begin{bmatrix} 1/3 & -2/3 \\ 2/3 & 0 \end{bmatrix}, \begin{bmatrix} 2/3 & -1/3 \\ -2/3 & 0 \end{bmatrix}, \begin{bmatrix} 2/3 & 2/3 \\ 1/3 & 0 \end{bmatrix} \right\}$

in $\mathbf{M}^{2,2}$; $(A, B) = a_{11}b_{11} + a_{12}b_{12} + a_{21}b_{21} + a_{22}b_{22}$

12. $\{2/\pi, (\pi/2) \cos x\}$ in $\mathbf{C}[0, \pi/2]$, $(\mathbf{f}, \mathbf{g}) = \int_0^{\pi/2} \mathbf{f}(x)\mathbf{g}(x)dx$

In exercises 13 through 16 apply the Gram-Schmidt process to find an *orthonormal* basis for the subspace of $\mathbf{C}[a, b]$ (under the inner product $(\mathbf{f}, \mathbf{g}) = \int_a^b \mathbf{f}(x)\mathbf{g}(x)dx$) spanned by the given set $\mathscr{S}$.

13. $\mathscr{S} = \{1, x, x^2, x^3\}$, $[a, b] = [-1, 1]$

14. $\mathscr{S} = \{1, x, x^2\}$, $[a, b] = [0, 1]$

15. $\mathscr{S} = \{1, e^x\}$, $[a, b] = [0, 1]$

16. $\mathscr{S} = \{1, e^{-x}\}$, $[a, b] = [0, 1]$

17. Prove (theorem 1b) that for all vectors $\mathbf{v}$ in an inner-product space, $\mathbf{V}$, we have $(\mathbf{0}, \mathbf{v}) = (\mathbf{v}, \mathbf{0}) = 0$.

18. Using the fact that $\mathscr{T}'$ of theorem 3 is an *orthogonal* basis for $\mathbf{V}$, show that the set $\mathscr{T}''$ of this theorem is an ortho*normal* basis for $\mathbf{V}$.

Applications of Linear Algebra 8

In this chapter we will present some of the many applications of linear algebra. Each of these will make use of material introduced in the previous chapters of this text. Specifically the prerequisite material is as follows:

8.1 Polynomial Curve Fitting—chapter 2.

8.2 Applications to Physics and Economics: Kirchhoff's Laws and Leontief Models—chapter 2.

8.3 Markov Chains—section 2.3 for the first part, section 6.1 for the (optional) second part.

8.4 Quadric Surfaces—sections 6.1 and 6.2.

8.5 Approximation of Continuous Functions—chapter 7, especially section 7.4.

8.1 Polynomial Curve Fitting

In this section we will obtain polynomial approximations to a given set of data points in two different ways. In both cases the construction leads to the problem of solving a system of linear equations.

Recall that a **polynomial** in the variable x is a function of the form

$$P(x) = a_0 + a_1 x + a_2 x^2 + \cdots + a_n x^n,$$

where the a_i are scalars and n is a nonnegative integer. If a_n is not equal to zero the **degree** of the polynomial is equal to n. Polynomials of degree one, two, and three are called **linear, quadratic,** and **cubic,** respectively. The graph of a linear polynomial is a line in the plane while that of a quadratic polynomial is a parabola.

The Interpolating Polynomial

We know that two distinct points determine a line and that we can obtain the equation of this line by using the techniques of chapter 1. Another way of arriving at this equation is illustrated as follows.

Since a nonvertical line in the xy-plane may be represented by a linear polynomial, $P(x) = a_0 + a_1x$, we can find the equation of the line through the points (x_0, y_0) and (x_1, y_1) by determining the coefficients a_0 and a_1. Each point must satisfy the desired equation, so we have

$$a_0 + a_1x_0 = y_0,$$

and

$$a_0 + a_1x_1 = y_1.$$

But this is just a system of two linear equations in the two unknowns a_0 and a_1. Moreover, since the coefficient matrix

$$\begin{bmatrix} 1 & x_0 \\ 1 & x_1 \end{bmatrix},$$

row reduces to the identity matrix if $x_0 \neq x_1$ (if the given points do not determine a vertical line), the system has a unique solution (theorem 6 of section 2.5). Solving it, we obtain a_0 and a_1 and hence the desired linear polynomial.

Following this lead, let us consider the situation in which we are given three points (x_0, y_0), (x_1, y_1), and (x_2, y_2), with x_0, x_1, and x_2 distinct. It is unlikely (although possible) that a line will pass through all three points. On the other hand, it is reasonable to believe that we might be able to construct a quadratic polynomial, $P(x) = a_0 + a_1x + a_2x^2$, through these points, since P has *three* undetermined coefficients. Now, requiring the given points to satisfy this polynomial equation, we obtain the linear system

$$a_0 + a_1x_0 + a_2x_0^2 = y_0$$
$$a_0 + a_1x_1 + a_2x_1^2 = y_1$$
$$a_0 + a_1x_2 + a_2x_2^2 = y_2.$$

This linear system may be written in the matrix form $V\mathbf{a} = \mathbf{y}$, where

$$V = \begin{bmatrix} 1 & x_0 & x_0^2 \\ 1 & x_1 & x_1^2 \\ 1 & x_2 & x_2^2 \end{bmatrix}, \quad \mathbf{a} = \begin{bmatrix} a_0 \\ a_1 \\ a_2 \end{bmatrix}, \quad \text{and} \quad \mathbf{y} = \begin{bmatrix} y_0 \\ y_1 \\ y_2 \end{bmatrix}.$$

As in the two-point case we can show that this 3×3 linear system has a unique solution (exercise 17). Solving it, we obtain $\mathbf{a} = (a_0, a_1, a_2)$ and hence the desired polynomial, $P(x)$.

Example 8.1

Find a parabola that passes through the points $(0, -5)$, $(1, -1)$, and $(2, 5)$.

We must find a quadratic polynomial, $P(x) = a_0 + a_1x + a_2x^2$, where a_0, a_1, and a_2 satisfy

$$a_0 + 0a_1 + 0a_2 = -5$$
$$a_0 + 1a_1 + 1a_2 = -1$$
$$a_0 + 2a_1 + 4a_2 = 5.$$

Solving this linear system, we obtain $a_0 = -5$, $a_1 = 3$, and $a_2 = 1$. Thus the parabola is given by $P(x) = -5 + 3x + x^2$ (or $y = x^2 + 3x - 5$).

We now turn to the general case. We wish to find a polynomial of the form

$$P(x) = a_0 + a_1x + a_2x^2 + \cdots + a_nx^n$$

that passes through the $n + 1$ points (x_0, y_0), (x_1, y_1), $\ldots$, (x_n, y_n), where all x_i are distinct. Since each point lies on the curve $y = P(x)$, we obtain the following system of $n + 1$ linear equations in the $n + 1$ unknowns $a_0, a_1, \ldots, a_n$:

$$a_0 + a_1x_0 + a_2x_0^2 + \cdots + a_nx_0^n = y_0$$
$$a_0 + a_1x_1 + a_2x_1^2 + \cdots + a_nx_1^n = y_1 \tag{1}$$
$$\vdots \qquad \vdots \qquad \vdots \qquad \vdots \qquad \vdots$$
$$a_0 + a_1x_n + a_2x_n^2 + \cdots + a_nx_n^n = y_n.$$

To determine the desired polynomial, we solve this system, obtaining the coefficients of $P(x)$ and hence the polynomial itself.

The following theorem summarizes this discussion and demonstrates the uniqueness of the solution of system (1). Notice that the same notation is used as in the 3×3 case, but we demonstrate the uniqueness in a different manner.

Theorem 1 Let (x_0, y_0), (x_1, y_1), $\ldots$, (x_n, y_n) be $n + 1$ points with all x_i distinct. Then there is a unique polynomial

$$P(x) = a_0 + a_1x + \cdots + a_nx^n$$

such that $P(x_i) = y_i$ for $i = 0, 1, \ldots, n$. The vector of coefficients $\mathbf{a} = (a_0, a_1, \ldots, a_n)$ satisfies the linear system

$$V\mathbf{a} = \mathbf{y},$$

where $\mathbf{y} = (y_0, y_1, \ldots, y_n)$ and

$$V = \begin{bmatrix} 1 & x_0 & x_0^2 \cdots x_0^n \\ 1 & x_1 & x_1^2 \cdots x_1^2 \\ \vdots & \vdots & \vdots \quad\quad \vdots \\ 1 & x_n & x_n^2 \cdots x_n^n \end{bmatrix}.$$

Proof   The only part of the theorem that still needs proof is that the system $V\mathbf{a} = \mathbf{y}$ has a solution for $\mathbf{a}$ and that the solution is unique. By theorem 6 of section 2.5 both of these will be demonstrated if we can show that the linear homogeneous system $V\mathbf{c} = \mathbf{0}$ has only the trivial solution, $\mathbf{c} = \mathbf{0}$.

Suppose $\mathbf{c}$ is a solution of $V\mathbf{c} = \mathbf{0}$. Performing the matrix-vector multiplication and equating each resulting component to zero yields the $n + 1$ equations

$$c_0 + c_1 x_i + c_2 x_i^2 + \cdots + c_n x_i^n = 0, \quad i = 0, 1, \ldots, n.$$

Letting $Q(x) = c_0 + c_1 x + \cdots + c_n x^n$, we have $Q(x_i) = 0$ for $i = 0$, $1, \ldots, n$. Since the x_i are distinct, this implies that $Q(x) = 0$ has at least $n + 1$ roots. But the degree of Q is less than or equal to n, so Q must be the zero polynomial, that is, $\mathbf{c} = (c_0, c_1, \ldots, c_n) = \mathbf{0}$, as desired. Thus $V\mathbf{a} = \mathbf{y}$ has a unique solution. But any polynomial, $P(x)$, of degree less than or equal to n that satisfies $P(x_i) = y_i$ must satisfy the equations of system (1). Consequently such a polynomial is unique. ∎

Definition   Let (x_0, y_0), (x_1, y_1), $\ldots$, (x_n, y_n) be given points for which all x_i are distinct. The unique polynomial

$$P(x) = a_0 + a_1 x + a_2 x^2 + \cdots + a_n x^n$$

that passes through these points is called the **interpolating polynomial** for these points. The coefficient matrix, V, of system (1) is known as a **Vandermonde matrix.**

Example 8.2

A table for e^x has the following entries

x	1.0	1.1	1.2	1.3
e^x	2.7183	3.0042	3.3201	3.6693

Use this data to find an approximation to e^x at $x = 1.15$ by means of a cubic interpolating polynomial.

Here (x_0, y_0), (x_1, y_1), (x_2, y_2), and (x_3, y_3) are $(1.0, 2.7183)$, $(1.1, 3.0042)$, $(1.2, 3.3201)$, and $(1.3, 3.6693)$, respectively. We shall find the cubic interpolating polynomial, $P(x) = a_0 + a_1 x + a_2 x^2 + a_3 x^3$, for these points and then evaluate it at $x = 1.15$.

For this data, system (1) is

$$a_0 + (1.0)a_1 + (1.0)^2 a_2 + (1.0)^3 a_3 = 2.7183$$
$$a_0 + (1.1)a_1 + (1.1)^2 a_2 + (1.1)^3 a_3 = 3.0042$$
$$a_0 + (1.2)a_1 + (1.2)^2 a_2 + (1.2)^3 a_3 = 3.3201$$
$$a_0 + (1.3)a_1 + (1.3)^2 a_2 + (1.3)^3 a_3 = 3.6693$$

Solving this linear system for $a_0, a_1, a_2,$ and a_3, we obtain $a_0 = 0.7832$, $a_1 = 1.7002$, $a_2 = 0.3152$, and $a_3 = 0.5501$. Therefore the desired approximation to $P(1.15)$ is 3.1582. (The correct value of $e^{1.15}$ to four decimal places is also 3.1582.)

Note: The interpolating polynomial, $P_n(x)$, for the $n + 1$ points (x_0, y_0), $(x_1, y_1), \ldots, (x_n, y_n)$ has degree less than or equal to n. Depending on the given data set, it may or may not have degree exactly equal to n. For example, in examples 8.1 and 8.2 the interpolating polynomials for the given sets of three $(n = 2)$ and four $(n = 3)$ points did have degrees two and three, respectively. On the other hand, the unique polynomial of the form $P(x) = a_0 + a_1x + a_2x^2$ that passes through the three $(n = 2)$ points (1, 10), (3, 10), and (5, 10) is just $P(x) = 10$, a constant function. Here the degree of the interpolating polynomial is zero, strictly less than n.

The Least Square Polynomial

The technique of polynomial curve fitting is often used to obtain a relationship between a dependent and independent variable, say, y and x, in an experimental situation. The experiment is usually performed a relatively large number of times resulting in many data points, (x_i, y_i). If we tried to determine the interpolating polynomial for these points, we would have to do quite a bit of computing. Moreover data obtained experimentally is subject to measurement errors, and these errors would be passed on to the approximating interpolating polynomial. It would be better in this situation to construct a relatively low degree polynomial that passes *near* the data points rather than through them. This is the idea behind *least square approximations.*

As an example consider the simplest case—the *least square linear polynomial* or *least square line.* Let $(x_0, y_0), (x_1, y_1), \ldots, (x_m, y_m)$ be a given set of points with all x_i distinct, and let d_k denote the vertical distance from the line $y = ax + b$ to the point (x_k, y_k) (see figure 8.1). That is, let $d_k = |y_k - (ax_k + b)|$. We wish to determine a and b such that $S = d_0^2 + d_1^2 + \cdots + d_m^2$ is minimized. In other words, we seek a linear polynomial $Q(x) = ax + b$ that minimizes the expression

$$S = (y_0 - (ax_0 + b))^2 + (y_1 - (ax_1 + b))^2 + \cdots + (y_m - (ax_m + b))^2.$$
(2)

Note: It may seem more logical to seek a polynomial that minimizes the sum of the vertical distances, or perhaps the sum of the perpendicular distances from each point to the line. However, these alternatives result

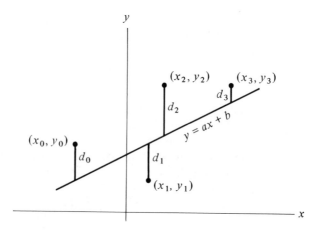

Figure 8.1

in a considerably more complicated analysis of the situation. Minimizing the expression S of equation (2) has a relatively simple as well as unique solution and works well in practical applications.

Instead of demonstrating how one can determine the least square line, we will go directly to the more general case of the nth-degree least square polynomial. Of course the least square line is just the special case of the latter, when $n = 1$.

Here we wish to determine a polynomial of the form

$$Q(x) = b_0 + b_1 x + b_2 x^2 + \cdots + b_n x^n$$

that minimizes the expression

$$S = (y_0 - Q(x_0))^2 + (y_1 - Q(x_1))^2 + \cdots + (y_m - Q(x_m))^2. \qquad (3)$$

Theorem 2 Let $(x_0, y_0), (x_1, y_1), \ldots, (x_m, y_m)$ be $m + 1$ points with all x_i distinct, and let n be a positive integer with $n \leq m$. Then there is a unique polynomial

$$Q(x) = b_0 + b_1 x + b_2 x^2 + \cdots + b_n x^n$$

that minimizes the expression S of equation (3). The vector of coefficients $\mathbf{b} = (b_0, b_1, \ldots, b_n)$ satisfies the linear system

$$(U^T U)\mathbf{b} = U^T \mathbf{y}, \qquad (4)$$

where $\mathbf{y} = (y_0, y_1, \ldots, y_m)$ and

$$U = \begin{bmatrix} 1 & x_0 & x_0^2 \cdots x_0^n \\ 1 & x_1 & x_1^2 \cdots x_1^n \\ \vdots & \vdots & \vdots \\ 1 & x_m & x_m^2 \cdots x_m^n \end{bmatrix}.$$

The proof of theorem 2 is similar to that of theorem 1 of section 8.5, so we will omit the proof.

Definition Given the points, (x_0, y_0), (x_1, y_1), $\ldots$, (x_m, y_m), the unique polynomial, $Q(x)$, of degree less than or equal to n that minimizes the expression S of equation (3) is called the nth-degree **least square polynomial** for these points. System (4) is called the set of **normal equations** for the given data.

Example 8.3

A student performs an experiment in an attempt to obtain an equation relating two physical quantities x and y. He performs the experiment six times, letting $x = $ 1, 2, 4, 6, 8, and 10 units. These experiments result in corresponding y values of 2, 2, 3, 4, 6, and 8 units, respectively. Find (a) the least square linear polynomial (least square line) and (b) the least square quadratic polynomial (least square parabola) for this data.

Using the notation of theorem 2, the given data is the set of six points (x_i, y_i), $i = 0, 1, \cdots, 5$: (1, 2), (2, 2), (4, 3), (6, 4), (8, 6), (10, 8). For part a we seek a polynomial of the form $Q_1(x) = b_0 + b_1 x$, while for part b we seek one of the form $Q_2(x) = b_0 + b_1 x + b_2 x^2$.

a. Here the vectors $\mathbf{b}$ and $\mathbf{y}$ are $\mathbf{b} = (b_0, b_1)$ and $\mathbf{y} = (2, 2, 3, 4, 6, 8)$, and the matrices U and U^T are

$$U = \begin{bmatrix} 1 & 1 \\ 1 & 2 \\ 1 & 4 \\ 1 & 6 \\ 1 & 8 \\ 1 & 10 \end{bmatrix}, \quad U^T = \begin{bmatrix} 1 & 1 & 1 & 1 & 1 & 1 \\ 1 & 2 & 4 & 6 & 8 & 10 \end{bmatrix}.$$

By computing $U^T U \mathbf{b}$ and $U^T \mathbf{y}$, we obtain system (4), $U^T U \mathbf{b} = U^T \mathbf{y}$, for part a:

$$6b_0 + 31b_1 = 25$$
$$31b_0 + 221b_1 = 170.$$

Solving this linear system, we see that $b_0 = 0.6986$ and $b_1 = 0.6712$ (approximately), so that the least square line for the data is

$$Q_1(x) = 0.6986 + 0.6712x.$$

b. Here we have $\mathbf{b} = (b_0, b_1, b_2)$, $\mathbf{y} = (2, 2, 3, 4, 6, 8)$, and

$$U = \begin{bmatrix} 1 & 1 & 1 \\ 1 & 2 & 4 \\ 1 & 4 & 16 \\ 1 & 6 & 36 \\ 1 & 8 & 64 \\ 1 & 10 & 100 \end{bmatrix}, \quad U^T = \begin{bmatrix} 1 & 1 & 1 & 1 & 1 & 1 \\ 1 & 2 & 4 & 6 & 8 & 10 \\ 1 & 4 & 16 & 36 & 64 & 100 \end{bmatrix}$$

Therefore system (4) for part b is

$$
\begin{aligned}
6b_0 + \quad 31b_1 + \quad 221b_2 &= \quad 25 \\
31b_0 + \quad 221b_1 + \quad 1801b_2 &= \quad 170 \\
221b_0 + 1801b_1 + 15665b_2 &= 1386,
\end{aligned}
$$

which has the approximate solution $b_0 = 1.836$, $b_1 = 0.0276$, and $b_2 = 0.0594$. Thus the least square parabola for this data is

$$Q_2(x) = 1.836 + 0.0276x + 0.0594x^2.$$

(In figure 8.2 we have plotted the given data of this example and the least square line and parabola for this data.)

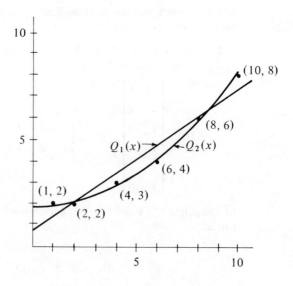

Figure 8.2

Note: Theorem 2 is a generalization of theorem 1. If in the former we have $m = n$, the matrix U is square and $U = V$, the Vandermonde matrix. Moreover, since V is invertible, the linear systems $V\mathbf{a} = \mathbf{y}$ and $V^T V\mathbf{a} = V^T \mathbf{y}$ are equivalent (just multiply both sides of the latter on the left by $(V^T)^{-1}$). Thus in the case that $m = n$ the interpolating polynomial and the least square polynomial are identical. Another way to see this last statement is to notice that, when $m = n$, the interpolating polynomial will result in a minimum value (namely, 0) for equation (3) since it passes *through* all the data points.

8.1 Exercises

In exercises 1 through 6 find a polynomial of the specified degree that passes through the given points.

1. $\{(0, -3), (1, 0), (2, 5)\}, n = 2$
2. $\{(-1, 4), (1, 2), (3, -4)\}, n = 2$
3. $\{(0, -1), (1, -1), (2, 3), (3, 17)\}, n = 3$
4. $\{(-1, 4), (0, 1), (1, 0), (3, 4)\}, n = 3$
5. $\{(-2, -15), (-1, -2), (0, 1), (1, 0), (2, 1)\}, n = 4$
6. $\{(0, -3), (1, -9), (2, -3), (3, 87), (4, 381)\}, n = 4$

A table of logarithms contains the following entries:

x	1.0	2.0	3.0	4.0
$\log_{10} x$	0.0000	0.3010	0.4771	0.6021

In exercises 7 and 8 approximate $\log_{10}(2.5)$ by constructing an interpolating polynomial.

7. Of degree 2 using the entries at $x = 1.0, 2.0$, and 3.0
8. Of degree 3 using all the entries

In exercises 9 through 12 find the least square polynomial of the specified degree for the given data points.

9. $\{(0, 0), (1, 0), (2, 1), (3, 3), (4, 5)\}, n = 1$
10. $\{(x, y) \mid y = x^3 \text{ and } x = 0, 1, 2, 3, 4, 5\}, n = 1$
11. $\{(x, y) \mid y = x^3 \text{ and } x = 0, 1, 2, 3, 4, 5\}, n = 2$
12. $\{(-2, 2), (-1, 1), (0, -1), (1, 0), (2, 3)\}, n = 2$

In exercises 13 and 14 determine the system of normal equations that occurs in the process of finding the least square polynomial of specified degree for the given data.

13. $\{(x, y) \mid y = x^4 - 8x^2, x = -3, -2, -1, 0, 1, 2, 3\}, n = 3$
14. $\{(0, 0), (1, 0), (2, 1), (3, 3), (4, 5), (5, 4), (6, 5)\}, n = 4$

A firm that manufactures widgets finds the daily consumer demand $d(x)$ for widgets as a function of their price x is as in the following table:

x	1	1.5	2	2.5	3
$d(x)$	200	180	150	100	25

In exercises 15 and 16, using least square polynomials, approximate the daily consumer demand.

15. By a linear function

16. By a quadratic function

17. Let

$$V = \begin{bmatrix} 1 & x_0 & x_0^2 \\ 1 & x_1 & x_1^2 \\ 1 & x_2 & x_2^2 \end{bmatrix},$$

where x_0, x_1, and x_2 are distinct numbers. Show that the row-reduced echelon form of V is the 3×3 identity matrix.

8.2 Applications to Physics and Economics: Kirchhoff's Laws and Leontief Models

In this section we will briefly examine two situations in which systems of linear equations arise in a natural way. Specifically we consider Kirchhoff's laws for electrical circuits and Leontief models for input-output analysis.

Kirchhoff's Laws

We will consider dc electrical circuits consisting of one or more *electromotive forces* (emf) and one or more *resistors*. Electromotive force, E, is measured in *volts* (denoted by V) and resistance, R, is measured in *ohms* (Ω). The relationship between emf, resistance, and *current*, I (measured in *amperes* or *amps*) in a simple circuit is given by *Ohm's Law*:

$$E = IR.$$

For example, in a circuit containing a simple emf of 6 volts and a single resistance of 30 ohms, as in figure 8.3, the current is calculated to be $6/30 = 1/5$ amp. The direction of the current is by convention assumed to be from $+$ to $-$ and is indicated by the arrow in figure 8.3.

 Assuming that the wires in a circuit have negligible resistance, the *voltage drop* across the resistor of figure 8.3 must equal the applied emf,

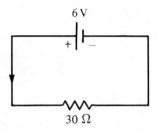

Figure 8.3

Ohm's law—since the current across the 30 Ω resistor is 1/5 amp, the voltage drop is $(1/5)(30) = 6$ V.) In somewhat more complicated circuits, such as that of figure 8.4, the sum of the voltage drops (across the resistors) must equal the *net* emf (the algebraic sum of the emfs). Here the net emf is 2 volts (8 V − 6 V), since the two emf's are in opposition to one another. Equating total voltage drop to net emf, we obtain

$$6I + 4I + 2I = 2$$

or $I = 1/6$ amp (flowing in the direction indicated).

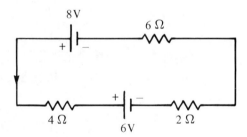

Figure 8.4

When a circuit consists of more than one *loop* (that is, *closed path*), we use Kirchhoff's laws to determine the current in any branch. In such circuits there will be points, called *nodes*, where three or more wires come together. For example, the circuit of figure 8.5 has three loops, *ABCD*, *ADEF*, and *ABCDEF*, and two nodes, *A* and *D*.

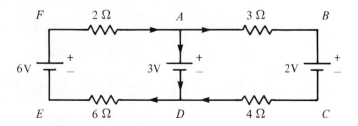

Figure 8.5

Kirchhoff's Laws

i. The algebraic sum of currents at any node in a circuit is zero.
ii. The algebraic sum of the voltage drops around any loop of the circuit is equal to the algebraic sum of the electromotive forces in that loop.

Note: It is not necessary to guess correctly the direction of the current in any given wire of a complicated circuit. An arbitrary assignment is made, and, if the current turns out to be negative, it simply means that the guess was wrong—the current flows in the opposite direction.

Example 8.4

Find the current in each wire of the circuit of figure 8.5.

Let I_1 be the current from D to E to F to A, let I_2 be the current from A to B to C to D, and let I_3 be the current from A to D.

Since A is a node, and I_1 has been assigned to flow into A while I_2 and I_3 flow out of A, we have by the first law,

$$I_1 - I_2 - I_3 = 0. \tag{1}$$

Since D is a node, and I_1 flows out of D while I_2 and I_3 flow into D, we have

$$I_2 + I_3 - I_1 = 0. \tag{2}$$

From loop $ABCD$, using the second law, we have

$$3I_2 + 4I_2 = 3 - 2,$$

or more simply

$$7I_2 = 1. \tag{3}$$

From loop $ADEF$ we have

$$2I_1 + 6I_1 = 6 - 3,$$

or more simply

$$8I_1 = 3. \tag{4}$$

From loop $ABCDEF$ we have

$$3I_2 + 4I_2 + 6I_1 + 2I_1 = 6 - 2,$$

or more simply

$$8I_1 + 7I_2 = 4. \tag{5}$$

Equations (1) through (5) then provide the following system of linear equations:

$$I_1 - I_2 - I_3 = 0$$
$$-I_1 + I_2 + I_3 = 0$$

$$7I_2 \quad = 1$$
$$8I_1 \qquad = 3$$
$$8I_1 + 7I_2 \quad = 4.$$

Following the methods of section 2.2, we have the unique solution $I_1 = 3/8$, $I_2 = 1/7$, $I_3 = 13/56$.

Input-Output Analysis: Closed Leontief Model

Input-output analysis is the study of the economics of situations involving production and consumption of various goods and services. The method of investigation described is due to the famous economist Wassily Leontief. We will consider two situations, first a closed model and then an open one.

A **closed economic model** is one in which each item (or value of service) produced by one segment of the economy is consumed by another segment of the same economy, in other words, there are no stored surpluses (or monetary equivalents) and no exports from the economy. We further assume that each segment of the economy requires a fixed fraction of the production of the other segments. The problem is to determine whether or not it is possible to have such a system with no "left-overs" and, if so, how much should each segment produce in order to achieve this state.

If we number the segments of the economy, say from 1 to n, the *input-output matrix* of the economy is defined to be the $n \times n$ matrix A in which each entry a_{ij} is the fraction of the total production of segment j needed to supply the needs of segment i. (Since all of any segment's production must be consumed, the sum of the entries in any column must be 1.)

Example 8.5

Assume the economy has three segments, farming, manufacturing, and service (such as repairing or transporting) and assume that 1/3 of the food produced by the farming segment is consumed by farmers themselves, the manufacturers consume 1/2 of it, and the remaining 1/6 is consumed by those offering services. Further assume that farmers require 1/2 of the manufactured goods, with the rest split evenly between the other two industries. Finally assume that all three segments have equal need of services.

By numbering the segments in order, we have

$$A = \begin{bmatrix} 1/3 & 1/2 & 1/3 \\ 1/2 & 1/4 & 1/3 \\ 1/6 & 1/4 & 1/3 \end{bmatrix}.$$

Let $\mathbf{x} = (x_1, x_2, x_3)$ be the dollar values associated with the output of each segment. Since the value produced by the farming segment must equal the sum of the amounts spent by farmers to fulfill their needs from the other segments (including other farmers), we have

$$x_1 = (1/3)x_1 + (1/2)x_2 + (1/3)x_3.$$

Similar equations for the expenditures by manufacturing and by service are

$$x_2 = (1/2)x_1 + (1/4)x_2 + (1/3)x_3,$$
$$x_3 = (1/6)x_1 + (1/4)x_2 + (1/3)x_3.$$

In matrix form these three equations can be written as

$$\mathbf{x} = A\mathbf{x},$$

and, since $\mathbf{x} = I\mathbf{x}$, we have the homogeneous linear system

$$(I - A)\mathbf{x} = \mathbf{0}.$$

Solving this homogeneous system, we obtain

$$x_1 = (5/3)s, \quad x_2 = (14/9)s, \quad x_3 = s.$$

Since these numbers are to represent the value of goods or services, only nonnegative solutions are realistic. Hence the system "balances" for any $s > 0$, and x_1, x_2, and x_3 as shown.

Returning to general considerations, let A be an input-output matrix, and let $\mathbf{x} = (x_1, \ldots, x_n)$ be the vector of values associated with the output of each segment. We wish to find the relative size of each segment of the economy.

Since the value produced by a given segment must equal the sum of the amounts spent by that segment to fulfill its needs from each of the other segments, we have the following system of equations:

$$x_1 = a_{11}x_1 + a_{12}x_2 + \cdots + a_{1n}x_n$$
$$x_2 = a_{21}x_1 + a_{22}x_2 + \cdots + a_{2n}x_n$$
$$\vdots \qquad \vdots \qquad \vdots \qquad \qquad \vdots$$
$$x_n = a_{n1}x_1 + a_{n2}x_2 + \cdots + a_{nn}x_n.$$

In matrix form this becomes

$$\mathbf{x} = A\mathbf{x},$$

and as in example 8.5 we have

$$(I - A)\mathbf{x} = \mathbf{0}.$$

This homogeneous system has the trivial solution (nobody produces anything!), but since each column of $I - A$ adds up to 0, the system has rank at most $n - 1$, so nontrivial solutions exist. To be realistic, we can only allow nonnegative values for the entries of $\mathbf{x}$ since they are dollar values of output. The following theorem, which we assume without proof, tells us that there are realistic, nontrivial solutions.

Theorem 1   Let A be the input-output matrix of a closed model. Then $(I - A)\mathbf{x} = \mathbf{0}$ has a solution in which all of its entries are positive.

Input-Output Analysis: Open Model

An **open economic model** is similar in structure to that of the closed model except that one extra segment called *consumers* is added. The consumer segment is assumed to have a certain demand for each of the goods or services produced by the other segments. These demands are subject to change, however, so we wish to analyze the problem of what to do if the demands change. The other assumptions remain exactly as in the closed case—each segment requires a fixed fraction of the total output of each of the other segment's total production, and there are no surpluses (above consumer demand) produced.

Again we define the *input-output matrix A* of the economy in terms of fractions of production but with a slight change. We let a_{ij} be the fraction of the production of segment j needed to supply the needs of segment i. Thus the matrix A is the same as that of a closed economy in case all consumer demands equal zero. Letting $\mathbf{x} = (x_1, \ldots, x_n)$ be the production vector, $A\mathbf{x}$ is the vector of internal consumption. For instance, the first row of $A\mathbf{x}$ is

$$a_{11}x_1 + a_{12}x_2 + \cdots + a_{1n}x_n.$$

This represents the fraction of product 1 consumed by segment 1, plus the fraction of product 1 consumed by segment 2, and so on. The resulting sum is the amount of product 1 consumed by all of the producing segments. Similarly the ith row of $A\mathbf{x}$ is the amount *produced* by the ith segment which is *consumed* by the producing segments.

Since there is an assumption that no surplus exists, the internal consumption plus the demand for each product must equal the total production of that product. Letting $\mathbf{d}$ be the vector of demands, with d_i being the demand for the ith product, we have for each i,

$$a_{i1}x_1 + a_{i2}x_2 + \cdots + a_{in}x_n + d_i = x_i.$$

In terms of matrices this system becomes

$$A\mathbf{x} + \mathbf{d} = \mathbf{x}.$$

Transposing $\mathbf{x}$, we have

$$(I - A)\mathbf{x} = \mathbf{d}$$

and, if $I - A$ is invertible,

$$\mathbf{x} = (I - A)^{-1}\mathbf{d}.$$

As in the case of the closed economy we can only accept vectors $\mathbf{x}$ for which each entry is nonnegative. The following theorem (assumed without proof) gives fairly general conditions under which a solution can be obtained.

Theorem 2   If a square matrix A has all positive entries, and the sum of the entries in each of its columns is less than one, then $(I - A)$ is invertible.

Example 8.6

In the economy of farming, manufacturing, and service industries of example 8.5, suppose there is also a consumer segment. Table 8.1 gives the dollar value of each product that is consumed by a given segment. That is, an entry in the ith row of table 8.1 gives the value (in millions of dollars) of the product associated with that row which is consumed by the industry associated with that column. If the anticipated consumer demands for five years later are farming, 20, manufacturing, 30, and service, 20, find the amount each segment needs to produce in order to balance the system.

	Farming	Manufacturing	Service	Consumer	Total
Farming	10	15	5	10	40
Manufacturing	14	7	7	21	49
Service	6	6	12	6	30

Table 8.1

The input-output matrix A for the economy is computed as follows: a_{ij} is that fraction of the total for the jth segment consumed by the ith segment (ignoring the consumer segment). For example the first column of A is $A^1 = (1/4, 3/8, 1/8)$ since farming consumes 10 of the total 40, manufacturing 15, and service 5. Computing the other entries, we have

$$A = \begin{bmatrix} 1/4 & 2/7 & 1/5 \\ 3/8 & 1/7 & 1/5 \\ 1/8 & 1/7 & 2/5 \end{bmatrix}.$$

Letting $\mathbf{d} = (20, 30, 20)$ be the vector of demands, we determine the solution to be

$$\mathbf{x} = (I - A)^{-1}\mathbf{d} = \begin{bmatrix} 3/4 & -2/7 & -1/5 \\ -3/8 & 6/7 & -1/5 \\ -1/8 & -1/7 & 3/5 \end{bmatrix}^{-1} \begin{bmatrix} 20 \\ 30 \\ 20 \end{bmatrix} = \begin{bmatrix} 5680/73 \\ 6230/73 \\ 5100/73 \end{bmatrix}.$$

Thus the approximate value of the farming segment should be $x_1 = 77.8$, manufacturing, $x_2 = 85.3$, and service, $x_3 = 69.9$.

8.2 Exercises

In exercises 1 and 2 let the circuit be given by figure 8.6. Calculate the current through each resistor using the given values of resistances and emf's.

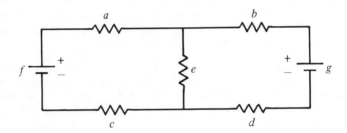

Figure 8.6

1. $a = 3\,\Omega, b = 4\,\Omega, c = 3\,\Omega, d = 2\,\Omega, e = 1\,\Omega, f = 6$ V, $g = 2$ V
2. $a = 4\,\Omega, b = 4\,\Omega, c = 5\,\Omega, d = 6\,\Omega, e = 4\,\Omega, f = 8$ V, $g = 10$ V

In exercises 3 and 4 let the circuit be given by Figure 8.7. Calculate the current in each branch of the circuit using the given values of resistances and emf's.

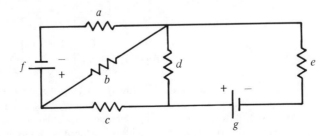

Figure 8.7

3. $a = 3\,\Omega, b = 4\,\Omega, c = 3\,\Omega, d = 2\,\Omega, e = 1\,\Omega, f = 6\,V, g = 2\,V$

4. $a = 4\,\Omega, b = 4\,\Omega, c = 5\,\Omega, d = 6\,\Omega, e = 4\,\Omega, f = 8\,V, g = 10\,V$

In exercises 5 and 6 the given matrix M is the input-output matrix of a closed economy of three segments A, B, and C. Find the value of production for each segment.

5.
$$M = \begin{bmatrix} 3/4 & 1/5 & 1/3 \\ 1/8 & 2/5 & 1/6 \\ 1/8 & 2/5 & 1/2 \end{bmatrix}$$

6.
$$M = \begin{bmatrix} 0.2 & 0.1 & 0.3 \\ 0.4 & 0.8 & 0.3 \\ 0.4 & 0.1 & 0.4 \end{bmatrix}$$

7. Assume a closed economy has three segments A, B, and C, with A requiring $1/2$ of its own production, and B and C splitting the rest evenly. All three segments require the same fraction of the production of B. Segment A requires $2/3$ of the production of C, while B and C require the same amount of C. Find the value of production for each segment.

8. The closed Leontief model problem is sometimes viewed as a problem in *pricing* rather than *production*. The assumption is that in a closed model each segment requires not only a fixed *fraction* of the production of the other segments but a fixed *amount* of the production. Viewed in this manner the problem is to decide how each segment should price its production in order to spend exactly what it earns. Explain the result of exercise 7 in this context.

In exercises 9 and 10 the given matrix M is the input-output matrix of an open economy of three segments A, B, and C, and the vector $\mathbf{d}$ is the predicted demand vector of the consumer segment. Find the production level for each segment needed to balance the economy.

9.
$$M = \begin{bmatrix} 1/3 & 1/5 & 1/6 \\ 1/4 & 2/5 & 1/3 \\ 1/4 & 1/5 & 1/6 \end{bmatrix}, \mathbf{d} = \begin{bmatrix} 3 \\ 5 \\ 1 \end{bmatrix}$$

10.
$$M = \begin{bmatrix} 0.2 & 0.1 & 0.5 \\ 0.3 & 0.2 & 0.3 \\ 0.1 & 0.1 & 0.1 \end{bmatrix}, \mathbf{d} = \begin{bmatrix} 20 \\ 15 \\ 5 \end{bmatrix}$$

11. Assume an open economy has three producing segments A, B, and C. Table 8.2 gives the current value of the production of the segment of each row that is consumed by the segment of each column. Calculate the production required of each industry needed to provide a balanced economy if the consumer demand rises to $\mathbf{d} = (15, 10, 6)$.

	A	*B*	*C*	Consumers
A	12	4	8	10
B	8	9	4	6
C	2	4	3	3

Table 8.2

8.3 Markov Chains

In this section we will demonstrate how multiplication of a vector by a matrix can be used to compute the probability of certain types of future events. Multiplication of matrices is the only requisite knowledge for most of the section. The latter part concerning equilibrium states presupposes a knowledge of eigenvalues and eigenvectors, section 6.1.

Markov Chains

We first need to recall a few facts about probabilities in general. The probability of a random event is a number between 0 and 1 (inclusive) that measures the likelihood that the event will occur. For example, the probability of rolling the number 1 with a single die is 1/6 since there is only one "successful" side out of the six sides of the die and all outcomes are equally likely. The probability of rolling an even number, however, is 1/2 since any of 2, 4, or 6 will be successful out of the six possible sides. Probability 0 implies that the event cannot occur, and probability 1 implies that the event must occur. As examples, the probability of rolling 7 with one die is 0, and the probability of rolling a number less than 10 with a single die is 1.

To compute probabilities of more complicated events, we use the following two facts. If two events are *independent*, the probability of both occurring is the *product* of the individual probabilities. For example, the probability of flipping a coin twice and obtaining heads both times is $(1/2)(1/2) = 1/4$. (The coin can't remember what it did before so the events are independent.) If two events are *disjoint* (that is, they cannot both occur), the probability of one *or* the other occurring is the *sum* of the individual probabilities. For example, the probability of rolling an even number on a die was shown to be $3/6 = 1/2$. This can be broken up into the three disjoint events—roll a 2, roll a 4, or roll a 6. Each of these has probability 1/6 and $1/6 + 1/6 + 1/6 = 1/2$.

We will now describe the setting for Markov chains. Repeated observations of a situation are made at regular intervals and the current state of the situation is recorded. As an example, suppose the weather is observed every 6 hours and the possible states of the weather are chosen to be fair (F), partly cloudy (P), cloudy (C), and rain (R). Based on past information, we assign to each state a probability of moving to some other state (or remaining the same). An important (and sometimes unrealistic) assumption is that these probabilities depend *only* on the current state and not on any previous ones. Continuing with the weather example, suppose the probabilities of moving from one weather state to another are as given in table 8.3.

		Next state			
		F	P	C	R
	F	0.6	0.3	0.1	0.0
Current state	P	0.2	0.4	0.2	0.2
	C	0.1	0.4	0.3	0.2
	R	0.0	0.2	0.3	0.5

Table 8.3

For instance, if it is cloudy now, the probability that it will be raining at the next observation is 0.2. Notice that each entry is nonnegative, (probabilities must lie between 0 and 1) and that the sum of the entries in each row is 1 (no matter what the initial state is, *some* state must occur at the next observation and the events are disjoint). Any such row is called a **probability vector,** a vector with nonnegative entries the sum of which are 1.

Our prediction for the weather will also be stated in terms of a probability vector, $\mathbf{x} = (x_1, x_2, x_3, x_4)$, where x_1 is the probability that the weather will be fair, x_2 is the probability of partly cloudy, and so on. Since *some* weather must occur, $x_1 + x_2 + x_3 + x_4 = 1$. For instance, if it is raining now, the prediction for 6 hours hence is (0.0, 0.2, 0.3, 0.5). What is the prediction for the following time period? Or more generally, if $\mathbf{x}$ is the prediction for one time period, what is the prediction for the following one?

To answer this question, we reason as follows: let $\mathbf{y} = (y_1, y_2, y_3, y_4)$ be the prediction of probabilities in the next time period. That is, y_1 is the probability that the weather will be fair, y_2 is the probability that it will be partly cloudy, and so on. Looking just at y_1, there are four different ways of getting fair weather in the next time period. It could be fair now and stay fair or it could change from one of three other states to fair. If it is fair, then the probability that it stays fair is 0.6. However, there is a probability of x_1 that it is fair, and moreover these events are independent. Thus the probability that it is fair and stays fair (FF) is $(x_1)(0.6)$. Similarly the probability that it is partly cloudy and changes to fair (PF) is $(x_2)(0.2)$. Listing all of the possibilities, we have

Probability of FF is $0.6x_1$,

Probability of PF is $0.2x_2$,

Probability of CF is $0.1x_3$,

Probability of RF is $0.0x_4$.

These four events are disjoint, so the probability that the weather is fair in the next time period is

$$y_1 = 0.6x_1 + 0.2x_2 + 0.1x_3 + 0.0x_4.$$

It should be noted that y_1 is exactly the dot product of the first column of table 8.3 with the vector $\mathbf{x}$. Carrying out a similar analysis for each of y_2, y_3, and y_4, we obtain y_i as the dot product of the ith column of the table with the vector $\mathbf{x}$. In order to make this conform to our usual convention for matrix multiplication, let T (for **transition matrix**) be the *transpose* of the entries in the table. Then this discussion yields the result that

$$\mathbf{y} = T\mathbf{x}. \tag{1}$$

Using equation (1), we can answer the question posed earlier. If it is raining now, what is the probability vector for two periods in the future? Letting $\mathbf{x} = (0, 0.2, 0.3, 0.5)$, the prediction for the first time period, we have

$$\mathbf{y} = T\mathbf{x} = \begin{bmatrix} 0.6 & 0.2 & 0.1 & 0.0 \\ 0.3 & 0.4 & 0.4 & 0.2 \\ 0.1 & 0.2 & 0.3 & 0.3 \\ 0.0 & 0.2 & 0.2 & 0.5 \end{bmatrix} \begin{bmatrix} 0.0 \\ 0.2 \\ 0.3 \\ 0.5 \end{bmatrix} = \begin{bmatrix} 0.07 \\ 0.30 \\ 0.28 \\ 0.35 \end{bmatrix}.$$

Since $\mathbf{x}$ is equal to $T(0, 0, 0, 1)$ we have another way of computing $\mathbf{y}$, $\mathbf{y} = T^2(0, 0, 0, 1)$. Similarly, if we wanted the probability vector for the following time period, equation (1) tells us that this prediction is given by

$$T\mathbf{y} = T(T^2(0, 0, 0, 1)) = T^3(0, 0, 0, 1).$$

Thus we see that, to go from any given state to any later state, we simply multiply the given vector by a power of T. The results of this specific example are generalized as the following procedure.

Markov chain procedure Suppose there are n possible states:

i. Prepare an $n \times n$ matrix where the ijth entry is the probability of moving from state i to state j.

ii. Let T be the transpose of this matrix.

iii. For any probability vector $\mathbf{x}$ in $\mathbf{R}^n$ and for any natural number k, the probability vector for k time periods in the future is given by $\mathbf{y} = T^k \mathbf{x}$.

Note: The vector $\mathbf{x}$ in the procedure is called an **initial vector** and is presented as a probability vector. If in fact we *know* the initial state, say, the ith state, then $\mathbf{x} = \mathbf{e}_i$, the ith elementary vector. Of course $\mathbf{e}_i$ is itself a probability vector; the probability of the ith state is 1 and of any other is 0. As an example, in the preceding discussion, if the initial state is cloudy, the probability vector is (0, 0, 1, 0).

Example 8.7

Suppose that a particular telephone wire is capable of supporting up to four calls at any one time and that the number of calls that will be in progress one minute later than at a given time depends only upon the number of calls in progress at the given time. Suppose further that the probability of moving from i calls at a given time to j calls one minute later is given in table 8.4. If at some particular time all possible numbers of calls are equally likely, what is the probability that the line will be unavailable (four calls in progress) three minutes later?

		j			
	0	1	2	3	4
0	0.3	0.3	0.2	0.2	0.0
1	0.2	0.3	0.3	0.1	0.1
i 2	0.1	0.2	0.4	0.2	0.1
3	0.1	0.1	0.2	0.3	0.3
4	0.0	0.0	0.3	0.3	0.4

Table 8.4

We follow the Markov chain procedure stated earlier. Step i is already done. The matrix T is the transpose of the given array,

$$T = \begin{bmatrix} 0.3 & 0.2 & 0.1 & 0.1 & 0.0 \\ 0.3 & 0.3 & 0.2 & 0.1 & 0.0 \\ 0.2 & 0.3 & 0.4 & 0.2 & 0.3 \\ 0.2 & 0.1 & 0.2 & 0.3 & 0.3 \\ 0.0 & 0.1 & 0.1 & 0.3 & 0.4 \end{bmatrix}.$$

Since all numbers of calls are equally likely, the initial vector is

$$\mathbf{x} = (0.2, 0.2, 0.2, 0.2, 0.2).$$

The probability vector for three minutes hence is then

$$\mathbf{y} = T^3\mathbf{x} = \begin{bmatrix} 0.3 & 0.2 & 0.1 & 0.1 & 0.0 \\ 0.3 & 0.3 & 0.2 & 0.1 & 0.0 \\ 0.2 & 0.3 & 0.4 & 0.2 & 0.3 \\ 0.2 & 0.1 & 0.2 & 0.3 & 0.3 \\ 0.0 & 0.1 & 0.1 & 0.3 & 0.4 \end{bmatrix}^3 \begin{bmatrix} 0.2 \\ 0.2 \\ 0.2 \\ 0.2 \\ 0.2 \end{bmatrix} = \begin{bmatrix} 0.128 \\ 0.174 \\ 0.292 \\ 0.222 \\ 0.184 \end{bmatrix}.$$

The probability of having four calls in progress at that time is the last coordinate of $\mathbf{y}$, $y_5 = 0.184$.

Equilibrium State

Definition   A probability vector $\mathbf{x}$ is an **equilibrium** or **steady state vector** if $\mathbf{x} = T\mathbf{x}$, where T is the transition matrix of a Markov chain.

In other words, an equilibrium vector is such that moving from one time period to the next results in no change. Every Markov chain has an equilibrium vector, and some have more than one. The following theorem shows how to determine them.

Theorem 1   Let T be the transition matrix of a Markov chain. Then $\lambda = 1$ is an eigenvalue of T. If $\mathbf{x}$ is an eigenvector corresponding to $\lambda = 1$ with non-negative components, and s is the sum of the components of $\mathbf{x}$, then $(1/s)\mathbf{x}$ is an equilibrium vector.

Proof   Let there be n states so that T is an $n \times n$ matrix. Since the sum of the entries in each column is 1, the sum of the entries in each column of $T - I$ is 0. This implies that, if each row is added to the last row, the result is a zero row, and hence the resulting matrix has determinant zero. But adding one row to another does not change the value of the determinant (theorem 4 of section 3.2), so the determinant of $T - I$ is 0. Since $T - I = T - 1I$, $\lambda = 1$ is an eigenvalue of T.

If $\mathbf{x}$ is an eigenvector corresponding to $\lambda = 1$, then $T\mathbf{x} = \mathbf{x}$. Assuming further that $\mathbf{x}$ has nonnegative components, and that $s = x_1 + \cdots + x_n$, the sum of the components of $\mathbf{x}$, we have $T((1/s)\mathbf{x}) = (1/s)T\mathbf{x} = (1/s)\mathbf{x}$. Now $(1/s)\mathbf{x}$ has nonnegative components, and the sum of the components is $(1/s)(x_1 + \cdots + x_n) = (1/s)s = 1$. Therefore $(1/s)\mathbf{x}$ is a probability vector, and consequently it is an equilibrium vector.   ∎

Example 8.8

Determine an equilibrium vector for a Markov chain with transition matrix

$$T = \begin{bmatrix} 1/3 & 1/3 & 1/3 \\ 1/2 & 1/6 & 0 \\ 1/6 & 1/2 & 2/3 \end{bmatrix}.$$

We first determine the eigenvectors of T corresponding to 1. To do so, we solve the homogeneous system $(T - I)\mathbf{x} = \mathbf{0}$ by row-reducing $T - I$,

$$T - I = \begin{bmatrix} -2/3 & 1/3 & 1/3 \\ 1/2 & -5/6 & 0 \\ 1/6 & 1/2 & -1/3 \end{bmatrix} \rightarrow \begin{bmatrix} 1 & 0 & -5/7 \\ 0 & 1 & -3/7 \\ 0 & 0 & 0 \end{bmatrix}.$$

From the row-reduced echelon form we see that $(5/7, 3/7, 1)$ is an eigenvector that corresponds to the eigenvalue 1. An (in fact, *the*) equilibrium vector is therefore

$$(1/(5/7 + 3/7 + 1))(5/7, 3/7, 1) = (1/3, 1/5, 7/15).$$

Often there is a *unique* equilibrium vector for a Markov chain, and succeeding stages of the chain give resulting vectors that are successively closer to this equilibrium vector and converge to it. The following theorem offers a very simple test for this situation. The proof is omitted.

Theorem 2 Let T be the transition matrix of a Markov chain. If some power of T has strictly positive entries, then there is a unique equilibrium vector $\mathbf{x}$, and for any probability vector $\mathbf{y}$, $T^n\mathbf{y}$ converges to $\mathbf{x}$ as n becomes large.

Example 8.9

Show that some power of the transition matrix T of example 8.7,

$$T = \begin{bmatrix} 0.3 & 0.2 & 0.1 & 0.1 & 0.0 \\ 0.3 & 0.3 & 0.2 & 0.1 & 0.0 \\ 0.2 & 0.3 & 0.4 & 0.2 & 0.3 \\ 0.2 & 0.1 & 0.2 & 0.3 & 0.3 \\ 0.0 & 0.1 & 0.1 & 0.3 & 0.4 \end{bmatrix},$$

has strictly positive entries so that there is a unique equilibrium vector. Find an approximation to the equilibrium vector by computing values of $T^n\mathbf{y}$ for some $\mathbf{y}$ and various values of n.

The second power of T, T^2, has strictly positive entries since

$$T^2 = \begin{bmatrix} 0.19 & 0.16 & 0.13 & 0.10 & 0.06 \\ 0.24 & 0.22 & 0.19 & 0.13 & 0.09 \\ 0.27 & 0.30 & 0.31 & 0.28 & 0.30 \\ 0.19 & 0.19 & 0.21 & 0.25 & 0.27 \\ 0.11 & 0.13 & 0.16 & 0.24 & 0.28 \end{bmatrix}.$$

Estimating the equilibrium vector by raising T to successively higher powers (with the aid of a computer), we see that T^7 and T^8 agree to the nearest hundredth.

$$T^7 = \begin{bmatrix} 0.125 & 0.123 & 0.123 & 0.122 & 0.121 \\ 0.170 & 0.170 & 0.169 & 0.167 & 0.170 \\ 0.295 & 0.295 & 0.295 & 0.295 & 0.295 \\ 0.224 & 0.224 & 0.224 & 0.225 & 0.226 \\ 0.188 & 0.188 & 0.189 & 0.190 & 0.192 \end{bmatrix},$$

$$T^8 = \begin{bmatrix} 0.123 & 0.123 & 0.123 & 0.122 & 0.122 \\ 0.170 & 0.169 & 0.169 & 0.168 & 0.168 \\ 0.295 & 0.295 & 0.295 & 0.295 & 0.295 \\ 0.224 & 0.224 & 0.224 & 0.225 & 0.225 \\ 0.189 & 0.189 & 0.189 & 0.190 & 0.190 \end{bmatrix}.$$

Letting $\mathbf{y} = (1, 0, 0, 0, 0)$, we take $T^8 \mathbf{y} = (0.123, 0.170, 0.295, 0.224, 0.189)$ as our approximation to the equilibrium vector.

Computational note: Estimating the equilibrium vector $\mathbf{x}$ by setting $\mathbf{x}$ approximately equal to $T^n \mathbf{y}$ for larger values of n and an arbitrary vector $\mathbf{y}$ (as in example 8.9) is commonly used in place of computing the actual equilibrium vector when the conditions of theorem 2 are met. By using the elementary vectors $\mathbf{e}_i$ for $\mathbf{y}$, we see that the columns of T^n must all equal $\mathbf{x}$ in the limit. Thus the procedure is to raise T to successively higher powers until all of the columns agree within the desired accuracy. Any column of this matrix is then a good approximation to the equilibrium vector.

8.3 Exercises

In exercises 1 through 10 use the information given in the table of probabilities for a Markov chain, table 8.5.

Next state

		1	2	3
	1	0.2	0.5	0.3
Current state	2	0.1	0.8	0.1
	3	0.2	0.4	0.4

Table 8.5

1. Find the probability of moving from state 3 to state 2.
2. Find the probability of moving from state 2 to state 3.
3. Find the transition matrix for the Markov chain.
4. Find the transition matrix for moving from the current state to two states in the future.
5. What is the initial vector if all states are equally likely?
6. What is the initial vector if state 1 is excluded but the other two are equally likely?
7. If all states are equally likely, find the probability vector for the next state.
8. If state 1 is excluded, but the other two are equally likely, find the probability vector for the next state.
9. If the initial state is state 1, find the probability of state 2 two time periods in the future.
10. If the initial state is state 1, find the probability of state 3 three time periods in the future.

In exercises 11 through 15 use the following information. Suppose a man has three modes of transportation to work: he can walk, drive his car, or take the bus. He never walks two days in a row nor drives two days in a row. If he walked the last time, he is twice as likely to drive as take the bus. In all other cases he shows no preference.

11. Prepare the table of probabilities for this Markov chain, and find the transition matrix.
12. If he drives on Monday, what is the probability that he will walk on Tuesday? That he will drive on Tuesday?
13. If he is equally likely to choose any mode of transportation on Monday, what is the prediction for all modes of transportation for Tuesday?
14. If he takes the bus on Monday, what is the probability that he will take the bus on Wednesday?
15. If he takes the bus on Monday, what is the probability that he will take the bus all week (five days in succession)?
16. Find all equilibrium vectors of the transition matrix

$$T = \begin{bmatrix} 1/2 & 1/3 & 0 \\ 1/2 & 1/3 & 0 \\ 0 & 1/3 & 1 \end{bmatrix}.$$

17. Find all equilibrium vectors of the transition matrix

$$T = \begin{bmatrix} 1/2 & 1/2 & 0 \\ 1/2 & 1/2 & 0 \\ 0 & 0 & 1 \end{bmatrix}.$$

18. Find all equilibrium vectors of the transition matrix of exercise 11.

19. If T is a transition matrix of a Markov chain, prove that the sum of the entries in each column of T^2 is also equal to 1.

20. Extend exercise 19 to T^n.

In exercises 21 and 22 use a computer program to estimate the components of each equilibrium vector to the nearest hundredth.

21. Find the equilibrium vector for the transition matrix of exercise 3.

22. Find the equilibrium vector for the transition matrix T obtained from table 8.3 on page 302.

8.4 Quadric Surfaces

A common topic in analytic geometry is the reduction of a quadratic equation in two variables—the equation of a *conic section*—to standard form. In this section we will discuss an extension of this idea. With the aid of the techniques of section 6.2 we will reduce a quadratic equation in three variables—that of a *quadric surface*—to certain standard forms.

Definition A **second-degree polynomial** (or **quadratic**) **equation** in x, y, and z is one of the form

$$ax^2 + by^2 + cz^2 + dxy + exz + fyz + gx + hy + iz = k, \tag{1}$$

where $a, b, \ldots, i, k$ are constants and $a, b, \ldots, f$ are not all zero. The graph of such an equation is called a **quadric** or **quadric surface.**

For example, the equation

$$x^2 + y^2 + z^2 - 2x + 4y = 4$$

is the special case of equation (1), with $a = b = c = 1$, $d = e = f = 0$, $g = -2$, $h = 4$, $i = 0$, and $k = 4$. (Its graph is a *sphere* in 3-space, with center at the point $(1, -2, 0)$ and radius equal to 3.)

There are nine basic types of *nondegenerate* quadric surfaces. These are listed below in *standard form*. We shall develop a *reduction procedure* to transform equation (1) to one of these standard forms. If it is impossible to reduce the equation to one of these forms, the corresponding surface is called a *degenerate quadric*. Graphs of the latter include planes, single points, and no points at all. It is not always clear from the original equation whether or not a quadric is nondegenerate, but, as we shall see, once the reduction is performed the given surface can be easily identified.

Standard Forms

i. **Ellipsoid** (figure 8.8):

$$\frac{x^2}{p^2} + \frac{y^2}{q^2} + \frac{z^2}{r^2} = 1 \quad (p, q, r > 0).$$

(A *sphere* is the special case of an ellipsoid for which $p = q = r$.)

ii. **Elliptic paraboloid** (figure 8.9):

$$z = \frac{x^2}{p^2} + \frac{y^2}{q^2} \quad (p, q > 0).$$

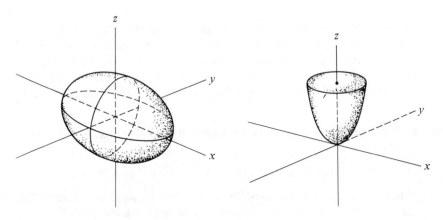

Figure 8.8   **Figure 8.9**

iii. **Hyperbolic paraboloid** (figure 8.10):

$$z = \frac{x^2}{p^2} - \frac{y^2}{q^2} \quad (p, q > 0).$$

iv. **Hyperboloid of one sheet** (figure 8.11):

$$\frac{x^2}{p^2} + \frac{y^2}{q^2} - \frac{z^2}{r^2} = 1 \quad (p, q, r > 0).$$

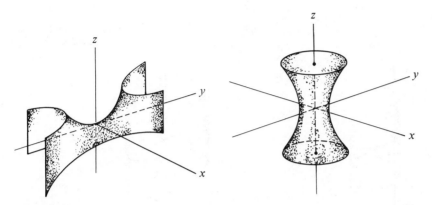

Figure 8.10   **Figure 8.11**

v. **Hyperboloid of two sheets** (figure 8.12):

$$\frac{x^2}{p^2} - \frac{y^2}{q^2} - \frac{z^2}{r^2} = 1 \quad (p, q, r > 0).$$

vi. **Cone** (figure 8.13):

$$\frac{x^2}{p^2} + \frac{y^2}{q^2} - \frac{z^2}{r^2} = 0 \quad (p, q, r > 0).$$

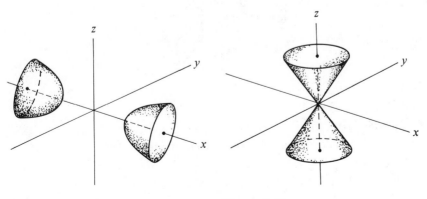

Figure 8.12   **Figure 8.13**

vii. **Parabolic cylinder** (figure 8.14):

$$x^2 = py + qz \quad (p, q \text{ not both zero}).$$

viii. **Elliptic cylinder** (figure 8.15):

$$\frac{x^2}{p^2} + \frac{y^2}{q^2} = 1 \quad (p, q > 0).$$

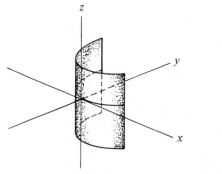
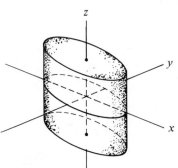

Figure 8.14   **Figure 8.15**

ix. **Hyperbolic cylinder** (figure 8.16):

$$\frac{x^2}{p^2} - \frac{y^2}{q^2} = 1 \quad (p, q > 0).$$

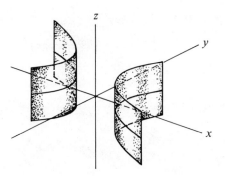

Figure 8.16

Example 8.10

Identify the quadrics whose equations are

a. $\dfrac{x^2}{4} + \dfrac{y^2}{9} + z^2 = 1,$

b. $9x^2 - y^2 - 16z^2 = 144,$

c. $x^2 - y^2 - z = 0.$

Equation a fits the standard form i with $p = 2$, $q = 3$, and $r = 1$; it represents an ellipsoid. Dividing equation b by 144, we obtain one of the standard form v with $p = 4$, $q = 12$, and $r = 3$; it represents a hyperboloid of two sheets. Lastly, by transposing the z term of equation c, we see that it takes the standard form iii with $p = 1$ and $q = 1$; it represents a hyperbolic paraboloid.

The Reduction Procedure

We will now derive a procedure for transforming (or *reducing*) those special cases of equation (1) that represent nondegenerate quadric surfaces to one of the given standard forms. We begin by writing equation (1) in the matrix form

$$\mathbf{x}^T A \mathbf{x} + B\mathbf{x} = k, \tag{2}$$

where A, B, and $\mathbf{x}$ are the matrices

$$A = \begin{bmatrix} a & d/2 & e/2 \\ d/2 & b & f/2 \\ e/2 & f/2 & c \end{bmatrix}, \quad B = \begin{bmatrix} g & h & i \end{bmatrix}, \quad \text{and} \quad \mathbf{x} = \begin{bmatrix} x \\ y \\ z \end{bmatrix}. \tag{3}$$

To see that equations (1) and (2) are equivalent, simply perform the operations indicated in equation (2), and equation (1) will result. Notice that A is a *symmetric* matrix.

Example 8.11

Find matrices A and B such that the equation $x^2 - 4y^2 - 2z^2 + 8xy - 6yz + 7x - 3y = 1$ takes the matrix form of equation (2).

Comparing the given equation to equation (1), we see that $a = 1$, $b = -4$, $c = -2$, $d = 8$, $e = 0$, $f = -6$, $g = 7$, $h = -3$, $i = 0$, and $k = 1$. Therefore

$$A = \begin{bmatrix} 1 & 4 & 0 \\ 4 & -4 & -3 \\ 0 & -3 & -2 \end{bmatrix} \quad \text{and} \quad B = \begin{bmatrix} 7 & -3 & 0 \end{bmatrix}.$$

The first step in the reduction procedure is the construction of a transformation that, when applied to equation (1), results in the elimination of the *cross-product* (the xy, xz, and yz) terms. In terms of equation (2) this is equivalent to diagonalizing the symmetric matrix A, since the off-diagonal entries of A are the ones that produce the coefficients of the cross-product terms.

Letting P be an orthogonal 3×3 matrix such that $D = P^T A P$ is diagonal (see section 6.2), we make the change of variable

$$\mathbf{x} = P\mathbf{x}'$$

and substitute into equation (2), obtaining

$$(P\mathbf{x}')^T A(P\mathbf{x}') + B(P\mathbf{x}') = k.$$

Now $(P\mathbf{x}')^T = (\mathbf{x}')^T P^T$, so this equation becomes

$$(\mathbf{x}')^T (P^T A P)\mathbf{x}' + BP\mathbf{x}' = k.$$

But $P^T A P = D$ (a diagonal matrix), so this last equation takes the form

$$(\mathbf{x}')^T D\mathbf{x}' + BP\mathbf{x}' = k. \tag{4}$$

Recalling from section 6.2 that the diagonal entries of D are just the eigenvalues of A, λ_1, λ_2, and λ_3, we need not actually perform the matrix multiplications $P^T A P$ to obtain D; we simply set

$$D = \begin{bmatrix} \lambda_1 & 0 & 0 \\ 0 & \lambda_2 & 0 \\ 0 & 0 & \lambda_3 \end{bmatrix}.$$

Obtaining BP requires more effort, since the procedure for determining P involves finding an orthonormal basis of eigenvectors of A. Letting P be a matrix with these eigenvectors as columns, we form the product BP and obtain a matrix we shall denote by

$$BP = \begin{bmatrix} a' & b' & c' \end{bmatrix}.$$

Finally, letting $\mathbf{x}' = (x', y', z')$, we see that equation (4) can be rewritten as the scalar equation

$$\lambda_1 (x')^2 + \lambda_2 (y')^2 + \lambda_3 (z')^2 + a'x' + b'y' + c'z' = k. \tag{5}$$

Notice that equation (5) contains no cross-product terms.

Note: If equation (1) contains no linear $(x, y, \text{ or } z)$ terms, then $B = 0$, and hence $BP = 0$. In this case we don't have to bother with P at all to identify the quadric surface, since equation (5) becomes simply

$$\lambda_1 (x')^2 + \lambda_2 (y')^2 + \lambda_3 (z')^2 = k$$

and depends only upon the eigenvalues of A. This is the case in the following example.

Example 8.12

Reduce the equation $5x^2 + 11y^2 + 2z^2 - 16xy + 20xz + 4yz = 18$ to standard form, and identify the quadric.

We first write the given equation in the form

$$\mathbf{x}^T A\mathbf{x} = 18,$$

where $\mathbf{x} = (x, y, z)$ and A is the matrix

$$A = \begin{bmatrix} 5 & -8 & 10 \\ -8 & 11 & 2 \\ 10 & 2 & 2 \end{bmatrix}.$$

To diagonalize A, we compute its eigenvalues, $\lambda_1 = 9, \lambda_2 = 18, \lambda_3 = -9$, and set

$$D = \begin{bmatrix} 9 & 0 & 0 \\ 0 & 18 & 0 \\ 0 & 0 & -9 \end{bmatrix}.$$

Then the transformed equation, $(\mathbf{x}')^T D \mathbf{x}' = 18$, is simply

$$9(x')^2 + 18(y')^2 - 9(z')^2 = 18.$$

Finally, dividing both sides by 18, we obtain the standard form

$$\frac{(x')^2}{2} + \frac{(y')^2}{1} - \frac{(z')^2}{2} = 1.$$

Thus the quadric is a hyperboloid of one sheet (see figure 8.17).

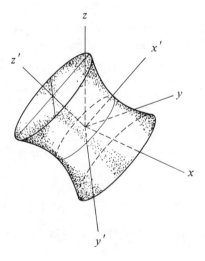

Figure 8.17

Note: Unless the eigenvalues of A (the diagonal entries of D) are arranged in the proper order, it may not be possible to obtain a final equation that is *exactly* the same as one of the listed standard forms. It may have some of the variables interchanged. This corresponds to a permutation of the coordinate axes relative to those of the given standard form. For instance, if in example 8.12 we take

$$D = \begin{bmatrix} 9 & 0 & 0 \\ 0 & -9 & 0 \\ 0 & 0 & 18 \end{bmatrix},$$

the resulting equation will be

$$\frac{(x')^2}{2} - \frac{(y')^2}{2} + (z')^2 = 1,$$

a hyperboloid of one sheet whose axis of symmetry is the y'-axis.

We now know how to transform equation (1) so that the cross-product terms are eliminated. In general we will also have to eliminate certain linear terms. This is accomplished with the aid of the process of *completing the square*. Recall that to complete the square in the expression $pu^2 + qu$, we proceed as follows:

$$\begin{aligned} pu^2 + qu &= p(u^2 + (q/p)u) \\ &= p(u^2 + (q/p)u + q^2/4p^2) - q^2/4p \\ &= p(u + q/2p)^2 - q^2/4p. \end{aligned}$$

The manner in which completing the square is used in the reduction process is illustrated in the next example.

Example 8.13

Reduce the equation $x^2 + 2y^2 - 2x + 8y - 4z = 3$ to standard form, and identify the quadric.

Notice that no cross-product terms appear in this equation. We complete the square in the quadratic variables, x and y, and transpose the z term to the right side:

$$(x^2 - 2x + 1) + 2(y^2 + 4y + 4) = 4z + 3 + 1 + 8,$$
$$(x - 1)^2 + 2(y + 2)^2 = 4z + 12.$$

We now divide both sides of this equation by the coefficient of z, obtaining

$$\frac{(x - 1)^2}{4} + \frac{(y + 2)^2}{2} = z + 3$$

Finally, we make the change of variables $x' = x - 1$, $y' = y + 2$, and $z' = z + 3$. This yields

$$\frac{(x')^2}{4} + \frac{(y')^2}{2} = z',$$

which is the standard form of an equation (in variables x', y', and z') of an elliptic paraboloid (figure 8.18).

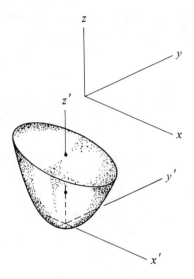

Figure 8.18

Note: Geometrically the change of variables of the type used in example 8.13, $x' = x - x_0$, $y' = y - y_0$, $z' = z - z_0$, *translates* the *xyz*- coordinate system into the $x'y'z'$-system. The origin of the $x'y'z'$ coordinate system will be the point (x_0, y_0, z_0) in the *xyz*-system. For example, the change of variables $x' = x - 1$, $y' = y + 2$, $z' = z + 3$ of example 8.13 results in an $x'y'z'$ origin at $(1, -2, -3)$ of the *xyz*-system (see figure 8.18).

Before giving another example, we will present this technique in the form of a procedure.

Procedure to reduce a quadratic equation in three variables to standard form   Given the equation

$$ax^2 + by^2 + cz^2 + dxy + exz + fyz + gx + hy + iz = k,$$

i. determine the matrices A and B described by (3),
ii. find an orthogonal matrix P and a diagonal matrix D such that $P^T A P = D$,
iii. make the change of variable $\mathbf{x} = P\mathbf{x}'$ by performing the indicated matrix operations in

$$(\mathbf{x}')^T D \mathbf{x}' + BP\mathbf{x}' = k$$

to eliminate the cross-product terms, and convert it to a scalar equation,

iv. complete the square, translate the coordinate system by making a change of variables, and rearrange terms in this scalar equation so that it is in standard form.

Example 8.14

Reduce the equation $2x^2 + 2y^2 + 2z^2 + 2xz + 4\sqrt{2}x + 12y = -10$ to standard form, and identify the quadric.

The given equation has the matrix form $\mathbf{x}^T A \mathbf{x} + B \mathbf{x} = -10$, where

$$A = \begin{bmatrix} 2 & 0 & 1 \\ 0 & 2 & 0 \\ 1 & 0 & 2 \end{bmatrix} \quad \text{and} \quad B = [4\sqrt{2} \quad 12 \quad 0].$$

Since the eigenvalues and corresponding normalized eigenvectors of A are

$$\lambda_1 = 1, \lambda_2 = 2, \lambda_3 = 3,$$

and

$$\mathbf{x}_1 = (-1/\sqrt{2}, 0, 1/\sqrt{2}), \mathbf{x}_2 = (0, 1, 0), \mathbf{x}_3 = (1/\sqrt{2}, 0, 1/\sqrt{2}),$$

we may take

$$P = \begin{bmatrix} -1/\sqrt{2} & 0 & 1/\sqrt{2} \\ 0 & 1 & 0 \\ 1/\sqrt{2} & 0 & 1/\sqrt{2} \end{bmatrix} \quad \text{and} \quad D = \begin{bmatrix} 1 & 0 & 0 \\ 0 & 2 & 0 \\ 0 & 0 & 3 \end{bmatrix}.$$

Thus, applying the transformation $\mathbf{x} = P\mathbf{x}'$, we obtain the equation

$$(\mathbf{x}')^T D \mathbf{x}' + BP\mathbf{x}' = -10,$$

or

$$(x')^2 + 2(y')^2 + 3(z')^2 - 4x' + 12y' + 4z' = -10.$$

We now complete the square in x', y', and z' and obtain

$$(x' - 2)^2 + 2(y' + 3)^2 + 3(z' + 2/3)^2 = 40/3.$$

Finally, letting $x'' = x' - 2$, $y'' = y' + 3$, and $z'' = z' + 2/3$ and dividing both sides of the last equation by $40/3$ yields the standard form

$$\frac{(x'')^2}{40/3} + \frac{(y'')^2}{20/3} + \frac{(z'')^2}{40/9} = 1.$$

The graph is an ellipsoid.

The techniques of this section can also be applied to the simpler case of a quadratic equation in two variables

$$ax^2 + bxy + cy^2 + dx + ey = f,$$

which is the equation of a *conic section*. The reduction of a conic section to standard form is explored in exercises 11 through 18.

8.4 Exercises

In exercises 1 through 4 the given equation contains no linear terms. Reduce it to standard form, and identify the quadric surface it represents.

1. $x^2 + y^2 + z^2 - 2xy - 2xz - 2yz = 2$
2. $3x^2 + 3y^2 + 3z^2 - 3xy - 4yz = 66$
3. $x^2 + y^2 + z^2 + 6xy + 8yz = 0$
4. $x^2 + 2\sqrt{2}xy = 4$

In exercises 5 and 6 the given equation contains no cross-product terms. Reduce it to standard form, and identify the quadric surface it represents.

5. $2x^2 + y^2 + z^2 - 4x + 8y + 6z = 9$
6. $2x^2 - 2y^2 - z^2 - 8x + 4y - 4z = 2$

In exercises 7 through 10 reduce the given equation to standard form, and identify the quadric surface it represents.

7. $2x^2 + y^2 + z^2 + 2\sqrt{6}xz + 2y + 2\sqrt{5}z = 2$
8. $x^2 + y^2 + z^2 + 2xy + 2xz + 2yz + \sqrt{6}y = 0$
9. $x^2 + y^2 - z^2 + 2xy - 2z = 3$
10. $3x^2 + 3y^2 + z^2 - 2xy + 12z = 7$

In exercises 11 through 14 a method for reducing quadratic equations in two variables—those of *conic sections*—to standard form is investigated.

11. Show that the equation $ax^2 + bxy + cy^2 + dx + ey = f$ can be written in the matrix form $\mathbf{x}^T A \mathbf{x} + B \mathbf{x} = f$, where

$$A = \begin{bmatrix} a & b/2 \\ b/2 & c \end{bmatrix}, \quad B = [d \ \ e], \quad \text{and} \quad \mathbf{x} = \begin{bmatrix} x \\ y \end{bmatrix}.$$

12. Use the result of exercise 11 to write the equation

$$3x^2 + 4xy - y^2 + 5x - 2y = 1$$

in matrix form.

13. Given the matrix equation $\mathbf{x}^T A \mathbf{x} + B \mathbf{x} = f$, let P be an orthogonal matrix such that $P^T A P = D$, a diagonal matrix. Show that the change of variable $\mathbf{x} = P \mathbf{x}'$ transforms the given equation into one of the form

$$(\mathbf{x}')^T D \mathbf{x}' + BP \mathbf{x}' = f.$$

14. Show that the matrix equation $(\mathbf{x}')^T D \mathbf{x}' + BP \mathbf{x}' = f$ of exercise 13 is equivalent to the scalar equation $a'(x')^2 + c'(y')^2 + d'x' + e'y' = f$, where a' and c' are the diagonal entries of D and d' and e' are the entries of BP. (Notice that there is no *cross-product*, that is, $x'y'$, term in this scalar equation.)

In exercises 15 and 16 use the results of exercises 11, 13, and 14 to transform the given equation into one in which no cross-product terms appear.

15. $x^2 + 4xy + y^2 + 5\sqrt{2}x + \sqrt{2}y = 0$

16. $x^2 - 4xy + 4y^2 + 10\sqrt{5}x = 15$

17–18. With the aid of the *completing-the-square* process, reduce the transformed equations obtained in exercises 15 and 16 to one of the following *standard forms* for a *nondegenerate conic section*:

(i) *ellipse,* $\dfrac{x^2}{p^2} + \dfrac{y^2}{q^2} = 1$ $(p, q > 0)$,

(ii) *parabola,* $y^2 = px$ $(p \neq 0)$,

(iii) *hyperbola,* $\dfrac{x^2}{p^2} - \dfrac{y^2}{q^2} = 1$ $(p, q > 0)$.

8.5 Approximation of Continuous Functions, Fourier Series

In section 8.1 we considered the problem of obtaining a polynomial approximation to a discrete set of data. Now we will turn to the related idea of approximation of continuous functions by other, generally simpler, functions. The key to the development of this material is the topic of inner-product spaces (section 7.4).

Least Square for Continuous Functions

There are many ways of approximating a continuous function on an interval by a polynomial. If $f(x)$ is the function, and $[a, b]$ is the interval, we could evaluate f at n distinct points in $[a, b]$ to obtain the data set $(x_i, f(x_i))$, $i = 1, 2, \ldots, n$, and then construct the interpolating or least square polynomial for this data (section 8.1). However, this approach only makes use of a small part of the given information. To involve all our knowledge of $f(x)$, we employ the following approach.

Recall that the least square polynomial, $p(x)$, for a finite set of data, $\{(x_i, y_i)\mid i = 1, 2, \ldots, n\}$, is the one that minimizes

$$S = (y_1 - p(x_1))^2 + (y_2 - p(x_2))^2 + \cdots + (y_n - p(x_n))^2.$$

By letting $y = (y_1, \ldots, y_n)$ and $z = (p(x_1), \ldots, p(x_n))$, and making use of the dot product for $\mathbf{R}^n$, we may write this equation as

$$S = (y - z) \cdot (y - z) = \|y - z\|^2.$$

This suggests one way to extend the idea of the least square approximation to a continuous function $\mathbf{f}(x)$: find a polynomial $\mathbf{p}(x)$ that minimizes

$$(\mathbf{f} - \mathbf{p}, \mathbf{f} - \mathbf{p}) = \|\mathbf{f} - \mathbf{p}\|^2,$$

where this inner product is the one on $C[a, b]$ (the vector space of functions continuous on the interval $[a, b]$) given by

$$(\mathbf{g}, \mathbf{h}) = \int_a^b \mathbf{g}(x)\mathbf{h}(x)dx \qquad (1)$$

(see section 7.4).

The following theorem provides the basis for the approximation techniques of this section. It will be proved at the end of the section.

Theorem 1   Let $\mathbf{f}(x)$ be a member of $C[a, b]$, and let $\mathbf{S}$ be a finite dimensional subspace of $C[a, b]$. Then there is an unique function $\mathbf{p}(x)$ in $\mathbf{S}$ that minimizes

$$\|\mathbf{f} - \mathbf{p}\|^2 = \int_a^b (\mathbf{f}(x) - \mathbf{p}(x))^2 dx.$$

Note: The function $\mathbf{p}$ of theorem 1 is called the **least square approximation** to $\mathbf{f}$ from $\mathbf{S}$. From a practical point of view the subspace $\mathbf{S}$ usually consists of polynomials of some sort.

Procedure for determining the least square approximation to a continuous function   Let $\mathbf{f}$ be a function in $C[a, b]$, and let $\mathbf{S}$ be a finite dimensional subspace of approximating functions. To construct the least square approximation to $\mathbf{f}$ from $\mathbf{S}$,

  i. find an orthonormal basis, $\{\mathbf{p}_1, \mathbf{p}_2, \ldots, \mathbf{p}_n\}$ for $\mathbf{S}$,
  ii. let $\mathbf{p} = c_1\mathbf{p}_1 + c_2\mathbf{p}_2 + \cdots + c_n\mathbf{p}_n$, where

$$c_i = (\mathbf{f}, \mathbf{p}_i) = \int_a^b \mathbf{f}(x)\mathbf{p}_i(x)dx.$$

Then $\mathbf{p}$ is the desired least square approximation.

Least Square Polynomial Approximation

Perhaps the simplest choice of the subspace $\mathbf{S}$ of theorem 1 is $\mathbf{P}_n[a, b]$, the vector space of all polynomials of degree less than or equal to n with domain restricted to $[a, b]$. In this case the desired approximating function is called the **least square polynomial** for $\mathbf{f}$ on $[a, b]$. To apply the procedure of theorem 1 to construct this polynomial, we first need an orthonormal basis for $\mathbf{P}_n[a, b]$. Since $\mathcal{B} = \{1, x, x^2, \ldots, x^n\}$ is a basis for this subspace, we can find an orthonormal basis by applying the Gram-Schmidt process of section 7.4 to $\mathscr{B}$.

Example 8.15

Find the least square quadratic polynomial approximating $\mathbf{f}(x) = x^3$ on $[-1, 1]$.

We apply the procedure of theorem 1. Here $S = \mathbf{P}_2[-1, 1]$, which has a basis $\mathscr{B} = \{1, x, x^2\}$. By applying the Gram-Schmidt process to $\mathscr{B}$, we obtain the orthonormal basis (see example 7.35) $\{\mathbf{p}_0, \mathbf{p}_1, \mathbf{p}_2\}$, where

$$\mathbf{p}_0(x) = \sqrt{1/2},$$
$$\mathbf{p}_1(x) = (\sqrt{3/2})x,$$

and

$$\mathbf{p}_2(x) = (1/2)(\sqrt{5/2})(3x^2 - 1).$$

Therefore the quadratic least square polynomial, $\mathbf{p}(x)$, for $\mathbf{f}(x) = x^3$ on $[-1, 1]$ is given by

$$\mathbf{p}(x) = c_0\mathbf{p}_0(x) + c_1\mathbf{p}_1(x) + c_2\mathbf{p}_2(x),$$

where

$$c_0 = (\mathbf{f}, \mathbf{p}_0) = \int_{-1}^{1} x^3(\sqrt{1/2})dx = 0$$

$$c_1 = (\mathbf{f}, \mathbf{p}_1) = \int_{-1}^{1} x^3(\sqrt{3/2})xdx = \sqrt{3/2} \int_{-1}^{1} x^4dx = (2/5)\sqrt{3/2}$$

and

$$c_2 = (\mathbf{f}, \mathbf{p}_2) = \int_{-1}^{1} x^3(1/2)(\sqrt{5/2})(3x^2 - 1)dx$$

$$= (1/2)\sqrt{5/2} \int_{-1}^{1} (3x^5 - x^3)dx = 0.$$

Thus

$$\mathbf{p}(x) = [(2/5)\sqrt{3/2}](\sqrt{3/2})x = (3/5)x.$$

(The given $\mathbf{f}(x)$ and approximating $\mathbf{p}(x)$ are shown in figure 8.19.)

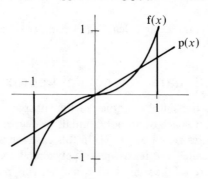

Figure 8.19

Note: As we can see from example 8.15, the least square approximation to $\mathbf{f}$ from $\mathbf{P}_2[a, b]$ does not need to have its degree exactly equal to 2.

The polynomials that form an orthonormal basis for $\mathbf{P}_n[-1, 1]$ are called *orthonormal Legendre* polynomials and are often listed in mathematical reference books. The next theorem demonstrates how we can construct a set of polynomials that are orthonormal on any interval $[a, b]$ in terms of the orthonormal Legendre polynomials.

Theorem 2   Let $\{\mathbf{q}_0(t), \mathbf{q}_1(t), \ldots, \mathbf{q}_n(t)\}$ be an orthonormal basis for $\mathbf{P}_n[-1, 1]$. Then $\mathscr{B} = \{\mathbf{p}_0(\mathbf{x}), \mathbf{p}_1(x), , \ldots, \mathbf{p}_n(x)\}$ is an orthonormal basis for $\mathbf{P}_n[a, b]$, where

$$\mathbf{p}_i(x) \quad \left(\sqrt{\frac{2}{b-a}}\right) \mathbf{q}_i\left(\frac{2x - (a + b)}{b - a}\right).$$

Proof   All we need do is to verify that $\mathscr{B}$ is an orthonormal *set* on $[a, b]$; that is,

$$\int_a^b (\mathbf{p}_i(x))^2 dx = 1$$

and

$$\int_a^b \mathbf{p}_i(x)\mathbf{p}_j(x)dx = 0, \quad \text{if} \quad i \neq j.$$

Then by lemma 3 of section 7.4 and theorem 7 of section 7.2, $\mathscr{B}$ is an orthonormal *basis* for $\mathbf{P}_n[a, b]$. We leave the details of this verification as an exercise.   ∎

Example 8.16

Find the least square linear polynomial, $\mathbf{p}(x)$, approximating

$$\mathbf{f}(x) = \sqrt{x}$$

on $[0, 1]$.

We again apply the procedure of theorem 1. The approximation we seek is of the form

$$\mathbf{p} = c_0\mathbf{p}_0 + c_1\mathbf{p}_1,$$

where $\{\mathbf{p}_0, \mathbf{p}_1\}$ is an orthonormal basis for $\mathbf{p}_1[0, 1]$,

$$c_0 = \int_0^1 \mathbf{f}(x)\mathbf{p}_0(x)dx$$

and

$$c_1 = \int_0^1 \mathbf{f}(x)\mathbf{p}_1(x)dx.$$

To find $\mathbf{p}_0$ and $\mathbf{p}_1$ we use theorem 2. Since an orthonormal basis for $\mathbf{p}_1[-1, 1]$ is $\{1/\sqrt{2}, (\sqrt{3/2})t\}$, and for this example $a = 0$ and $b = 1$, we obtain

$$\mathbf{p}_0(x) = \sqrt{2}(1/\sqrt{2}) = 1$$

and

$$\mathbf{p}_1(x) = \sqrt{2}(\sqrt{3/2})(2x - 1) = \sqrt{3}(2x - 1).$$

Consequently

$$c_0 = \int_0^1 \sqrt{x}\, dx = 2/3,$$

$$c_1 = \int_0^1 \sqrt{x}\, \sqrt{3}(2x - 1)dx$$

$$= \sqrt{3}\int_0^1 (2x^{3/2} - x^{1/2})dx = 2\sqrt{3}/15,$$

yielding

$$\mathbf{p}(x) = c_0\mathbf{p}_0(x) + c_1\mathbf{p}_1(x) = (4/5)x + 4/15.$$

(The given $\mathbf{f}(x)$ and approximating $\mathbf{p}(x)$ are shown in figure 8.20)

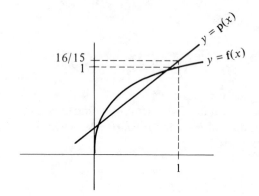

Figure 8.20

Fourier Series

Another type of relatively simple function used for approximation purposes is the *trigonometric polynomial*.

Definition   A **trigonometric polynomial** is a function of the form

$$\mathbf{p}(x) = a_0 + a_1 \cos x + a_2 \cos 2x + \cdots + a_n \cos nx$$
$$+ b_1 \sin x + b_2 \sin 2x + \cdots + b_n \sin nx.$$

If a_n and b_n are not both zero, we say that $\mathbf{p}(x)$ has **order** n.

It can be shown that $\mathbf{T}_n[a, b]$, the set of all trigonometric polynomials on $[a, b]$ of order less than or equal to n, is a subspace of $\mathbf{C}[a, b]$. If we choose the subspace $\mathbf{S}$ of theorem 1 to be $\mathbf{T}_n[-\pi, \pi]$, the resulting approximation is called the **Fourier series of order** n for the given function.

To obtain the Fourier series of order n for a continuous function $\mathbf{f}(x)$, we apply the procedure of theorem 1 with $\mathbf{S} = \mathbf{T}_n[-\pi, \pi]$. This subspace has the orthogonal basis (exercise 17)

$$\mathscr{B} = \{1, \cos x, \cos 2x, \ldots, \cos nx, \sin x, \sin 2x, \ldots, \sin nx\}.$$

Thus, to obtain an orthonormal basis for $\mathbf{T}_n[-\pi, \pi]$, we need only normalize the vectors in $\mathscr{B}$. In other words, we seek constants $c_0, c_1, \ldots, c_n$, $d_1, \ldots, d_n$ so that

$$(c_0, c_0) = \int_{-\pi}^{\pi} c_0^2 \, dx = 1,$$

$$(c_k \cos kx, c_k \cos kx) = \int_{-\pi}^{\pi} c_k^2 \cos^2 kx \, dx = 1 \quad (k = 1, 2, \ldots, n),$$

and

$$(d_k \sin kx, d_k \sin kx) = \int_{-\pi}^{\pi} d_k^2 \sin^2 kx \, dx = 1 \quad (k = 1, 2, \ldots, n).$$

Performing the indicated integrations and solving for the c_k and d_k yields

$$c_0 = 1/\sqrt{2\pi},$$

$$c_k = d_k = 1/\sqrt{\pi} \quad \text{for } k = 1, 2, \ldots, n.$$

Now that we have an orthonormal basis for $\mathbf{T}_n[-\pi \ \pi]$, we can complete the procedure of theorem 1 and obtain the following theorem.

Theorem 3   Let $\mathbf{f}(x)$ be a continuous function on $[-\pi, \pi]$. Then the Fourier series for $\mathbf{f}$ (the least square trigonometric polynomial approximating $\mathbf{f}$ on $[-\pi, \pi]$) of order n is

$$\mathbf{p}(x) = a_0 + a_1 \cos x + a_2 \cos 2x + \cdots + a_n \cos nx$$
$$+ b_1 \sin x + b_2 \sin 2x + \cdots + b_n \sin nx,$$

where

$$a_0 = (1/2\pi) \int_{-\pi}^{\pi} \mathbf{f}(x) \, dx,$$

$$a_k = (1/\pi) \int_{-\pi}^{\pi} \mathbf{f}(x) \cos kx \, dx \quad \text{for } k = 1, 2, \ldots, n,$$

and

$$b_k = (1/\pi) \int_{-\pi}^{\pi} \mathbf{f}(x) \sin kx \, dx \quad \text{for } k = 1, 2, \ldots, n.$$

Example 8.17

Find the Fourier series of order 2 for $\mathbf{f}(x) = x$.

Applying theorem 3, the desired function has the form

$$\mathbf{p}(x) = a_0 + a_1 \cos x + a_2 \cos 2x + b_1 \sin x + b_2 \sin 2x,$$

where

$$a_0 = (1/2\pi) \int_{-\pi}^{\pi} x \, dx = 0,$$

$$a_1 = (1/\pi) \int_{-\pi}^{\pi} x \cos x \, dx = 0,$$

$$a_2 = (1/\pi) \int_{-\pi}^{\pi} x \cos 2x \, dx = 0,$$

$$b_1 = (1/\pi) \int_{-\pi}^{\pi} x \sin x \, dx = 2,$$

$$b_2 = (1/\pi) \int_{-\pi}^{\pi} x \sin 2x \, dx = -1.$$

Therefore $\mathbf{p}(x) = 2 \sin x - \sin 2x$.

(The given $\mathbf{f}(x)$ and the approximating $\mathbf{p}(x)$ are shown in figure 8.21.)

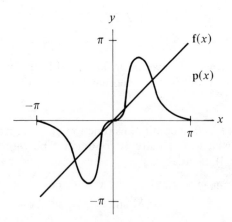

Figure 8.21

Note: If a function $\mathbf{f}$ is continuous on $[-\pi, \pi]$, there is a Fourier series of order n for $\mathbf{f}$ for every natural number n. If this sequence of Fourier series is denoted by $\{\mathbf{p}_n\}$, then $\|\mathbf{f} - \mathbf{p}_n\| \to 0$ as $n \to \infty$. Therefore the infinite series

$$a_0 + \sum_{k=1}^{\infty} (a_k \cos kx + b_k \sin kx)$$

converges to the function f. This series is called the **Fourier series** (without the qualifier *of order n*) for **f**.

We close this section with the proof of theorem 1. This proof will also justify the procedure immediately following the statement of the theorem.

Proof of theorem 1 We first show existence of the least square approximation. Let $\mathscr{B} = \{\mathbf{p}_1, \mathbf{p}_2, \ldots, \mathbf{p}_n\}$ be an orthonormal basis for **S**. Then any vector in **S** can be written

$$\mathbf{p} = c_1\mathbf{p}_1 + c_2\mathbf{p}_2 + \cdots + c_n\mathbf{p}_n.$$

We wish to find **p** in $C[a, b]$ that minimizes

$$\|\mathbf{f} - \mathbf{p}\|^2 = (\mathbf{f} - \mathbf{p}, \mathbf{f} - \mathbf{p})$$
$$= (\mathbf{f} - (c_1\mathbf{p}_1 + c_2\mathbf{p}_2 + \cdots + c_n\mathbf{p}_n),$$
$$\mathbf{f} - (c_1\mathbf{p}_1 + c_2\mathbf{p}_2 + \cdots + c_n\mathbf{p}_n)).$$

But the $\mathbf{p}_i$ form an orthonormal set, so $(\mathbf{p}_i, \mathbf{p}_i) = 1$ and $(\mathbf{p}_i, \mathbf{p}_j) = 0$ if $i \neq j$. This implies that

$$\|\mathbf{f} - \mathbf{p}\|^2 = (c_1, c_1) + (c_2, c_2) + \cdots + (c_n, c_n) - 2c_1(\mathbf{f}, \mathbf{p}_1)$$
$$- 2c_2(\mathbf{f}, \mathbf{p}_2) - \cdots - 2c_n(\mathbf{f}, \mathbf{p}_n) + (\mathbf{f}, \mathbf{f}).$$

We now let $\mathbf{c} = (c_1, c_2, \ldots, c_n)$, $\mathbf{d} = (d_1, d_2, \ldots, d_n)$, where $d_i = (\mathbf{f}, \mathbf{p}_i)$ and $(\mathbf{f}, \mathbf{f}) = e$. Then, using the dot product in $\mathbf{R}^n$, we can write

$$\|\mathbf{f} - \mathbf{p}\|^2 = \mathbf{c} \cdot \mathbf{c} - 2\mathbf{c} \cdot \mathbf{d} + e. \tag{2}$$

Since the desired function **p** is completely determined by the n-vector **c**, we seek **c** that minimizes $\|\mathbf{f} - \mathbf{p}\|^2$. If we can find a vector **r** so that the change of variable

$$\mathbf{c} = \mathbf{c}' + \mathbf{r} \tag{3}$$

results in an equation of the form

$$\|\mathbf{f} - \mathbf{p}\|^2 = \mathbf{c}' \cdot \mathbf{c}' + e', \tag{4}$$

and e' depends only on the fixed choice of **r**, not on the variable **c**′, the resulting function **p** will minimize $\|\mathbf{f} - \mathbf{p}\|^2$. This is because $\mathbf{c}' \cdot \mathbf{c}' \geq 0$ is minimized if $\mathbf{c}' = \mathbf{0}$. Thus, if we can determine such an **r**, we see from equation (4) that $\mathbf{c} = \mathbf{0} + \mathbf{r} = \mathbf{r}$ will minimize $\|\mathbf{f} - \mathbf{p}\|^2$.

To find the desired $\mathbf{r}$, we substitute equation (3) into equation (2). This results in

$$\|\mathbf{f} - \mathbf{p}\|^2 = (\mathbf{c}' + \mathbf{r}) \cdot (\mathbf{c}' + \mathbf{r}) - 2(\mathbf{c}' + \mathbf{r}) \cdot \mathbf{d} + e$$
$$= \mathbf{c}' \cdot \mathbf{c}' + 2\mathbf{c}' \cdot (\mathbf{r} - \mathbf{d}) + (\mathbf{r} \cdot \mathbf{r} - 2\mathbf{r} \cdot \mathbf{d} + e).$$

Choosing $\mathbf{r} = \mathbf{d}$, we obtain equation (4) with

$$e' = \mathbf{r} \cdot \mathbf{r} - 2\mathbf{r} \cdot \mathbf{d} + e = -\mathbf{d} \cdot \mathbf{d} + e.$$

Thus, taking $\mathbf{c} = \mathbf{d}$, that is,

$$\mathbf{c}_i = (\mathbf{f}, \mathbf{p}_i),$$

results in a minimum for $\|\mathbf{f} - \mathbf{p}\|^2$.

To show uniqueness, suppose $\mathbf{q}$ is a vector in $\mathbf{S}$ with

$$\|\mathbf{f} - \mathbf{q}\|^2 = \|\mathbf{f} - \mathbf{p}\|^2,$$

where

$$\mathbf{p} = (\mathbf{f}, \mathbf{p}_1)\mathbf{p}_1 + (\mathbf{f}, \mathbf{p}_2)\mathbf{p}_2 + \cdots + (\mathbf{f}, \mathbf{p}_n)\mathbf{p}_n.$$

Writing $\mathbf{q}$ as a linear combination of the basis vectors

$$\mathbf{q} = a_1\mathbf{p}_1 + a_2\mathbf{p}_2 + \cdots + a_n\mathbf{p}_n,$$

we let $\mathbf{a} = (a_1, \ldots, a_n)$ and make the change of variable of equation (3) (now with $\mathbf{r} = \mathbf{d}$),

$$\mathbf{a} = \mathbf{a}' + \mathbf{d}.$$

Using equation (4) and the fact that the minimum $\mathbf{p}$ was obtained with $\mathbf{c}' = \mathbf{0}$, we have

$$e' = \|\mathbf{f} - \mathbf{p}\|^2 = \|\mathbf{f} - \mathbf{q}\|^2 = \mathbf{a}' \cdot \mathbf{a}' + e'.$$

This implies that $\mathbf{a}' = \mathbf{0}$, and hence $\mathbf{a} = \mathbf{d}$. Thus $\mathbf{q} = \mathbf{p}$ as desired. ∎

8.5 Exercises

In exercises 1 through 4 find the *linear* least square polynomial approximating the given function on the indicated interval.

1. $f(x) = x^4$ on $[-1, 1]$ 2. $f(x) = x^3 + 3x$ on $[-1, 1]$
3. $f(x) = 1/x$ on $[1, 2]$ 4. $f(x) = 1/\sqrt{x}$ on $[1, 4]$

5–8. Find the *quadratic* least square polynomial approximating the functions of exercises 1 through 4 on the indicated intervals.

9–10. Find the *cubic* least square polynomial approximating the functions of exercises 1 and 2 on the indicated intervals.

In exercises 11 through 16 find the Fourier series of order n (for the indicated n) for the given function.

11. $f(x) = x + 1$ $(n = 1)$ **12.** $f(x) = x^2$ $(n = 1)$

13. $f(x) = x^3 + 1$ $(n = 1)$ **14.** $f(x) = e^x$ $(n = 1)$

15. $f(x) = x - 1$ $(n = 2)$ **16.** $f(x) = \sin 3x$ $(n = 2)$

17. Show that $\{1, \cos x, \cos 2x, \cdots, \cos nx, \sin x, \sin 2x, \cdots, \sin nx\}$ is an orthogonal basis for $T_n[-\pi, \pi]$.

Linear Programming

9

From a mathematical standpoint the subject of linear programming deals with the problem of optimizing a linear function subject to the constraints imposed by a system of linear inequalities. After introducing the topic in the first section, we will demonstrate the basic solution technique in section 9.2 and then extend this technique to a more general situation in section 9.3. The proofs of the major theorems will be delayed until section 9.4. Adequate preparation for this chapter is given by chapter 2.

9.1   Introduction and Terminology

In this section we see how linear programming problems might arise and how very simple ones can be solved geometrically. We also introduce some of the basic terminology and a standard form for linear programming problems.

A Simple Example

Suppose a manufacturer produces two types of liquids, X and Y. Because of past sales experience the market researcher estimates that at least twice as much Y as X is needed. The manufacturing capacity of the plant will only allow for a total of 9 units to be manufactured. If each unit of liquid X results in a profit of $2, while the profit for each unit of Y is $1, how much of each should be produced to maximize the profit?

Let x be the amount of X to be produced, and let y be the amount of Y. Then the condition that there be at least twice as much Y as X becomes

$$2x \leq y.$$

The manufacturing capacity of 9 units implies that

$$x + y \leq 9.$$

Let $z = z(x, y)$ be the profit associated with the production schedule of producing x units of X and y units of Y. Then the *profit function* or *objective function* is given by

$$z = 2x + y.$$

There is also a further condition implied but not stated in the problem. The variables must be nonnegative since we cannot produce negative amounts of a product. Thus we have

$$x \geq 0, \quad y \geq 0.$$

Collecting all of this information together, we may state the given problem mathematically as follows: maximize $z = 2x + y$ subject to the constraints

$$2x - y \leq 0$$
$$x + y \leq 9$$
$$x \geq 0, \quad y \geq 0.$$

In the next section we will develop machinery (the *simplex algorithm*) for solving such problems in general, but we can solve this particular one geometrically. First we sketch the set of points in $\mathbf{R}^2$ that satisfy the set of constraints (shaded part of figure 9.1). This region is called the *feasible region*. If a point is in the region, it satisfies all of the constraints and is called *feasible*; if it is not in the region, it violates at least one of them and is called *infeasible*.

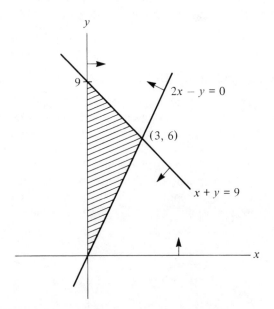

Figure 9.1

Different points in the feasible region give different values of the objective function. For example, the origin $(0, 0)$ is feasible and gives a profit of $2(0) + (0) = 0$. Similarly $(1, 5)$ is feasible and gives a profit of

$2(1) + (5) = 7$, which is better than 0. We seek the point or points of the feasible region that yield the maximal profit.

Now, corresponding to a *fixed* value of z, the set of solutions to the equation $z = 2x + y$ is the line (called the *objective line*) with slope -2 and y-intercept z. In other words, all points along the line $y = -2x + z$ correspond to the same z value. Remember that we are only interested in those points that lie in the feasible region, and we want z to be as large as possible (figure 9.2).

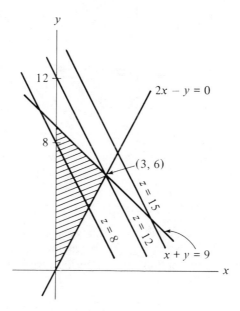

Figure 9.2

It is clear from figure 9.2 that the optimal value occurs when $z = 12$, since moving the objective line up increases its y-intercept, the value of z, but moving beyond the point (3, 6) gives a line that does not intersect the feasible region at all.

Note: In the preceding discussion we obtained a unique solution; the maximum value $z = 12$ occurs only at the point (3, 6). If, however, the objective function is changed to $z = 2x + 2y$, all of the points in the feasible region that lie on the edge determined by $x + y = 9$ will be optimal with value $z = 18$ (figure 9.3). Thus the point at which the optimal value occurs need not be unique.

Moreover, if the feasible region is *unbounded*, as would be the case in the preceding discussion if the constraint $x + y \leq 9$ were removed, an optimal value may not exist. That is, it is possible that no matter which point is chosen, a "better" one always exists.

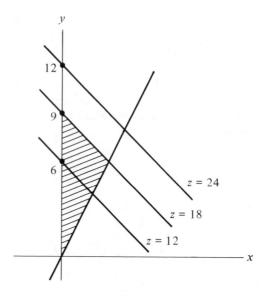

Figure 9.3

The General Problem

The previous example is a simple case of the general *linear programming problem*. We will describe this problem mathematically after introducing the following notation for comparing certain vectors.

Definition   Let $\mathbf{u}$ and $\mathbf{v}$ be vectors in $\mathbf{R}^m$. We say $\mathbf{u}$ **is less than** $\mathbf{v}$ and write $\mathbf{u} < \mathbf{v}$ if $u_i < v_i$ for each $i = 1, \ldots, m$. Likewise $\mathbf{u} \leq \mathbf{v}$ means $u_i \leq v_i$ for each i. Similar definitions hold for $\mathbf{u} > \mathbf{v}$ and $\mathbf{u} \geq \mathbf{v}$. The vector $\mathbf{v}$ is **positive** if $\mathbf{v} > \mathbf{0}$ and **nonnegative** if $\mathbf{v} \geq \mathbf{0}$.

Example 9.1

Let $\mathbf{u} = (1, 2, -3), \mathbf{v} = (2, 4, 1), \mathbf{w} = (2, 0, 1)$. Then $\mathbf{u} < \mathbf{v}$ and $\mathbf{w} \leq \mathbf{v}$, but no special relationship exists between $\mathbf{u}$ and $\mathbf{w}$. It is also true that $\mathbf{v} > \mathbf{0}$ and $\mathbf{w} \geq \mathbf{0}$.

Given a system of inequalities,

$$a_{11}x_1 + a_{12}x_2 + \cdots + a_{1n}x_n \leq b_1$$
$$a_{21}x_1 + a_{22}x_2 + \cdots + a_{2n}x_n \leq b_2$$
$$\vdots \qquad \vdots \qquad \qquad \vdots \qquad \vdots$$
$$a_{m1}x_1 + a_{m2}x_2 + \cdots + a_{mn}x_n \leq b_m,$$

we can use the vector inequality notation to write the system as

$$A\mathbf{x} \leq \mathbf{b}$$

where, as was the case with linear equations,

$$A = \begin{bmatrix} a_{11} & \cdots & a_{1n} \\ \vdots & & \vdots \\ a_{m1} & \cdots & a_{mn} \end{bmatrix}, \quad \mathbf{b} = \begin{bmatrix} b_1 \\ \vdots \\ b_m \end{bmatrix}, \quad \text{and} \quad \mathbf{x} = \begin{bmatrix} x_1 \\ \vdots \\ x_n \end{bmatrix}.$$

Example 9.2

Graph the set of solutions of the system of inequalities

$$A\mathbf{x} \leq \mathbf{b}, \quad \mathbf{x} \geq \mathbf{0},$$

where

$$A = \begin{bmatrix} 2 & -1 \\ 1 & 1 \end{bmatrix} \quad \text{and} \quad \mathbf{b} = \begin{bmatrix} 0 \\ 9 \end{bmatrix}.$$

The matrix inequality,

$$A\mathbf{x} = \begin{bmatrix} 2 & -1 \\ 1 & 1 \end{bmatrix} \begin{bmatrix} x \\ y \end{bmatrix} \leq \begin{bmatrix} 0 \\ 9 \end{bmatrix}, \quad \mathbf{x} \geq \mathbf{0},$$

is equivalent to the system of inequalities

$$\begin{aligned} 2x - y &\leq 0 \\ x + y &\leq 9 \\ x \geq 0, \quad y &\geq 0. \end{aligned}$$

This is the same set of inequalities as was derived at the beginning of this section. Thus the set of solutions is just that of figure 9.1.

Consider the system of linear inequalities

$$A\mathbf{x} \leq \mathbf{b}, \quad \mathbf{b} \geq \mathbf{0}, \quad \mathbf{x} \geq \mathbf{0}.$$

The set of all solutions to this system is called the **feasible region.** Notice that $\mathbf{0}$ is always in the feasible region of this system since $A\mathbf{0} = \mathbf{0} \leq \mathbf{b}$, and usually there are infinitely many feasible points. As in the example presented earlier, we shall be seeking solutions that *optimize* (*maximize* or *minimize*) the value z of a linear function

$$z = \mathbf{c} \cdot \mathbf{x} = c_1 x_1 + c_2 x_2 + \cdots + c_n x_n,$$

where the c_i are scalars. The function $z = \mathbf{c} \cdot \mathbf{x}$ is called the **objective function.**

Definition A linear programming problem is in **standard form** if it is stated in the following way: optimize the function $z = \mathbf{c} \cdot \mathbf{x}$, subject to the conditions

$$A\mathbf{x} \leq \mathbf{b}, \quad \mathbf{b} \geq \mathbf{0}, \text{ and } \mathbf{x} \geq \mathbf{0}.$$

To apply our knowledge of linear *equations* to the study of linear *inequalities*, we convert the system of linear inequalities to a system of linear equations by adding a nonnegative variable, called a **slack variable**, to each constraint. For example, the inequality

$$2x + 3y \leq 5$$

is equivalent to the equation

$$2x + 3y + z = 5,$$

with the restriction that z be nonnegative. Notice that $z = 5 - (2x + 3y)$ is the "slack" or difference between $2x + 3y$ and 5. There is a trade-off in this process—in order to work with equations instead of inequalities, the number of variables increases by one for each inequality.

The mathematical description of this conversion process is as follows. Let

$$A\mathbf{x} \leq \mathbf{b}, \quad \mathbf{b} \geq \mathbf{0}, \quad \mathbf{x} \geq \mathbf{0},$$

be a system of m linear inequalities in n unknowns. The corresponding system of equations is

$$A'\mathbf{x}' = \mathbf{b}, \quad \mathbf{b} \geq \mathbf{0}, \quad \mathbf{x}' \geq \mathbf{0},$$

where A' is the $m \times (n + m)$ matrix whose first n columns are those of A and last m columns are those of the identity matrix of order m. The first n components of the vector $\mathbf{x}'$ correspond to those of $\mathbf{x}$, and the last m components are the slack variables of the system.

Example 9.3

Convert the following system of linear inequalities to the corresponding system of linear equations.

$$2x_1 - x_2 + x_3 \leq 5$$
$$x_2 - 2x_3 \leq 2$$
$$x_1 - x_2 + 2x_3 \leq 4$$
$$x_i \geq 0, \quad i = 1, 2, 3.$$

The resulting system in equation form is

$$2x_1 - x_2 + x_3 + x_4 \qquad\qquad = 5$$
$$x_2 - 2x_3 \qquad + x_5 \qquad = 2$$
$$x_1 - x_2 + 2x_3 \qquad\qquad + x_6 = 4$$
$$x_i \geq 0, \quad i = 1, \ldots 6.$$

The matrices A and A' are given by

$$A = \begin{bmatrix} 2 & -1 & 1 \\ 0 & 1 & -2 \\ 1 & -1 & 2 \end{bmatrix}, \quad A' = \begin{bmatrix} 2 & -1 & 1 & 1 & 0 & 0 \\ 0 & 1 & -2 & 0 & 1 & 0 \\ 1 & -1 & 2 & 0 & 0 & 1 \end{bmatrix}.$$

Note: We are not interested in the *values* of the slack variables. Thus in a linear programming problem with objective function $z = \mathbf{c} \cdot \mathbf{x}$, in this conversion process we take as the new objective function $z' = \mathbf{c}' \cdot \mathbf{x}'$, where $c'_j = c_j$ for $j = 1, \ldots, n$ and $c'_j = 0$ for $j = n + 1, \ldots, n + m$.

A system of linear equations that corresponds to the linear inequalities of a linear programming problem in standard form is always *row-reduced* in the following sense.

Definition   A matrix of rank r is **row-reduced** if it contains r nonzero rows, the rest are zero, and there exists a subset of r columns that are distinct elementary columns.

Note: A matrix in row-reduced echelon form is of course row-reduced in this sense. The columns that contain the leading ones provide the set of elementary columns.

Example 9.4

Which of the following are row-reduced?

$$B = \begin{bmatrix} 1 & 3 & 1 & 0 \\ -1 & 2 & 0 & 1 \\ 0 & 0 & 0 & 0 \end{bmatrix}, \quad C = \begin{bmatrix} 2 & 0 & 3 & 0 \\ 1 & 1 & 2 & 0 \\ 3 & 0 & 4 & 1 \end{bmatrix},$$

$$D = \begin{bmatrix} 2 & 0 & 0 & 2 & 1 \\ 1 & 1 & 0 & 1 & 0 \\ -1 & 0 & 1 & 4 & 0 \end{bmatrix}.$$

The matrix B is row-reduced, since $r = 2$ and $B^3 = \mathbf{e}_1$ and $B^4 = \mathbf{e}_2$. Also D is row-reduced, since $r = 3$ and $D^5 = \mathbf{e}_1$, $D^2 = \mathbf{e}_2$, and $D^3 = \mathbf{e}_3$. However, C is not row-reduced, since $r = 3$, but there is no column equal to $\mathbf{e}_1$.

For an $m \times n$ matrix of rank m, being row-reduced is tantamount to possessing a subset of m columns that may be rearranged to form an identity matrix. It is sometimes convenient to speak of an *embedded identity matrix* when referring to the set of elementary columns. For example, in the matrix A' of a linear system of equations that corresponds to a system of linear inequalities, the last m columns form an embedded identity matrix.

As in the case of a system in row-reduced *echelon* form, it is easy to identify a particular solution, called a **basic solution**. To obtain the corresponding basic solution, let the variables that correspond to the elementary columns assume the corresponding **b** column value (these are called the **basic variables** of the solution), and let all others be zero (these are the **nonbasic variables**). Basic solutions correspond to the vertices of the feasible region in $\mathbf{R}^n$.

Example 9.5

Find a basic solution to the following system of equations, and identify the basic and nonbasic variables.

$$\begin{aligned} 2x_2 + x_3 \quad - \quad x_5 &= 2 \\ x_1 - \quad x_2 \qquad\quad + 3x_5 &= 3 \\ x_2 \quad + x_4 \qquad &= 4. \end{aligned}$$

Forming the corresponding augmented matrix, we see that it is already in row-reduced form,

$$\begin{bmatrix} 0 & 2 & 1 & 0 & -1 & | & 2 \\ 1 & -1 & 0 & 0 & 3 & | & 3 \\ 0 & 1 & 0 & 1 & 0 & | & 4 \end{bmatrix}.$$

A particular solution is then obtained by letting $x_3 = 2$ (because $A^3 = e_1$), $x_1 = 3$ (because $A^1 = e_2$), $x_4 = 4$ (because $A^4 = e_3$), and $x_2 = x_5 = 0$. The basic variables are x_1, x_3, and x_4, and the nonbasic variables are x_2 and x_5.

9.1 Exercises

In exercises 1 through 4 sketch the feasible region in $\mathbf{R}^2$ described by the constraints.

1. $x - y \le 1$
   $x + 2y \le 8$
   $x, y \ge 0$

2. $x - y \le 2$
   $3x - y \le 6$
   $x, y \ge 0$

3. $x + y \le 2$
   $-x + 3y \le 9$
   $2x - y \le 6$
   $x, y \ge 0$

4. $x - y \le 1$
   $2x + 2y \le 5$
   $2x - y \ge 6$
   $x \ge 1, y \le 2$

5. Maximize the objective function $z = 2x + y$ on the feasible region of exercise 1, or show that no maximum exists.

6. Maximize the objective function $z = 3x - y$ on the feasible region of exercise 2, or show that no maximum exists.

7. Minimize the function $z = x - 3y$ on the feasible region of exercise 3, or show that no minimum exists.

8–10. Convert each system of constraints in exercises 1 through 3 into equation form.

In exercises 11 through 14 decide whether or not the system of linear equations is row-reduced. If it is row-reduced, identify the columns that yield an embedded identity matrix, and find the corresponding basic solution.

11. $\begin{aligned} 2x_1 \quad\quad\ + x_3 - 3x_4 \quad\quad\ &= 2 \\ x_1 + x_2 \quad\quad\ - x_4 \quad\quad\ &= 1 \\ 3x_1 \quad\quad\quad\ + 2x_4 + x_5 &= 4 \end{aligned}$

12. $\begin{aligned} x_1 \quad\quad\ - x_4 &= 1 \\ x_2 \quad\ + x_4 &= 3 \\ x_3 + 2x_4 &= 5 \end{aligned}$

13. $\begin{aligned} x_1 \quad\ + x_3 \quad\quad\quad\ &= 2 \\ -x_1 + x_2 \quad\ + x_4 \quad\ &= 6 \\ 2x_1 \quad\quad\quad\quad\ + x_5 &= 1 \end{aligned}$

(*Note:* two possible answers)

14. $\begin{aligned} 2x_2 - x_3 + 2x_4 + x_5 + 2x_6 &= 2 \\ x_1 - x_2 \quad\quad\quad\quad\ - 3x_6 &= 2 \end{aligned}$

In exercises 15 and 16 convert each to a linear programming problem in standard form and then to equation form. You do *not* need to try to solve them.

15. A man wishes to invest a portion of $50,000 in savings, stocks, and property. He feels that no more than 1/5 of the total amount should be in savings and that no more than 2/5 in savings and stocks together. If he estimates an 8% return from savings, 12% from real estate, and 15% from stocks, how should he invest his money?

16. A toy manufacturer has available 200 lbs of wood, 400 lbs of moldable plastic, and 50 lbs of metal to produce toys A, B, C, and D, each of which requires some or all of the materials for component parts. Table 9.1 gives the requirements for each toy, the price of each toy, and the price of the materials used in production of the toys. How many of each toy should the manufacturer produce to maximize profit?

	Wood	Plastic	Metal	$/toy
A	0.2	0.3	0.1	5
B	0	0.5	0.2	4
C	0	0.9	0.1	3
D	1	0	0.1	6
$/lb	1	0.5	2	

Table 9.1

9.2 The Simplex Algorithm

In this section we introduce a general procedure for solving linear programming problems. Now that the translation from inequalities to equations has been described (section 9.1), we will assume that the linear programming problem is in equation form and row-reduced.

Let the linear programming problem be given as follows: optimize the linear function

$$z = \mathbf{c} \cdot \mathbf{x}$$

subject to the system of constraints

$$A\mathbf{x} = \mathbf{b}, \quad \mathbf{b} \geq \mathbf{0}, \quad \mathbf{x} \geq \mathbf{0}, \tag{1}$$

where A has an embedded identity matrix.

Let $\mathbf{c}^*$ be the $1 \times m$ matrix that consists of the components of $\mathbf{c}$ associated with the basic variables in the current basic solution. Then the value associated with the current solution is $\mathbf{c}^*\mathbf{b}$ (the nonbasic variables are zero). The m-vector $\mathbf{c}^*A - \mathbf{c}$ is called the vector of **simplex indicators.** (Here $\mathbf{c}$ is viewed as a $1 \times n$ matrix so that the operations are defined.) We call the jth component of this vector, $\mathbf{c}^*A^j - c^j$, the jth **simplex indicator.**

Associated with system (1), we prepare the array shown in figure 9.4, called the **tableau,** associated with the current solution. For convenience tableaus will sometimes be presented with the row of variables at the top and the basic variables and the vector $\mathbf{c}^*$ on the left (see, for example, figure 9.5).

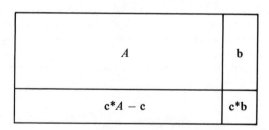

Figure 9.4

Example 9.6

Let $A\mathbf{x} = \mathbf{b}, \mathbf{b} \geq \mathbf{0}, \mathbf{x} \geq \mathbf{0}$ be the system of example 9.5,

$$
\begin{aligned}
2x_2 + x_3 \quad\quad - x_5 &= 2 \\
x_1 - x_2 \quad\quad\quad + 3x_5 &= 3 \\
x_2 \quad + x_4 \quad\quad &= 4,
\end{aligned}
$$

with objective function $z = 2x_1 - x_2 + 4x_3 - 2x_5$. Prepare the corresponding tableau.

Since the basic solution is $x_3 = 2$, $x_1 = 3$, and $x_4 = 4$, we have $\mathbf{c}^* = (c_3, c_1, c_4) = (4, 2, 0)$. The tableau is given in figure 9.5.

	$\mathbf{c}^*$	x_1	x_2	x_3	x_4	x_5	$\mathbf{b}$
x_3	4	0	2	1	0	-1	2
x_1	2	1	-1	0	0	3	3
x_4	0	0	1	0	1	0	4
		0	7	0	0	4	14

Figure 9.5

Notice in figure 9.5 the second simplex indicator has been computed $\mathbf{c}^*A^2 - c_2 = (4, 2, 0) \cdot (2, -1, 1) - (-1) = 7$. The right-most bottom row entry is $\mathbf{c}^*\mathbf{b} = 14$, the value associated with the basic solution $x_3 = 2$, $x_1 = 3$, and $x_4 = 4$.

Conditions for Terminating the Algorithm

The next theorem implies that the solution of example 9.6 is maximal. In other words, there is no solution with value greater than 14, which occurs when $x_3 = 2$, $x_1 = 3$, $x_4 = 4$, and $x_2 = x_5 = 0$. The proof of this and subsequent theorems will be delayed until section 9.4.

Theorem 1 Let a linear programming problem be given as follows: optimize $z = \mathbf{c} \cdot \mathbf{x}$, subject to the constraints

$$A\mathbf{x} = \mathbf{b}, \quad \mathbf{b} \geq 0, \quad \mathbf{x} \geq 0,$$

where A has an embedded identity matrix. If no simplex indicator of the associated tableau is negative (that is, $\mathbf{c}^*A - \mathbf{c} \geq 0$), then the corresponding basic solution is maximal. If no simplex indicator is positive, the corresponding basic solution is minimal.

Theorem 1 gives us conditions under which we have solved the problem and can terminate the procedure. The next theorem also gives us conditions under which we can cease our search for a different reason— no optimal solution exists.

Theorem 2   Let a linear programming problem be given as follows: optimize $z = \mathbf{c} \cdot \mathbf{x}$, subject to the constraints

$$A\mathbf{x} = \mathbf{b}, \quad \mathbf{b} \geq 0, \quad \mathbf{x} \geq 0,$$

where A has an embedded identity matrix. Let j be a column number such that the simplex indicator is negative ($c^*A^j - c_j < 0$). If every element of A^j is less than or equal to zero ($A^j \leq 0$), then no maximal solution exists. If everything is the same except that the simplex indicator is positive, then no minimal solution exists.

Example 9.7

Show that the following linear programming problem has no maximal solution. Subject to the constraints

$$
\begin{array}{rcl}
x_1 \quad + 2x_3 - x_4 \quad\quad + x_6 &=& 2 \\
- x_3 \quad\quad + x_5 + 3x_6 &=& 1 \\
x_2 + x_3 - 2x_4 \quad\quad + x_6 &=& 4 \\
x_j \geq 0, j = 1, \ldots, 6
\end{array}
$$

maximize the function

$$z = x_1 + 3x_2 - x_3 + 3x_4 - 2x_5.$$

We prepare the corresponding tableau, figure 9.6, using $\mathbf{c} = (1, 3, -1, 3, -2, 0)$ and $\mathbf{c^*} = (1, -2, 3)$ since the basic variables are $x_1, x_5,$ and x_2.

	c*	x_1	x_2	x_3	x_4	x_5	x_6	b
x_1	1	1	0	2	-1	0	1	2
x_5	-2	0	0	-1	0	1	3	1
x_2	3	0	1	1	-2	0	1	4
		0	0	8	-10	0	-2	12

Figure 9.6

We have $c^*A^4 - c_4 = -10$ and $A^4 = (-1, 0\ -2) \leq 0$, so by theorem 2 no maximal solution exists.

Finding a Better Solution

We now have a criterion for optimality (theorem 1) and a criterion under which no optimal solution exists (theorem 2). Next we will introduce a procedure that takes us from a basic solution that meets neither criterion to a new one that might meet one of them. This process is called **pivoting** and is just the row-reduction technique of chapter 2. Once the pivot is chosen, divide that row by the pivot value (to obtain a 1 in the pivot position), and use the pivot row to obtain zeros in the rest of the pivot column.

Example 9.8

In the following tableau (that of example 9.7 without the bottom row) pivot on the 1,3 position:

1	0	②	−1	0	1	2
0	0	−1	0	1	3	1
0	1	1	−2	0	1	4

We first divide the top row by the circled entry in the 1, 3 position to obtain

1/2	0	1	−1/2	0	1/2	1
0	0	−1	0	1	3	1
0	1	1	−2	0	1	4

Now add the top row to the second and the negative of the top row to the third to create zeros in the rest of the third column. This yields the new tableau:

1/2	0	1	−1/2	0	1/2	1
1/2	0	0	−1/2	1	7/2	2
−1/2	1	0	−3/2	0	1/2	3

The new tableau is row-reduced and corresponds to the basic solution $x_3 = 1$, $x_5 = 2$, $x_2 = 3$, and $x_1 = x_4 = x_6 = 0$. The new c^* vector is $c^* = (-1, -2, 3)$. Completing the tableau to check for optimality, we have the tableau of figure 9.7.

	c^*	x_1	x_2	x_3	x_4	x_5	x_6	b
x_3	−1	1/2	0	1	−1/2	0	1/2	1
x_5	−2	1/2	0	0	−1/2	1	7/2	2
x_2	3	−1/2	1	0	−3/2	0	1/2	3
		−4	0	0	−6	0	−6	4

Figure 9.7

By theorem 1 this new basic solution is minimal.

Notice that in example 9.8 we could have computed the bottom row of the new tableau (figure 9.7) by continuing the pivot procedure to the bottom row. In other words, the bottom row of the tableau of figure 9.6 was

$$[0 \quad 0 \quad 8 \quad -10 \quad 0 \quad -2 \mid 12].$$

If we add -8 times the pivot row to it, we obtain

$$[-4 \quad 0 \quad 0 \quad -6 \quad 0 \quad -6 \mid 4],$$

which is the bottom row of the new tableau of figure 9.7. This technique will always work, and we state it as the next theorem.

Theorem 3 The bottom row of a new tableau may be computed from the preceding tableau by using the pivot row and row operations to obtain a zero in the pivot column position of the bottom row.

In this pivoting process we must check two conditions. The simplex indicator (bottom row entry) will determine whether the value of the solution will increase or decrease in the next tableau. That is, if the simplex indicator is *negative*, that column can be chosen to *increase* the value in the next tableau; if it is *positive*, that column will cause the value to *decrease*. However, care must also be taken to assure that the next solution remains *feasible*; that is, that all entries in the new **b** column are nonnegative.

Once an appropriate column has been chosen, a row is chosen as follows. Compute the quotients of the **b** column entries divided by the corresponding *positive* entries in the desired column. Choose a row that gives the smallest such quotient, because it will lead to a new solution that is feasible. In example 9.8 we chose to pivot on the 1,3 position by the following procedure. To minimize we first chose the 3rd column, since the simplex indicator of this column, 8, is greater than zero. Then, we computed the quotients $b_1/a_{13} = 2/2 = 1$ and $b_3/a_{33} = 4/1 = 4$ (b_2/a_{23} was ignored because $a_{23} \leq 0$), using the smaller of the two quotients (the first one) as the basis for choosing the first row. The justification for the process just described is the following theorem.

Theorem 4 Let a linear programming problem be given as follows: optimize $z = \mathbf{c} \cdot \mathbf{x}$, subject to the constraints

$$A\mathbf{x} = \mathbf{b}, \quad \mathbf{b} \geq 0, \quad \mathbf{x} \geq 0,$$

where A has an embedded identity matrix. Let j be a column number such that the simplex indicator is negative, that is, $\mathbf{c}*A^j - c_j < 0$. If a_{ij} is positive for at least one row number $i = 1, \ldots, m$, let k be a row number such that $a_{kj} > 0$ and $b_k/a_{kj} \leq b_i/a_{ij}$ for all $i = 1, \ldots, m$ with $a_{ij} > 0$. Then pivoting on the k, j position leads to a new basic solution with value greater than or equal to the former. If everything is the same except that the simplex indicator is positive, the new solution will have value less than or equal to the former.

The simplex algorithm Combining theorems 1 through 4, we have the following procedure for *maximizing* a linear programming problem. To *minimize*, consideration of negative simplex indicators should be replaced by that of *positive* simplex indicators.

i. Convert the given linear programming problem to the form: Optimize $z = \mathbf{c} \cdot \mathbf{x}$ subject to the constraints

$$A\mathbf{x} = \mathbf{b}, \quad \mathbf{x} \geq \mathbf{0}, \quad \mathbf{b} \geq \mathbf{0},$$

where A has an embedded identity matrix.
ii. Prepare the first tableau.
iii. If no simplex indicator is negative, the solution is maximal, and the algorithm terminates.
iv. If some simplex indicator is negative, and no entry in that column is positive, there is no maximal solution, and the algorithm terminates.
v. If at least one simplex indicator is negative, and each such column has a positive entry, choose any such column j, and pivot on the k, j position, where $b_k/a_{kj} = \min \{b_i/a_{ij} | a_{ij} > 0\}$.
vi. Repeat step v until either step iii or step iv is satisfied.

Because of theorem 3 we only need to construct $\mathbf{c}^*$ for the first tableau. For each succeeding tableau the vector of simplex indicators $\mathbf{c}^*A - \mathbf{c}$ is obtained by pivoting. Furthermore we do not need to worry about which variables are basic and which are nonbasic until the final tableau. For these reasons, tableaux subsequent to the first will not include the columns that indicate the basic variables or $\mathbf{c}^*$.

Example 9.9

Use the simplex algorithm to solve the motivating example of section 9.1. Maximize the function $z = 2x + y$, subject to the constraints

$$2x - y \leq 0$$
$$x + y \leq 9$$
$$x, y \geq 0.$$

We first convert it to equation form by use of slack variables s and t as follows: maximize the function $z = 2x + y$, subject to the constraints

$$2x - y + s \quad\quad = 0$$
$$x + y \quad\quad + t = 9,$$
$$x, y, s, t \geq 0.$$

We now have the problem in the desired form, and we proceed to prepare the first tableau with $\mathbf{c} = (2, 1, 0, 0)$ and $\mathbf{c^*} = (0, 0)$, since the embedded identity matrix columns correspond to the slack variables s and t.

	$\mathbf{c^*}$	x	y	s	t	$\mathbf{b}$
s	0	2	-1	1	0	0
t	0	1	1	0	1	9
		-2	-1	0	0	0

Here we could choose either column 1 or column 2 as the pivot column since the simplex indicators are negative for these columns (-2 and -1, respectively). We choose column 1. Both entries in this column are positive, so to determine the pivot row, we need to form both quotients $0/2 = 0$ and $9/1 = 9$. Since $0 < 9$, the pivot row is row 1. Pivoting on the 1,1 position we have

1	$-1/2$	$1/2$	0	0
0	$3/2$	$-1/2$	1	9
0	-2	1	0	0

Now the only negative simplex indicator is in the second column, so this will be the pivot column. There is only one positive entry in the second column ($3/2 > 0$), so the new pivot row is the second row. Pivoting on the 2,2 position yields

1	0	$5/6$	$1/3$	3
0	1	$-1/3$	$2/3$	6
0	0	$1/3$	$4/3$	12

Since all of the simplex indicators are nonnegative, the resulting basic solution $x = 3$, $y = 6$, and $s = t = 0$ is maximal, and the value of the objective function at this point is 12. Of course this is the result that was obtained geometrically in section 9.1.

Note: From now on we will proceed directly from one tableau to the next without explanation, only circling the pivot entry at each stage. For example, the simplex algorithm for example 9.9 would be presented as shown in figure 9.8.

	c*	x	y	s	t	b
s	0	②	-1	1	0	0
t	0	1	1	0	1	9
		-2	-1	0	0	0
		1	-1/2	1/2	0	0
		0	③/②	-1/2	1	9
		0	-2	1	0	0
		1	0	5/6	1/3	3
		0	1	-1/3	2/3	6
		0	0	1/3	4/3	12

Figure 9.8

Example 9.10

Minimize $z = 3x_1 - x_2 - 2x_3 + x_4$, subject to the constraints

$$2x_1 - 4x_2 - x_3 + x_4 \leq 8$$
$$x_1 + x_2 + 2x_3 - 3x_4 \leq 7$$
$$x_1 - x_2 - x_3 + x_4 \leq 3$$
$$x_j \geq 0, \quad j = 1, \ldots, 4.$$

We first convert the problem to equation form: minimize $z = 3x_1 - x_2 - 2x_3 + x_4 + 0x_5 + 0x_6 + 0x_7$, subject to

$$2x_1 - 4x_2 - x_3 + x_4 + x_5 \qquad\qquad = 8$$
$$x_1 + x_2 + 2x_3 - 3x_4 \qquad + x_6 \qquad = 7$$
$$x_1 - x_2 - x_3 + x_4 \qquad\qquad + x_7 = 3$$
$$x_j \geq 0, \quad j = 1, \ldots, 7.$$

We then prepare the first tableau, figure 9.9, and use it to start the

	c*	x_1	x_2	x_3	x_4	x_5	x_6	x_7	b
x_5	0	2	-4	-1	1	1	0	0	8
x_6	0	1	①	2	-3	0	1	0	7
x_7	0	1	-1	-1	1	0	0	1	3
		-3	1	2	-1	0	0	0	0
		6	0	7	-11	1	4	0	36
		1	1	2	-3	0	1	0	7
		2	0	1	-2	0	1	1	10
		-4	0	0	2	0	-1	0	-7

Figure 9.9

simplex algorithm. The pivot used is circled in the tableau. From the fourth column we see that no minimal solution exists.

Example 9.11

A factory makes three products, products A, B, and C. The company has available each week 200 hours of lathe time, 100 hours of milling time, 100 hours of grinding time, and 80 hours of packing time. The time requirements for production of one unit of each product are given in table 9.2.

	Lathe	Mill	Grind	Pack
A	3	1	2	1
B	1	2	1	2
C	4	2	1	2

Table 9.2

If each unit of A yields a profit of \$20, each unit of B yields a profit of \$16, and each unit of C yields a profit of \$24, find a production schedule for the week that maximizes profit.

Let x_1 be the number of units of A to be produced, x_2 the number of B, and x_3 the number of C. Knowing that the total lathe time cannot exceed 200, we have

$$3x_1 + x_2 + 4x_3 \leq 200.$$

Similarly the milling, grinding, and packing limitations yield the constraints

$$x_1 + 2x_2 + 2x_3 \leq 100$$
$$2x_1 + x_2 + x_3 \leq 100$$
$$x_1 + 2x_2 + 2x_3 \leq 80.$$

The profit function is $z = 20x_1 + 16x_2 + 24x_3$, and obviously each variable must be nonnegative. Putting all these conditions together, we have the mathematical problem: maximize $z = 20x_1 + 16x_2 + 24x_3$, subject to the constraints

$$3x_1 + x_2 + 4x_3 \leq 200$$
$$x_1 + 2x_2 + 2x_3 \leq 100$$
$$2x_1 + x_2 + x_3 \leq 100$$
$$x_1 + 2x_2 + 2x_3 \leq 80$$
$$x_j \geq 0, \quad j = 1, 2, 3.$$

We restate the problem in equation form: maximize

$$z = 20x_1 + 16x_2 + 24x_3 + 0x_4 + 0x_5 + 0x_6 + 0x_7,$$

subject to the constraints

$$
\begin{aligned}
3x_1 + x_2 + 4x_3 + x_4 \qquad\qquad\qquad\qquad &= 200\\
x_1 + 2x_2 + 2x_3 \qquad + x_5 \qquad\qquad\quad &= 100\\
2x_1 + x_2 + x_3 \qquad\qquad + x_6 \quad\quad &= 100\\
x_1 + 2x_2 + 2x_3 \qquad\qquad\qquad + x_7 &= 80\\
x_j \ge 0, \quad j = 1, \cdots, 7.&
\end{aligned}
$$

We prepare the first tableau (Figure 9.10) and use it to start the simplex algorithm. The pivot used at each iteration is circled.

	c^*	x_1	x_2	x_3	x_4	x_5	x_6	x_7	b
x_4	0	3	1	4	1	0	0	0	200
x_5	0	1	2	2	0	1	0	0	100
x_6	0	2	1	1	0	0	1	0	100
x_7	0	1	2	②	0	0	0	1	80
		-20	-16	-24	0	0	0	0	0
		1	-3	0	1	0	0	-2	40
		0	0	0	0	1	0	-1	20
		③/②	0	0	0	0	1	$-1/2$	60
		1/2	1	1	0	0	0	1/2	40
		-8	8	0	0	0	0	12	960
		0	-3	0	1	0	$-2/3$	$-5/3$	0
		0	0	0	0	1	0	-1	20
		1	0	0	0	0	2/3	$-1/3$	40
		0	1	1	0	0	$-1/3$	2/3	20
		0	8	0	0	0	16/3	28/3	1280

Figure 9.10

Since no simplex indicator is negative, the solution is maximal. The conclusion is that the company should produce 40 units of A, 20 units of C, and produce 0 units of B. The associated profit is $1280 per week.

9.2 Exercises

In exercises 1 and 2 complete the first tableau for the given linear programming problem.

1. Objective function: $z = 3x_1 - 2x_2 + 2x_4 - x_5$

	c^*	x_1	x_2	x_3	x_4	x_5	b
x_3	0	3	1	1	0	0	1
x_4	2	-1	2	0	1	0	5
x_5	-1	4	5	0	0	1	9

2. Objective function: $z = 2x_1 + x_2 - 3x_3 + x_4$

c^*	x_1	x_2	x_3	x_4	x_5	x_6	x_7	x_8	b
	2	1	1	0	3	0	-2	0	10
	1	0	-1	0	1	0	1	1	9
	1	0	1	0	2	1	2	0	7
	3	0	2	1	0	0	-1	0	2

In exercises 3 through 6 prepare the first tableau for each linear programming problem.

3. Maximize $z = x_1 - 2x_2 + x_3 + 2x_5$, subject to

$$3x_1 + x_2 + 2x_3 \qquad\qquad = 2$$
$$x_1 \qquad - 3x_3 \qquad + x_5 = 5$$
$$2x_1 \qquad - x_3 + x_4 \qquad = 8$$
$$x_j \geq 0, \quad j = 1, \cdots, 5.$$

4. Minimize $z = 2x_1 - x_2 + x_3$, subject to

$$x_1 + x_2 + x_3 \leq 10$$
$$2x_1 - x_2 - x_3 \leq 5$$
$$x_1 + 3x_2 - 4x_3 \leq 1$$
$$x_j \geq 0, \quad j = 1, 2, 3.$$

5. Minimize $z = x_1 - 2x_2 + x_3 + 4x_4$, subject to

$$2x_1 + x_2 + x_3 + 3x_4 \leq 12$$
$$x_1 - 3x_2 - x_3 + x_4 \leq 8$$
$$x_j \geq 0, \quad j = 1, \cdots, 4.$$

6. Maximize $z = x_1 - x_2 + 2x_3$, subject to

$$x_1 - 3x_2 - x_3 \leq 6$$

$$x_1 + 3x_2 - 4x_3 \leq 8$$
$$x_1 - x_2 - 3x_3 \leq 10$$
$$x_j \geq 0, \quad j = 1, 2, 3.$$

In exercises 7 through 10 explain why each tableau is the final tableau for some linear programming problem. If the solution is optimal, give the optimal solution and corresponding value of the objective function.

7. Maximize:

x_1	x_2	x_3	x_4	x_5	b
-3	1	0	-2	0	2
-1	0	0	3	1	4
-2	0	1	1	0	1
2	0	0	3	0	5

8. Same as exercise 7 only *minimize*.

9. Minimize:

x_1	x_2	x_3	x_4	x_5	x_6	b
0	1	1	-3	1	0	8
1	2	0	-2	-3	4	11
0	-2	0	-1	-2	-5	-8

10. Same as exercise 9 only *maximize*.

11–14. Solve completely each of exercises 3 through 6.

15–16. Solve exercises 15 and 16 of section 9.1.

9.3 The Two-Phase Problem

In this section we show how other types of problems can be put in the form needed for the simplex algorithm.

The Two-Phase Problem

So far we have only considered problems where the constraints are of the type

$$A\mathbf{x} \leq \mathbf{b}, \quad \mathbf{b} \geq \mathbf{0}, \quad \text{and} \quad \mathbf{x} \geq \mathbf{0}.$$

What if we have constraints of the type $A_i \cdot \mathbf{x} \geq b_i$ for $b_i \geq 0$? To handle this situation we introduce a slack variable, as before, but this time we *subtract* it. That is, $A_i \cdot \mathbf{x} \geq b_i$ is equivalent to the equation $A_i \cdot \mathbf{x} - x_{n+1} = b_i$, where $x_{n+1} \geq 0$ (and $\mathbf{x} = (x_1, \ldots, x_n)$). Unfortunately a problem arises here. We no longer have an embedded identity matrix to start the solution process. The same problem arises if we have equations as well as inequalities in the system.

Example 9.12

Convert the following problem to equation form:

$$3x_1 - 2x_2 + 5x_3 \leq 14$$
$$x_1 - x_2 - 2x_3 \geq 1$$
$$x_2 - x_3 = 5$$
$$x_j \geq 0, \quad j = 1, 2, 3.$$

We add a slack variable to the first equation and subtract one from the second to obtain the system:

$$3x_1 - 2x_2 + 5x_3 + x_4 \qquad = 14$$
$$x_1 - x_2 - 2x_3 \qquad - x_5 = 1$$
$$x_2 - x_3 \qquad = 5$$
$$x_j \geq 0, \quad j = 1, \cdots, 5.$$

In example 9.12 we have only one elementary column (the fourth) and the simplex algorithm cannot be initiated. There is, however, a device that allows us to use the same method. As needed, we add new variables, called **artificial variables,** to create an embedded identity matrix. In this case we add one artificial variable to each of the second and third constraints to obtain

$$3x_1 - 2x_2 + 5x_3 + x_4 \qquad = 14$$
$$x_1 - x_2 - 2x_3 \qquad - x_5 + x_6 \qquad = 1$$
$$x_2 - x_3 \qquad + x_7 = 5$$
$$x_j \geq 0, \quad j = 1, \ldots, 7.$$

Now we have an embedded identity matrix: the entries of the fourth, sixth, and seventh columns.

Unfortunately, the new system is *not* equivalent to the original. In the last equation, for example, if x_7 is strictly positive, then $x_2 - x_3 < 5$ instead of $x_2 - x_3 = 5$. If $x_7 = 0$, however, we have $x_2 - x_3 + 0 = 5$ which is equivalent to the original. The idea then is to use the basic solution that has been "rigged" ($x_4 = 14$, $x_6 = 1$, $x_7 = 5$) to start the simplex algorithm and then try to move to a basic solution in which the artificial variables are *not* basic. To this end we introduce an objective function that is zero if the artificial variables are all zero and positive otherwise. This objective function is taken to be the sum of the artificial variables. In the case of example 9.12 this is $z = x_6 + x_7$. We then use the simplex method to *minimize* this function. If the minimum value is zero, we have a solution—in fact a *basic* solution—to the original problem. If the minimum value is positive, the original system can have *no* solutions, that is, the constraints are inconsistent.

Example 9.13

Find a basic solution to the following system or show that none exists.

$$3x_1 - 2x_2 + 5x_3 \le 14$$
$$x_1 - x_2 - 2x_3 \ge 1$$
$$x_2 - x_3 = 5$$
$$x_j \ge 0, \quad j = 1, 2, 3.$$

Including the slack variables x_4 and x_5 and artificial variables x_6 and x_7, we minimize the function $z = x_6 + x_7$. The tableaus are given in Figure 9.11.

	c*	x_1	x_2	x_3	x_4	x_5	x_6	x_7	b
x_4	0	3	−2	5	1	0	0	0	14
x_6	1	①	−1	−2	0	−1	1	0	1
x_7	1	0	1	−1	0	0	0	1	5
		1	0	−3	0	−1	0	0	6
		0	1	11	1	3	−3	0	11
		1	−1	−2	0	−1	1	0	1
		0	①	−1	0	0	0	1	5
		0	1	−1	0	0	−1	0	5
		0	0	12	1	3	−3	−1	6
		1	0	−3	0	−1	1	1	6
		0	1	−1	0	0	0	1	5
		0	0	0	0	0	−1	−1	0

Figure 9.11

From the final tableau of figure 9.11 we see that x_6 and x_7 are non-basic variables, so the basic solution that corresponds to the final tableau is a solution to the original problem. That is, $x_4 = 6$, $x_1 = 6$, $x_2 = 5$, and $x_3 = x_5 = 0$ is a basic solution to the original problem. Ignoring variables x_6 and x_7, the final tableau corresponds to the following system of equations,

$$
\begin{aligned}
12x_3 + x_4 + 3x_5 &= 6 \\
x_1 - 3x_3 \quad\quad - x_5 &= 6 \\
x_2 - x_3 \quad\quad\quad &= 5.
\end{aligned} \tag{1}
$$

The process just described is called the **first phase** of the linear programming problem, locating a basic feasible solution. The **second phase** is taking the result of the first phase to construct the first tableau with the *original* objective function. From then on the problem is optimized in the usual manner.

Example 9.14

Maximize the function $z = x_1 - 2x_2 + 2x_3$, subject to the constraints

$$
\begin{aligned}
3x_1 - 2x_2 + 5x_3 &\leq 14 \\
x_1 - x_2 - 2x_3 &\geq 1 \\
x_2 - x_3 &= 5 \\
x_j \geq 0, \quad j &= 1, 2, 3.
\end{aligned}
$$

Since the constraints are exactly those of example 9.12, the first phase has already been completed, and we start from there. The first tableau of the second phase is constructed from system (1) of example 9.13 and the objective function $z = x_1 - 2x_2 + 2x_3 + 0x_4 + 0x_5$. Notice that except for the bottom row, the columns of the first tableau will be exactly those of the final tableau of the first stage, without the columns that correspond to the artificial variables. The tableaus of the second phase are shown in figure 9.12.

	c*	x_1	x_2	x_3	x_4	x_5	b
x_4	0	0	0	12	1	③	6
x_1	1	1	0	−3	0	−1	6
x_2	−2	0	1	−1	0	0	5
		0	0	−3	0	−1	−4
		0	0	4	1/3	1	2
		1	0	1	1/3	0	8
		0	1	−1	0	0	5
		0	0	1	1/3	0	−2

Figure 9.12

Since no simplex indicators are negative, we have the maximal solution $x_1 = 8$, $x_2 = 5$, $x_5 = 2$, and $x_3 = x_4 = 0$.

We summarize the preceding discussion with the following procedure.

The two-phase procedure

i. Add artificial variables to create an embedded identity matrix.
ii. Use the simplex algorithm to minimize this linear programming problem where the objective function is the sum of the artificial variables.
iii. If the minimum is not zero, there is no solution to the system of constraints.
iv. If the minimum is zero, use the resulting basic solution, and delete the columns that correspond to artificial variables to create a first tableau with the original objective function.
v. Use the simplex algorithm to optimize the resulting linear programming problem.

Note: In some kinds of problems, knowing whether or not a set of constraints has *any* solution is the only important question. In this case only the first phase is necessary.

Example 9.15

Show that there is no solution to the set of constraints

$$4x_1 + 2x_2 - x_3 \leq 5$$
$$2x_1 + x_2 + x_3 \geq 8$$
$$-2x_1 - x_2 + 2x_3 \leq 2$$
$$x_j \geq 0, \quad j = 1, 2, 3.$$

Adding slack variables to the first and third constraints and subtracting one from the second, we obtain the equivalent set of equations,

$$4x_1 + 2x_2 - x_3 + x_4 \qquad\qquad = 5$$
$$2x_1 + x_2 + x_3 \qquad - x_5 \qquad = 8$$
$$-2x_1 - x_2 + 2x_3 \qquad\qquad + x_6 = 2$$
$$x_j \geq 0, \quad j = 1, \dots, 6.$$

Adding an artificial variable to the second constraint, we obtain the (nonequivalent) first-phase problem: minimize $z = x_7$, subject to the constraints

$$4x_1 + 2x_2 - x_3 + x_4 \qquad\qquad\qquad = 5$$
$$2x_1 + x_2 + x_3 \qquad - x_5 \qquad + x_7 = 8$$
$$-2x_1 - x_2 + 2x_3 \qquad\qquad + x_6 \qquad = 2$$
$$x_j \geq 0, \quad j = 1, \dots, 7.$$

The tableaus for the simplex algorithm are shown in figure 9.13. Since this solution is minimal, and the artificial variable is still in the corresponding basic solution ($x_1 = 2$, $x_3 = 3$, $x_7 = 1$), there is no solution to the original problem.

	c^*	x_1	x_2	x_3	x_4	x_5	x_6	x_7	b
x_4	0	④	2	−1	1	0	0	0	5
x_7	1	2	1	1	0	−1	0	1	8
x_6	0	−2	−1	2	0	0	1	0	2
		2	1	1	0	−1	0	0	8
		1	1/2	−1/4	1/4	0	0	0	5/4
		0	0	3/2	−1/2	−1	0	1	11/2
		0	0	③/②	1/2	0	1	0	9/2
		0	0	3/2	−1/2	−1	0	0	11/2
		1	1/2	0	1/3	0	1/6	0	2
		0	0	0	−1	−1	−1	1	1
		0	0	1	1/3	0	2/3	0	3
		0	0	0	−1	−1	−1	0	1

Figure 9.13

9.3 Exercises

In exercises 1 through 4 convert each problem to the form $A\mathbf{x} = \mathbf{b}$, $\mathbf{b} \geq 0$, $\mathbf{x} \geq 0$, where A has an embedded identity matrix, by use of slack and/or artificial variables.

1.
$$2x - 3y - z \leq 5$$
$$y + 2z \geq 8$$
$$x - 2z = 3$$
$$x, y, z \geq 0$$

2.
$$x_1 + 2x_2 + x_4 \geq 6$$
$$x_2 + x_3 + 4x_4 \geq 12$$
$$x_1 - 2x_3 + x_4 \leq 4$$
$$x_j \geq 0, \quad j = 1, \ldots, 4$$

3.
$$2x - y = 7$$
$$x + 2y \leq 18$$
$$3x - 5y \leq -3$$
$$x, y \geq 0$$

4.
$$x_1 + 2x_2 - x_3 = 5$$
$$4x_2 - x_3 + x_4 \geq 8$$
$$x_2 - 3x_3 + x_4 \leq -2$$
$$x_j \geq 0, \quad j = 1, \ldots, 4$$

5–8. Find a basic solution to each of exercises 1 through 4 or show that none exists by using the simplex algorithm with objective function given by the sum of the artificial variables.

In exercises 9 through 12 use the results of exercises 5 through 8 and the simplex algorithm to optimize the given objective function subject to the constraints imposed in exercises 1 through 4.

9. Maximize $z = x + 3y - z$

10. Minimize $z = x_1 + 2x_3 + x_4$

11. Minimize $z = 2x - 3y$

12. Minimize $z = x_1 + x_2 + x_3 + x_4$

In exercises 13 and 14 use the two-phase method to solve each linear programming problem.

13. Maximize the function $z = 2x_1 + x_2 + 2x_4$, subject to

$$\begin{align}
x_1 + 2x_2 - x_3 &\geq 5 \\
x_2 + x_3 + x_4 &\leq 12 \\
2x_1 - x_2 - 3x_3 - 2x_4 &\leq -10 \\
x_j \geq 0, \quad j = 1, \ldots, 4.
\end{align}$$

14. Maximize the function $z = x_1 - x_2 + x_3 - x_4$, subject to

$$\begin{align}
x_2 + 3x_3 + x_4 &\leq 20 \\
x_1 - 3x_2 + x_3 - 2x_4 &\geq 5 \\
x_2 - x_3 + 3x_4 &\geq 6 \\
x_j \geq 0, \quad j = 1, \ldots, 4.
\end{align}$$

15. *The diet problem:* a company makes dog food from a mixture of four different ingredients A, B, C, D that contains three basic nutrients P, Q, R in various quantities. The chart of figure 9.14 gives the amount of each nutrient in one pound of each ingredient. The column at the right gives the price per pound (in cents) of each ingredient, and the row at the bottom gives the minimum daily requirement of each nutrient per pound of final product. Find the dog food mix that supplies all of the needs at the cheapest price.

	P	Q	R	
A	3	2	3	40
B	3	1	1	20
C	0	4	4	10
D	1	2	3	20
	2	2	3	

Figure 9.14

16. *The transportation problem:* a company has a particular raw material stored at warehouses W_1, W_2, W_3 that is required by factories F_1, F_2. The chart of figure 9.15 gives the number of miles (in hundreds) between each warehouse and each factory. The column at the right gives the quantity of material available at each warehouse, and the row at the bottom gives the amount needed by each factory. Find a shipping formula that supplies the needs of each factory and minimizes the total shipping distance.

	F_1	F_2	
W_1	4	8	6
W_2	2	1	8
W_3	0	3	5
	7	4	

Figure 9.15

9.4   Theory of Linear Programming

We now examine the proofs of Theorems 1-4 of Section 9.2 which describe the simplex algorithm. We consider only the problem of *maximizing*. The problem of minimizing requires only minor modifications and will be omitted.

Theorem 1   Let a linear programming problem be given as follows: maximize $z = \mathbf{c} \cdot \mathbf{x}$, subject to the constraints

$$A\mathbf{x} = \mathbf{b}, \quad \mathbf{b} \geq \mathbf{0}, \quad \mathbf{x} \geq \mathbf{0},$$

where A has an embedded identity matrix. If no simplex indicator is negative (that is, $\mathbf{c}^*A - \mathbf{c} \geq \mathbf{0}$) then the corresponding basic solution is maximal.

Proof   As usual, assume that A is an $m \times n$ matrix. By renumbering the variables, if necessary, we may assume that the embedded identity matrix occurs in the first m columns and in their natural order. $A\mathbf{x} = \mathbf{b}$ has the form

$$
\begin{aligned}
x_1 \quad\quad &+ a_{1,m+1}x_{m+1} + \cdots + a_{1,n}x_n = b_1 \\
x_2 \quad &+ a_{2,m+1}x_{m+1} + \cdots + a_{2,n}x_n = b_2 \\
&\quad\vdots \\
x_m &+ a_{m,m+1}x_{m+1} + \cdots + a_{m,n}x_n = b_m.
\end{aligned}
\tag{1}
$$

The corresponding basic solution is obtained by letting $x_i = b_i$ for $i = 1, \ldots, m$ and $x_{m+1} = x_{m+2} = \cdots = x_n = 0$. Then with the truncated cost vector $\mathbf{c}^* = (c_1, \ldots, c_m)$ the value of the current basic solution is

$$
\begin{aligned}
z &= c_1 b_1 + c_2 b_2 + \cdots + c_m b_m + c_{m+1}(0) + \cdots + c_n(0) \\
&= c_1 b_1 + c_2 b_2 + \cdots + c_m b_m = \mathbf{c}^*\mathbf{b}.
\end{aligned}
$$

From the theory of linear equation of chapter 2 we have the *general* solution to the system of linear equations (1) as a parameterized family by letting $x_{m+1} = s_1, x_{m+2} = s_2, \ldots, x_n = s_{n-m}$, and then

$$
\begin{aligned}
x_1 &= b_1 - a_{1,m+1}s_1 - \cdots - a_{1,n}s_{n-m} \\
x_2 &= b_2 - a_{2,m+1}s_1 - \cdots - a_{2,n}s_{n-m} \\
&\vdots \\
x_m &= b_m - a_{m,m+1}s_1 - \cdots - a_{m,n}s_{n-m}. \\
x_{m+1} &= \phantom{b_m - a_{m,m+1}} s_1 \\
&\vdots \\
x_n &= \phantom{b_m - a_{m,m+1}s_1 - \cdots -} s_{n-m}.
\end{aligned}
\tag{2}
$$

The value associated with this general solution $\mathbf{x}$ is given by

$$
\begin{aligned}
z = \mathbf{c} \cdot \mathbf{x} &= c_1 x_1 + c_2 x_2 + \cdots + c_n x_n \\
&= c_1 b_1 - c_1 a_{1,m+1}s_1 - \cdots - c_1 a_{1,n}s_{n-m} \\
& + c_2 b_2 - c_2 a_{2,m+1}s_1 - \cdots - c_2 a_{2,n}s_{n-m} \\
& \vdots \\
& + c_m b_m - c_m a_{m,m+1}s_1 - \cdots - c_m a_{m,n}s_{n-m} \\
& + c_{m+1}s_1 + \cdots + \phantom{a_{m,m}} c_n s_{n-m}.
\end{aligned}
$$

Simplifying by columns, we have

$$
z = \mathbf{c}^* \mathbf{b} - (\mathbf{c}^* A^{m+1} - c_{m+1})s_1 - (\mathbf{c}^* A^{m+2} - c_{m+2})s_2 - \cdots - (\mathbf{c}^* A^n - c_n)s_{n-m}.
$$

Since $\mathbf{c}^* A^{m+j} - c_{m+j} \geq 0$ and $s_j \geq 0$ for each $j = 1, \ldots, n - m$, the value of z is maximal when $s_1 = s_2 = \cdots = s_{n-m} = 0$. Thus the corresponding basic solution is maximal. ◾

An important special case of the final expression for z is the following.

Corollary Let $A\mathbf{x} = \mathbf{b}$, $\mathbf{b} \geq 0$, $\mathbf{x} \geq 0$, and $z = \mathbf{c} \cdot \mathbf{x}$ be as in theorem 1. Let j be the number of a nonbasic variable, and let $\mathbf{x}$ be the solution obtained from the basic solution by letting $x_j = s$ and the other nonbasic variables be zero. Then the value of the solution $\mathbf{x}$ is given by

$$
z = \mathbf{c}^* \mathbf{b} - (\mathbf{c}^* A^j - c_j)s.
$$

Thus we see that, if $\mathbf{c}^* A^j - c_j < 0$, a larger value of z results in the new solution (if $s > 0$). That is, if the simplex indicator is negative, the new solution will be better for the purpose of maximizing. In fact the equation $z = \mathbf{c}^* \mathbf{b} - (\mathbf{c}^* A^j - c_j)s$ expresses the objective function value

z as a *linear* function of s. In other words, $-(c*A^j - c_j)$ measures the *rate of increase* (or *decrease* if $c*A^j - c_j$ is positive) in the value of z as a function of s. If there are no restrictions on the size of s, the value of z is unbounded, and no maximum occurs. This is the situation of theorem 2. If s is restricted, we make s as large as the restrictions will allow. This yields a new basic solution and is the situation of theorem 4. We restate both theorems and prove them simultaneously.

Let a linear programming problem be given as follows: optimize $z = c \cdot x$ subject to the constraints

$$Ax = b, \quad b \geq 0, \quad x \geq 0,$$

where A has an embedded identity matrix. Let j be a column number such that the simplex indicator is negative, that is, $c*A^j - c_j < 0$.

Theorem 2   If no element of A^j is positive ($A^j \leq 0$), then no maximal solution exists.

Theorem 4   If a_{ij} is positive for at least one row number $i = 1, \ldots, m$, let k be a row number such that $a_{kj} > 0$ and $b_k/a_{kj} \leq b_i/a_{ij}$ for all $i = 1, \ldots, m$ such that $a_{ij} > 0$. Then pivoting on the k,j position leads to a new basic solution with value greater than or equal to the former.

Proof   From system (2) the solution x obtained from the basic solution $x_1 = b_1$, $\ldots, x_m = b_m$, $x_{m+1} = \cdots = x_n = 0$ by letting nonbasic variable $x_j = s$ is given by

$$x_1 = b_1 - a_{1j}s$$
$$x_2 = b_2 - a_{2j}s$$
$$\vdots$$
$$x_m = b_m - a_{mj}s \qquad (3)$$
$$x_j = \qquad s$$
$$x_i = 0 \quad \text{if} \quad i \neq 1, \ldots, m \quad \text{and} \quad i \neq j.$$

From this we see that x is a solution only if $s \geq 0$ (since $x_j = s$) and, for $i = 1, \ldots, m$, $b_i - a_{ij}s \geq 0$. If $a_{ij} \leq 0$, then $b_i - a_{ij}s \geq b_i \geq 0$, so this condition is satisfied for any value of $s \geq 0$. If $A^j \leq 0$, that is, if *every* $a_{ij} \leq 0$, we have no constraints on s (except $s \geq 0$), and by the corollary to theorem 1, theorem 2 is proved. If, however, $a_{ij} > 0$ for some i, the condition $b_i - a_{ij}s \geq 0$ implies that $s \leq b_i/a_{ij}$. Thus s must be chosen less than or equal to b_i/a_{ij} for each $i = 1, \ldots, m$ such that $a_{ij} > 0$. By choosing $s = \min \{b_i/a_{ij} | a_{ij} > 0\}$, we obtain a new solution. Theorem 4 will be proved when it is shown that this new solution is the result of pivoting on the k,j position, where $s = b_k/a_{kj}$. To this end we put system (1) in augmented matrix form (emphasizing the pivot column j and pivot row k),

$$\begin{bmatrix} 1 & & & \cdots a_{1j} \cdots & b_1 \\ & \ddots & & \vdots & \vdots \\ & & 1 & \textcircled{a_{kj}} \cdots & b_k \\ & & & \ddots & \vdots & \vdots \\ & & 1 \cdots & a_{mj} \cdots & b_m \end{bmatrix}. \tag{4}$$

Pivoting on the k,j position, we obtain the new matrix

$$\begin{bmatrix} 1 & -a_{1j}/a_{kj} & & \cdots 0 \cdots & b_1 - a_{1j}(b_k/a_{kj}) \\ & \ddots & & \vdots & \vdots \\ & 1/a_{kj} & & \cdots 1 \cdots & b_k/a_{kj} \\ & \vdots & & \vdots & \vdots \\ & -a_{mj}/a_{kj} & 1 \cdots & 0 \cdots & b_m - a_{mj}(b_k/a_{kj}) \end{bmatrix}. \tag{5}$$

Since $s = b_k/a_{kj}$, the corresponding basic solution is that of system (3) as desired. ∎

The final theorem needed for the simplex algorithm states that a new bottom row may be computed by continuing row operations to the bottom row.

Theorem 3 The bottom row of a new tableau may be computed from the preceding tableau by using the pivot row and row operations to obtain a zero in the pivot column position of the bottom row.

Proof Again we assume that the current basic solution is in terms of the first m variables and in their natural order. Then matrix (4) of the preceding proof is the "heart" of the simplex tableau for the current solution and matrix (5) is that of the new tableau. The bottom row of the first tableau is given by $\mathbf{c}^*A - \mathbf{c}$ for the first n entries and $\mathbf{c}^*\mathbf{b}$ for the last, where $\mathbf{c}^* = (c_1, \ldots, c_m)$. Then the entry in the jth (the pivot column) position is $\mathbf{c}^*A^j - c_j$. To obtain a zero in that position by use of the pivot row A_k, we must add $-[(\mathbf{c}^*A^j - c_j)/a_{kj}]A_k$ to the bottom row. The result is that the first n positions in the bottom row will be given by

$$(\mathbf{c}^*A - \mathbf{c}) - [(\mathbf{c}^*A^j - c_j)/a_{kj}]A_k$$

and the last position by

$$\mathbf{c}^*\mathbf{b} - [(\mathbf{c}^*A^j - c_j)/a_{kj}]b_k.$$

To complete the proof, we must show that this is also the bottom row if we construct the bottom row from matrix (5), $\mathbf{c}^{*\prime}A' - \mathbf{c}$, where $\mathbf{c}^{*\prime}$ is the $1 \times m$ matrix of components of $\mathbf{c}$ that corresponds to basic variables in the new solution and A' is the new tableau. In the new solution x_j has replaced x_k, so $\mathbf{c}^{*\prime} = (c_1, \ldots, c_{k-1}, c_j, c_{k+1}, \ldots, c_n)$. The matrix A' from matrix (5) is given by its rows as $A'_i = A_i - \dfrac{a_{ij}}{a_{kj}} A_k$ for $i \neq k$ and $A'_k = \dfrac{1}{a_{kj}} A_k$. We can then compute $\mathbf{c}^{*\prime}A' - \mathbf{c}$ by rows as follows:

$$\mathbf{c}^{*\prime}A' - \mathbf{c} = c_1 \left(A_1 - \frac{a_{1j}}{a_{kj}} A_k \right) + \cdots + c_j \left(\frac{1}{a_{kj}} A_k \right)$$

$$+ \cdots + c_m (A_m - \frac{a_{mj}}{a_{kj}} A_k) - \mathbf{c}$$

$$= (c_1 A_1 + \cdots + c_m A_m - c_k A_k)$$

$$- \left(\frac{c_1 a_{1j}}{a_{kj}} + \cdots - \frac{c_j}{a_{kj}} + \cdots + \frac{c_m a_{mj}}{a_{kj}} \right) A_k - \mathbf{c}$$

$$= \mathbf{c}^* A - [(c_1 a_{1j} + \cdots + c_m a_{mj} - c_j)/a_{kj}] A_k - \mathbf{c}$$

$$= \mathbf{c}^* A - [(\mathbf{c}^* A^j - c_j)/a_{kj}] A_k - \mathbf{c}$$

$$= (\mathbf{c}^* A - \mathbf{c}) - [(\mathbf{c}^* A^j - c_j)/a_{kj}] A_k.$$

Thus the first n positions in the bottom row are correct, and we only need check the last position. This entry is just the value of the new basic solution, and by the corollary to theorem 1 with $s = b_k/a_{kj}$ this position must be

$$\mathbf{c}^* \mathbf{b} - (\mathbf{c}^* A^j - c_j)(b_k/a_{kj}).$$

Since this agrees with the row operation result, the proof is complete. ∎

One natural question that remains is whether or not the process must terminate. In other words, if we are maximizing the objective function, are we guaranteed that eventually some tableau must satisfy the conditions of theorem 1 or theorem 2? The answer is, "Not quite, but close enough." Although examples have been constructed that *cycle* (in other words, repeat the same sequence of basic solutions ad infinitum without locating the optimal value), none have arisen by accident. Thus in practical terms it is not necessary to take such situations into account. In one important case (the most common situation) it is easy to prove that the process must terminate as the next theorem shows.

Theorem 5 If in each tableau of the simplex algorithm the basic solution **b** is strictly positive, then the process terminates in a finite number of steps.

Proof Assume the problem is to maximize the objective function. Then at any stage in the algorithm we chose a column j for which $\mathbf{c}*A^j - c_j < 0$ and let $s = \min \{b_i/a_{ij}|a_{ij} > 0\}$. Since $\mathbf{b} > \mathbf{0}$, each $b_i > 0$, and thus $s > 0$. By the corollary to theorem 1 the value of the next solution is $\mathbf{c}*\mathbf{b} - (\mathbf{c}*A^j - c_j)s$ which is strictly greater than the value of the current solution, $\mathbf{c}*\mathbf{b}$. That is, the value associated with each basic solution *strictly increases* with each iteration of the method, and thus no solution can be repeated or the value would drop back to its former level, a contradiction. The proof is then complete if we know that there are only finitely many basic solutions. This follows from the fact that any basic solution arises as a transformation of the original system $A\mathbf{x} = \mathbf{b}$ into a new row-equivalent one with elementary columns that correspond to the basic solution variables. Since A has only finitely many columns, and since any such row-reduced form is unique, there are only finitely many such matrices. Thus there are only finitely many basic solutions. ∎

9.4 Exercises

1. Each of the theorems in this section was stated as a maximizing problem. Prove the analogous results for minimizing problems by proving the following statement. A solution $\mathbf{x}_0$ maximizes a linear function $z = \mathbf{c} \cdot \mathbf{x}$ subject to some set of constraints if and only if it minimizes the linear function $z' = (-\mathbf{c}) \cdot \mathbf{x}$ subject to the same set of constraints.

2. At some stage in the operation of the simplex algorithm there may be several simplex indicators that will improve the value of the objective function if chosen for the pivot column. Show that the best choice is the simplex indicator

$$\mathbf{c}*A^j - c_j \text{ such that } |(\mathbf{c}*A^j - c_j)q_j|$$

   is greatest where

$$q_j = \min \{b_i/a_{ij}|a_{ij} > 0\}.$$

3. Theorem 5 guarantees that if $\mathbf{b} > \mathbf{0}$ at each stage, the simplex algorithm will terminate in a finite number of iterations. If $\mathbf{b}$ has some zero entries, this need not be the case. Demonstrate this fact by showing that the following problem cycles if the simplex indicator of proper sign and maximal absolute value are always chosen. If a choice between pivot rows occurs, always choose the first one. (The example is due to Beale.)
   Minimize the linear function

$$z = -0.75x_1 + 150x_2 - 0.02x_3 + 6x_4,$$

subject to the constraints

$$0.25x_1 - 60x_2 - 0.04x_3 + 9x_4 \le 0$$
$$0.50x_1 - 90x_2 - 0.02x_3 + 3x_4 \le 0$$
$$x_3 \qquad\qquad \le 1$$
$$x_j \ge 0, \quad j = 1, \cdots, 4.$$

4. Find the optimal solution to the problem in exercise 3 by row-reducing the equational form and obtaining a matrix with embedded identity matrix in columns 1, 3, and 4. Form the corresponding tableau and proceed in the usual manner.

Appendix: Computer Programs

The interactive programs in this appendix can be used to reduce or even eliminate the computations inherent in various linear algebra problems. Programs similar to these are often included in the software package associated with a computer system. If this is the case, it is not necessary to bother with the ones given here. If not, these programs can be entered once by anyone with minimal computer experience and then *used by everyone* in the class and all subsequent classes.

Once a program is entered onto a computer system, its use is self-explanatory and requires *no computer experience* to operate. All that is required is the ability to "log on" and access the program. The interactive mode allows the input to be entered at the terminal in a very natural manner, and the response is immediate.

The programs given here have one distinct advantage over the canned programs of commercial software packages. They allow coefficients that are fractions to be entered as quotients of integers rather than as decimal equivalents. Moreover answers are given in the same form, provided the denominators are not too large. That is, if $x = 1/7$, that is the way the answer will be given rather than $x = 0.142857$. Since this feature requires a sizable percentage of each program, we also indicate a "bare-bones" alternative to each.

It should be noted that these programs are designed for instructional purposes rather than for efficiency or universality. For example, the language chosen is BASIC since it lends itself most easily to the matrix algebra of this text and can be easily read by someone with no computer background. As another example, the programs assume that a coefficient smaller than 0.00001 is zero. This is no limitation for the kinds of problems contained in this book but would be disastrous if a problem were carefully written to demonstrate the shortcoming.

Row-Reduction to Row-Reduced Echelon Form

This program is by far the most important, and there is pedagogical justification for allowing only this one throughout the course. Except for matrix arithmetic problems nearly every question that requires computation can be completely answered or substantially simplified by row-reducing an appropriate matrix.

Omitting lines 70–100 and 130–250 suppresses the ability to enter fractions; omitting lines 520–670 and 720–800 does the same for output.

```
10 DIM B(10,15),C(15)
20 DIM E(10,15),F(10,15)
30 PRINT ""
40 PRINT "ENTER THE DIMENSIONS OF THE MATRIX AS M,N"
50 INPUT M,N
60 PRINT ""
70 PRINT "DO YOU WANT TO ENTER FRACTIONS?  YES OR NO"
80 INPUT A$
90 PRINT ""
100 IF A$="YES" THEN GO TO 140
110 PRINT "ENTER THE MATRIX BY ROWS (USE LINE FEED
    BETWEEN ROWS)
120 MAT INPUT B(M,N)
130 GO TO 260
140 PRINT "ENTER MATRIX OF NUMERATORS (BY ROWS, USE LINE
    FEED BETWEEN ROWS)
150 MAT INPUT B(M,N)
160 PRINT ""
170 PRINT "ENTER MATRIX OF DENOMINATORS BY ROWS"
180 MAT INPUT F(M,N)
190 FOR I=1 TO M
200 FOR J=1 TO N
210 IF ABS(F(I,J))<.00001 THEN PRINT "INVALID DENOMINATOR"
220 IF ABS(F(I,J))<.00001 THEN GO TO 820
230 LET B(I,J)=B(I,J)/F(I,J)
240 NEXT J
250 NEXT I
260 LET J=1
270 FOR I=1 TO M
280 LET K=I-1
290 LET K=K+1
300 IF ABS(B(K,J))>.00001 THEN GO TO 350
310 IF K < M THEN GO TO 290
320 LET J=J+1
330 IF J>N THEN GO TO 520
340 GO TO 280
350 LET D=B(K,J)
360 FOR J1=J TO N
370 LET C(J1)=B(I,J1)
380 LET B(I,J1)=B(K,J1)
390 LET B(K,J1)=C(J1)
400 LET B(I,J1)=B(I,J1)/D
410 NEXT J1
420 FOR I1=1 TO M
430 IF I1=I THEN GO TO 490
440 LET D=B(I1,J)
450 FOR J1=J TO N
460 LET B(I1,J1)=B(I1,J1)-B(I,J1)*D
470 IF ABS(B(I1,J1))<.00001 THEN LET B(I1,J1)=0
480 NEXT J1
490 NEXT I1
500 LET J=J+1
```

```
510 NEXT I
520 LET T=1
530 FOR I=1 TO M
540 FOR J=1 TO N
550 LET K=1
560 LET G=ABS(B(I,J))
570 IF G*K-INT(G*K)<.00001 THEN GO TO 630
580 IF G*K-INT(G*K)>.99999 THEN GO TO 630
590 LET K=K+1
600 IF K>10000 THEN GO TO 680
610 LET T=2
620 GO TO 570
630 LET E(I,J)=B(I,J)*K
640 LET F(I,J)=K
650 NEXT J
660 NEXT I
670 IF T=2 THEN GO TO 730
680 PRINT ""
690 PRINT "THE ROW-REDUCED ECHELON FORM IS:"
700 PRINT ""
710 MAT PRINT B(M,N);
720 GO TO 810
730 PRINT ""
740 PRINT "THE ROW-REDUCED ECHELON FORM IS THE ENTRY-BY-
    ENTRY"
750 PRINT "QUOTIENT OF THE FOLLOWING TWO MATRICES:"
760 PRINT ""
770 MAT PRINT E(M,N);
780 PRINT ""
790 PRINT ""
800 MAT PRINT F(M,N);
810 PRINT ""
820 END
```

Matrix Multiplication

This program multiplies two matrices (if the multiplication is defined) and allows a new matrix to be entered for products of more than two once the first operation is completed. To suppress the ability to enter fractions, delete lines 90–400, 440, and 990–1020. To do the same for output, delete lines 610–760 and 810–880.

```
10 DIM A(15,15),B(15,15)
20 DIM C(15,15),E(15,15),F(15,15)
30 PRINT ""
40 PRINT "GIVE THE DIMENSIONS OF A AS M,N"
50 INPUT M,N
60 PRINT "GIVE THE DIMENSIONS OF B AS M,N"
70 INPUT M1,N1
80 IF N<>M1 THEN GO TO 1040
90 PRINT
```

```
100 PRINT "IF ONLY  A  HAS FRACTIONS, TYPE  A , IF ONLY
    B  , TYPE  B"
110 PRINT "IF BOTH HAVE FRACTIONS, TYPE  AB ,  IF NEITHER,
    TYPE  N"
120 INPUT F$
130 IF F$="N" THEN GO TO 410
140 IF F$="B" THEN GO TO 410
150 PRINT ""
160 PRINT "ENTER MATRIX OF NUMERATORS OF A (BY ROWS, USE
    LINE FEED)
170 MAT INPUT A(M,N)
180 PRINT ""
190 PRINT "ENTER MATRIX OF DENOMINATORS OF A"
200 MAT INPUT C(M,N)
210 FOR I=1 TO M
220 FOR J=1 TO N
230 IF ABS(C(I,J))<.000001  THEN PRINT "INVALID DENOMINATOR"
240 LET A(I,J)=A(I,J)/C(I,J)
250 NEXT J
260 NEXT I
270 IF F$="A" THEN GO TO 450
280 PRINT""
290 PRINT "ENTER MATRIX OF NUMERATORS OF  B  (BY ROWS, USE
    LINE FEED)"
300 MAT INPUT B(M1,N1)
310 PRINT ""
320 PRINT "ENTER MATRIX OF DENOMINATORS OF  B"
330 MAT INPUT C(M1,N1)
340 FOR I=1 TO M1
350 FOR J=1 TO N1
360 IF ABS(C(I,J))<.00001 THEN PRINT "INVALID DENOMINATOR"
370 LET B(I,J)=B(I,J)/C(I,J)
380 NEXT J
390 NEXT I
400 GO TO 480
410 PRINT ""
420 PRINT "ENTER A BY ROWS (USE LINE FEED BETWEEN ROWS)"
430 MAT INPUT A(M,N)
440 IF F$="B"  THEN GO TO 280
450 PRINT ""
460 PRINT "ENTER B BY ROWS (USE LINE FEED BETWEEN ROWS)"
470 MAT INPUT B(N,N1)
480 FOR I=1TO M
490 FOR J=1 TO N1
500 LET C(I,J)=0
510 FOR K=1 TO N
520 LET C(I,J)=C(I,J)+A(I,K)*B(K,J)
530 NEXT K
540 NEXT J
550 NEXT I
560 FOR I=1 TO M
```

```
570 FOR J=1 TO N1
580 LET A(I,J)=C(I,J)
590 NEXT J
600 NEXT I
610 LET T=1
620 FOR I=1 TO M
630 FOR J=1 TO N1
640 LET K=1
650 LET G =ABS(C(I,J))
660 IF G*K-INT(G*K)<.0001 THEN GO TO 720
670 IF G*K-INT(G*K)>.99999 THEN GO TO 720
680 LET K=K+1
690 IF K>10000 THEN GO TO 770
700 LET T=2
710 GO TO 660
720 LET E(I,J)=C(I,J)*K
730 LET F(I,J)=K
740 NEXT J
750 NEXT I
760 IF T=2 THEN GO TO 810
770 PRINT ""
780 PRINT "THE PRODUCT IS:"
790 MAT PRINT C(M,N1);
800 GO TO 880
810 PRINT ""
820 PRINT "THE PRODUCT IS THE ENTRY-BY-ENTRY QUOTIENT"
830 PRINT "OF THE FOLLOWING TWO MATRICES"
840 PRINT ""
850 MAT PRINT E(M,N1);
860 PRINT ""
870 MAT PRINT F(M,N1);
880 PRINT ""
890 PRINT "DO YOU WANT A FURTHER PRODUCT?   YES OR NO"
900 INPUT G$
910 IF G$="NO" THEN GO TO 1060
920 PRINT ""
930 PRINT "GIVE THE DIMENSIONS OF THE NEW  B   MATRIX AS M,N"
940 INPUT M2,N2
950 IF N1<>M2 THEN GO TO 1040
960 LET N=N1
970 LET M1=M2
980 LET N1=N2
990 PRINT ""
1000 PRINT "DO YOU WANT TO ENTER FRACTIONS?  YES OR NO"
1010 INPUT A$
1020 IF A$="YES" THEN GO TO 280
1030 GO TO 450
1040 PRINT ""
1050 PRINT "THE PRODUCT IS NOT DEFINED."
1060 PRINT ""
1070 END
```

Determinant

This program computes the determinant of a square matrix. To suppress the ability to enter fractions, delete lines 70–100 and 130–250. To do the same for output, delete lines 550–660, 690–700, and 720–750.

```
10 DIM B(10,15),C(15)
20 PRINT ""
30 PRINT "ENTER THE ORDER OF THE MATRIX."
40 INPUT M
50 LET N=M
60 PRINT ""
70 PRINT "DO YOU WANT TO ENTER FRACTIONS? YES OR NO"
80 INPUT A$
90 PRINT ""
100 IF A$="YES" THEN GO TO 140
110 PRINT "ENTER THE MATRIX BY ROWS (USE LINE FEED BETWEEN
    ROWS"
120 MAT INPUT B(M,N)
130 GO TO 260
140 PRINT "ENTER THE MATRIX OF NUMERATORS BY ROWS (USE LINE
    FEED BETWEEN ROWS)"
150 MAT INPUTB(M,M)
160 PRINT""
170 PRINT "ENTER MATRIX OF DENOMINATORS BY ROWS"
180 MAT INPUT F(M,M)
190 FOR I=1 TO M
200 FOR J=1 TO M
210 IF ABS(F(I,J))<.00001 THEN PRINT "INVALID DENOMINATOR"
220 IF ABS(F(I,J))<.00001 THEN GO TO 760
230 LET B(I,J)=B(I,J)/F(I,J)
240 NEXT J
250 NEXT I
260 LET J=1
270 LET D1=1
280 FOR I=1 TO M
290 LET K=I-1
300 LET K=K+1
310 IF B(K,J)<>0 THEN GO TO 360
320 IF K < M THEN GO TO 300
330 LET J=J+1
340 IF J>N THEN GO TO 550
350 GO TO 290
360 LET D=B(K,J)
370 LET D1=D1*D
380 IF K>I THEN LET D1=-D1
390 FOR J1=J TO N
400 LET C(J1)=B(I,J1)
410 LET B(I,J1)=B(K,J1)
420 LET B(K,J1)=C(J1)
430 LET B(I,J1)=B(I,J1)/D
440 NEXT J1
```

```
450 FOR I1=I+1 TO M
460 IF I1=I THEN GO TO 520
470 LET D=B(I1,J)
480 FOR J1=J TO N
490 LET B(I1,J1)=B(I1,J1)-B(I,J1)*D
500 IF ABS(B(I1,J1))<.00001 THEN LET B(I1,J1)=0
510 NEXT J1
520 NEXT I1
530 LET J=J+1
540 NEXT I
550 LET T=1
560 LET D1=D1*B(M,M)
570 IF ABS(D1)<.00001 THEN LET D1=0
580 LET K=1
590 LET G=ABS(D1)
600 IF G*K-INT(G*K)<.00001 THEN GO TO 660
610 IF G*K-INT(G*K)>.99999 THEN GO TO 660
620 LET K=K+1
630 IF K>10000 THEN GO TO 670
640 LET T=2
650 GO TO 600
660 LET D2=D1*K
670 PRINT ""
680 PRINT "THE DETERMINANT IS"
690 IF K>10000 THEN GO TO 710
700 IF T=2 THEN GO TO 730
710 PRINT D1
720 GO TO 760
730 PRINT D2
740 PRINT "_____"
750 PRINT K
760 PRINT ""
770 END
```

Matrix Inverse

This program determines whether or not a square matrix is invertible and gives its inverse if it is. To suppress the ability to enter fractions, delete lines 90–120 and 150–270. To do the same for output, delete lines 760–910 and 560–1030.

```
10 DIM B(10,20),C(15)
20 DIM H(10,15)
30 DIM E(10,15),F(10,15)
40 PRINT ""
50 PRINT "ENTER THE ORDER OF THE MATRIX."
60 INPUT M
70 LET N=M
80 PRINT ""
90 PRINT "DO YOU WANT TO ENTER FRACTIONS?  YES OR NO"
100 INPUT A$
```

```
110 PRINT ""
120 IF A$="YES" THEN GO TO 160
130 PRINT "ENTER THE MATRIX BY ROWS (USE LINE FEED
    BETWEEN ROWS)"
140 MAT INPUT H(M,M)
150 GO TO 280
160 PRINT "ENTER MATRIX OF NUMERATORS (BY ROWS, USE LINE
    FEED BETWEEN ROWS)"
170 MAT INPUT H(M,M)
180 PRINT ""
190 PRINT "ENTER MATRIX OF DENOMINATORS BY ROWS"
200 MAT INPUT F(M,M)
210 FOR I=1 TO M
220 FOR J=1 TO M
230 IF ABS(F(I,J))<.00001 THEN PRINT "INVALID DENOMINATOR"
240 IF ABS(F(I,J))<.00001 THEN GO TO 1050
250 LET H(I,J)=H(I,J)/F(I,J)
260 NEXT J
270 NEXT I
280 LET N=2*M
290 FOR I=1 TO M
300 FOR J=1 TO N
310 IF J<=M THEN LET B(I,J)=H(I,J)
320 IF J>M THEN LET B(I,J)=0
330 IF J=I+M THEN LET B(I,J)=1
340 NEXT J
350 NEXT I
360 LET J=1
370 FOR I=1 TO M
380 LET K=I-1
390 LET K=K+1
400 IF ABS(B(K,J))>.00001 THEN GO TO 450
410 IF K < M THEN GO TO 390
420 LET J=J+1
430 IF J>N THEN GO TO 690
440 GO TO 380
450 LET D=B(K,J)
460 FOR J1=J TO N
470 LET C(J1)=B(I,J1)
480 LET B(I,J1)=B(K,J1)
490 LET B(K,J1)=C(J1)
500 LET B(I,J1)=B(I,J1)/D
510 NEXT J1
520 FOR I1=1 TO M
530 IF I1=I THEN GO TO 590
540 LET D=B(I1,J)
550 FOR J1=J TO N
560 LET B(I1,J1)=B(I1,J1)-B(I,J1)*D
570 IF ABS(B(I1,J1))<.00001 THEN LET B(I1,J1)=0
580 NEXT J1
590 NEXT I1
600 LET J=J+1
610 NEXT I
```

```
620 LET T=0
630 FOR J=1 TO M
640 LET T=T+B(M,J)
650 NEXT J
660 IF ABS(T)>.00001 THEN GO TO 700
670 PRINT ""
680 PRINT "THE MATRIX IS NOT INVERTIBLE."
690 GO TO 1050
700 FOR I=1 TO M
710 FOR J=1 TO M
720 LET B(I,J)=B(I,J+M)
730 NEXT J
740 NEXT I
750 LET N=M
760 LET T=1
770 FOR I=1 TO M
780 FOR J=1 TO N
790 LET K=1
800 LET G=ABS(B(I,J))
810 IF G*K-INT(G*K)<.00001 THEN GO TO 870
820 IF G*K-INT(G*K)>.99999 THEN GO TO 870
830 LET K=K+1
840 IF K>10000 THEN GO TO 920
850 LET T=2
860 GO TO 810
870 LET E(I,J)=B(I,J)*K
880 LET F(I,J)=K
890 NEXT J
900 NEXT I
910 IF T=2 THEN GO TO 970
920 PRINT ""
930 PRINT "THE INVERSE IS:"
940 PRINT ""
950 MAT PRINT B(M,N);
960 GO TO 1040
970 PRINT ""
980 PRINT "THE INVERSE IS THE QUOTIENT OF THE FOLLOWING
    MATRICES:"
990 PRINT ""
1000 MAT PRINT E(M,N);
1010 PRINT ""
1020 PRINT ""
1030 MAT PRINT F(M,N);
1040 PRINT ""
1050 END
```

Matrix Powers

This program raises a square matrix to a sequence of positive integer powers. To suppress the ability to enter fractions, delete lines 120–280. To do the same for output, delete lines 530–760 and 800–890.

```
10 DIM A(15,15),B(15,15)
20 DIM C(15,15),F(15,15)
30 DIM D(15,15)
40 PRINT ""
50 PRINT "ENTER THE ORDER OF  A "
60 INPUT N
70 LET M=N
80 LET N1=N
90 PRINT "ENTER THE EXPONENT"
100 INPUT K2
110 LET K1=1
120 PRINT " IF  A  HAS FRACTIONS , TYPE  A , IF NOT TYPE N"
130 INPUT F$
140 IF F$="N" THEN GO TO 290
150 IF F$="N" THEN GO TO 290
160 PRINT ""
170 PRINT "ENTER MATRIX OF NUMERATORS OF A (BY ROWS, USE
    LINE FEED)
180 MAT INPUT A(M,N)
190 PRINT ""
200 PRINT "ENTER MATRIX OF DENOMINATORS OF A"
210 MAT INPUT C(M,N)
220 FOR I=1 TO M
230 FOR J=1 TO N
240 IF ABS(C(I,J))<.000001  THEN PRINT "INVALID DENOMINATOR"
250 LET A(I,J)=A(I,J)/C(I,J)
260 NEXT J
270 NEXT I
280 GO TO 320
290 PRINT ""
300 PRINT "ENTER A BY ROWS (USE LINE FEED BETWEEN ROWS)"
310 MAT INPUT A(M,N)
320 FOR I=1 TO N
330 FOR J=1 TO N
340 LET B(I,J)=A(I,J)
350 NEXT J
360 NEXT I
370 FOR I=1TO M
380 FOR J=1 TO N1
390 LET C(I,J)=0
400 FOR K=1 TO N
410 LET C(I,J)=C(I,J)+A(I,K)*B(K,J)
420 NEXT K
430 NEXT J
440 NEXT I
450 IF K1=K2 THEN GO TO 540
460 LET K1=K1+1
470 IF K1=K2 THEN GO TO 540
480 FOR I=1 TO N
490 FOR J=1 TO N
500 LET B(I,J)=C(I,J)
510 NEXT J
520 NEXT I
```

```
530 GO TO 370
540 LET T=1
550 LET S=1
560 FOR I=1 TO M
570 FOR J=1 TO N1
580 LET K=1
590 LET G =ABS(C(I,J))
600 IF G*K-INT(G*K)<.00001 THEN GO TO 720
610 IF G*K-INT(G*K)>.99999 THEN GO TO 720
620 LET K=K+1
630 IF K>10000 THEN GO TO 770
640 LET T=2
650 IF S=2 THEN GO TO 600
660 PRINT "THE RESULT IS NOT INTEGRAL. IF YOU WANT DECIMALS
    TYPE   D"
670 INPUT G$
680 IF G$="D" THEN GO TO 770
690 IF G$="D" THEN GO TO 770
700 LET S=2
710 GO TO 600
720 LET D(I,J)=C(I,J)*K
730 LET F(I,J)=K
740 NEXT J
750 NEXT I
760 IF T=2 THEN GO TO 810
770 PRINT ""
780 PRINT "A   TO THE ";K2;" IS:"
790 MAT PRINT C(M,N1);
800 GO TO 900
810 PRINT ""
820 PRINT "A    TO THE ";K2;" IS THE ENTRY-BY-ENTRY QUOTIENT"
830 PRINT "OF THE FOLLOWING TWO MATRICES"
840 PRINT ""
850 MAT PRINT D(M,N1);
860 PRINT ""
870 MAT PRINT F(M,N1);
880 GO TO 900
890 PRINT ""
900 PRINT " "
910 PRINT "DO YOU WANT A DIFFERENT POWER?   YES OR NO"
920 INPUT A$
930 IF A$="NO" THEN GO TO 1030
940 IF A$="NO" THEN GO TO 1030
950 PRINT "ENTER THE EXPONENT"
960 INPUT K3
970 IF K3<=K2 THEN GO TO 1000
980 LET K2=K3
990 GO TO 480
1000 LET K2=K3
1010 LET K1=1
1020 GO TO 320
1030 PRINT ""
1040 END
```

Simplex Algorithm

This program allows a sequence of pivot operations to be performed on a matrix. Deciding which pivots to use and interpreting the results are left to the operator. To suppress the ability to enter fractions, delete lines 70–100 and 130–250. To do the same for output, delete lines 450–600 and 650–730.

```
10 DIM B(10,15),C(15)
20 DIM E(10,15),F(10,15)
30 PRINT ""
40 PRINT "ENTER THE DIMENSIONS OF THE MATRIX AS M,N"
50 INPUT M,N
60 PRINT ""
70 PRINT "DO YOU WANT TO ENTER FRACTIONS?  YES OR NO"
80 INPUT A$
90 PRINT ""
100 IF A$="YES" THEN GO TO 140
110 PRINT "ENTER THE MATRIX BY ROWS (USE LINE FEED BETWEEN
    ROWS)
120 MAT INPUT B(M,N)
130 GO TO 260
140 PRINT "ENTER MATRIX OF NUMERATORS (BY ROWS, USE LINE
    FEED BETWEEN ROWS)
150 MAT INPUT B(M,N)
160 PRINT ""
170 PRINT "ENTER MATRIX OF DENOMINATORS BY ROWS"
180 MAT INPUT F(M,N)
190 FOR I=1 TO M
200 FOR J=1 TO N
210 IF ABS(F(I,J))<.00001 THEN PRINT "INVALID DENOMINATOR"
220 IF ABS(F(I,J))<.00001 THEN GO TO 750
230 LET B(I,J)=B(I,J)/F(I,J)
240 NEXT J
250 NEXT I
260 PRINT ""
270 PRINT "CHOSE THE PIVOT ROW I AND COLUMN J AS  I,J"
280 INPUT I,J
290 LET D=B(I,J)
300 IF D^2>.00001 THEN GO TO 340
310 PRINT ""
320 PRINT "THE PIVOT ENTRY IS  0 .  CHOOSE ANOTHER."
330 GO TO 260
340 FOR J1=1 TO N
350 LET B(I,J1)=B(I,J1)/D
360 NEXT J1
370 FOR I1=1 TO M
380 IF I1=I THEN GO TO 440
390 LET D=B(I1,J)
400 FOR J1 = 1 TO N
410 LET B(I1,J1)=B(I1,J1)-B(I,J1)*D
420 IF ABS(B(I1,J1))<.00001 THEN LET B(I1,J1)=0
430 NEXT J1
```

```
440 NEXT I1
450 LET T=1
460 FOR I=1 TO M
470 FOR J=1 TO N
480 LET K=1
490 LET G=ABS(B(I,J))
500 IF G*K-INT(G*K)<.00001 THEN GO TO 560
510 IF G*K-INT(G*K)>.99999 THEN GO TO 560
520 LET K=K+1
530 IF K>10000 THEN GO TO 610
540 LET T=2
550 GO TO 500
560 LET E(I,J)=B(I,J)*K
570 LET F(I,J)=K
580 NEXT J
590 NEXT I
600 IF T=2 THEN GO TO 660
610 PRINT ""
620 PRINT "THE NEW MATRIX IS:"
630 PRINT ""
640 MAT PRINT B(M,N);
650 GO TO 740
660 PRINT ""
670 PRINT "THE NEW MATRIX IS THE ENTRY-BY-ENTRY"
680 PRINT "QUOTIENT OF THE FOLLOWING TWO MATRICES:"
690 PRINT ""
700 MAT PRINT E(M,N);
710 PRINT ""
720 PRINT ""
730 MAT PRINT F(M,N);
740 PRINT ""
760 PRINT "DO YOU WANT TO PIVOT AGAIN?   YES OR NO"
770 INPUT A$
780 IF A$="NO" THEN GO TO 810
790 IF A$="NO" THEN GO TO 810
800 GO TO 260
810 PRINT ""
820 END
```

Answers to Odd-Numbered Exercises

Section 1.1 (page 10)

1. $(-4, -8, 12)$ **3.** $(5, 1, 1)$ **5.** $(-3, -2, 2)$ **7.** $(-9, -3, 1)$

9. $\sqrt{14}$ **11.** $\sqrt{17}$ **13.** $\sqrt{14} - \sqrt{5}$ **15.** $(3/5, -4/5)$

17. $(4, 2)$ **19.** $2\mathbf{i} - 3\mathbf{k}$ **21.** $\sqrt{14}$

27. If $r < 0$, no solutions. If $r = 0$, simply point $\mathbf{p}$. If $r > 0$:
in $\mathbf{R}^2$, a circle centered at $\mathbf{p}$ with radius r;
in $\mathbf{R}^3$, a sphere centered at $\mathbf{p}$ with radius r.

29. $(2/3)\mathbf{p} + (1/3)\mathbf{q}$ **31.** yes

Section 1.2 (page 20)

1. 4   **3.** 4   **5.** $(11, -7, -10)$ **7.** $(-22, 14, 20)$

9. obtuse   **11.** $-8/\sqrt{145}$ **13.** of form $c(-4, 7, -1)$

15. $\sqrt{14}/2$

19. The quantity $\mathbf{v} \cdot \mathbf{w}$ is a *scalar*, and dot product is only defined for two vectors.

Section 1.3 (page 32)

1. $(3, 1, 2) \cdot (\mathbf{x} - (2, -1, -4)) = 0$
$3x + y + 2z = -3$

3. $(2, 1, 3) \cdot (\mathbf{x} - \mathbf{0}) = 0$
$2x + y + 3z = 0$

5. $(3, 6, 1) \cdot (\mathbf{x} - (0, 2, 0)) = 0$
$3x + 6y + z = 12$

7. $(2, 3, 0) \cdot (\mathbf{x} - (1/2, 0, 0)) = 0$
$2x + 3y = 1$

9. $(2, 1) \cdot (\mathbf{x} - (-1, 2)) = 0$
$2x + y = 0$

11. $(1, 1) \cdot (\mathbf{x} - (-1, 2)) = 0$
$x + y = 1$

13. $(2, -1) \cdot (\mathbf{x} - (-2, 5)) = 0$
$2x - y = -9$

15. $\mathbf{x}(t) = (2, 1, -3) + t(1, 2, 2)$
$x = 2 + t$
$y = 1 + 2t$
$z = -3 + 2t$

17. $\mathbf{x}(t) = (2, -3, 1) + t(1, 0, 0)$
$x = 2 + t$
$y = -3$
$z = 1$

19. $\mathbf{x}(t) = (1 - t)(2, 0, -2) + t(1, 4, 2)$
$x = \quad 2 - t$
$y = \quad\quad 4t$
$z = -2 + 4t$

21. $\mathbf{x}(t) = (2, 4, 5) + t(1, -1, -2)$
$x = 2 + t$
$y = 4 - t$
$z = 5 - 2t$

23. $\mathbf{x}(t) = (1, -1) + t(1, -3)$
$x = \quad 1 + t$
$y = -1 - 3t$

25. $(3, 1) \cdot (\mathbf{x} - (2, 3)) = 0$

27. $\sqrt{6}/2$

29. $9\sqrt{5}/5$

31. $(13/5, 1, -7/5)$

37. $2/5$

39. 0 (the planes are perpendicular)

Section 1.4 (page 42)

1. $(1, 2, 5, 1)$ **3.** $(12, -6, 14, 40)$ **5.** 2

7. $6\sqrt{11}$ **9.** $2\sqrt{47}$

11. $\mathbf{u}_1$ and $\mathbf{u}_3$, $\mathbf{u}_2$ and $\mathbf{u}_3$, $\mathbf{u}_2$ and $\mathbf{u}_4$, $\mathbf{u}_3$ and $\mathbf{u}_4$

15. $(2/\sqrt{31}, 1/\sqrt{31}, -1/\sqrt{31}, 0, 3/\sqrt{31}, 4/\sqrt{31})$

17. $\mathbf{x}(t) = (1 - t)(-1, 0, 3, 2) + t(-1, 0, 4, 5)$ (two point form)
$\mathbf{x}(t) = (-1, 0, 3, 2) + t(0, 0, 1, 3)$ (point-parallel form)
$x_1 = -1, x_2 = 0, x_3 = 3 + t, x_4 = 2 + 3t$ (parametric equations)

19. $(1, -1, 1, -1, 1) \cdot (\mathbf{x} - (3, 4, 5, 6, 7)) = 0$ (point-normal form)
$x_1 - x_2 + x_3 - x_4 + x_5 = 5$ (standard form)

21. $(2, -3, 0, 1, -1, 0) \cdot (\mathbf{x} - (1, 0, 0, 0, 0, 0)) = 0$

23. product of variables   **25.** products of variables

31. $\mathbf{x} = (3, 1, 0, 2, 1) + r(-1, 0, 4, 0, -1) + s(-4, 1, 1, 1, 0) + t(-3, 1, 0, -1, -1)$

35. $(2, 0, 2, -2)$

Section 2.1 (page 52)

1. yes   **3.** no   **5.** no

7. $\quad y + z = 6$ **11.** $(1 + 4s, s)$
$x + y - z = 0$
$-x + y + z = 3$

13. $(2 - 3t, -1 - s + 2t, s, t)$ **15.** point

17. line   **19.** -1

Section 2.2 (page 68)

1. $\begin{bmatrix} 2 & -3 & 1 & | & 0 \\ 1 & 0 & -2 & | & 1 \\ 0 & -4 & 1 & | & -1 \end{bmatrix}$ **3.** no

5. yes   **7.** no

9. $x_1 = 0,$
   $x_2 = 2,$
   $x_3 = -1$

11. $(4 - 2s - 3t, s, t)$

13. $(-3/4, -5/4, 13/4)$

15. $(2s, -1/3 + (5/3)s, s)$

17. no solution

19. $(1/2 + s, 1 + 2s - t, s, t)$

31. $k = 2$; no

33. Answers vary with computer system but are approximately those of exercises 13 through 20.

Section 2.3 (page 81)

1. $\begin{bmatrix} -1 & 4 & -1 \\ 5 & 0 & 5 \\ -1 & 1 & -1 \end{bmatrix}$

3. $\begin{bmatrix} -2 & -12 & 8 \\ -5 & -5 & 0 \\ 3 & 2 & -7 \end{bmatrix}$

5. $\begin{bmatrix} -1 - \lambda & 0 & 1 \\ 2 & -1 - \lambda & 3 \\ 0 & 1 & -2 - \lambda \end{bmatrix}$

7. $\begin{bmatrix} -1 & 2 & 0 \\ 0 & -1 & 1 \\ 1 & 3 & -2 \end{bmatrix}$

9. $\begin{bmatrix} -1 & -4 & 3 \\ -6 & 7 & -3 \\ 5 & 1 & 0 \end{bmatrix}$

11. $\begin{bmatrix} -1 & -6 & 5 \\ -4 & 7 & 1 \\ 3 & -3 & 0 \end{bmatrix}$

13. $\begin{bmatrix} 2 & 1 \\ -1 & 0 \\ 1/2 & -2 \end{bmatrix}$

15. $A^1 = (2, 1)$
    $A^3 = (1/2, -2)$
    $A_2 = (1, 0, -2)$

17. $AB = \begin{bmatrix} 7 & -1 & 1 \\ 0 & -4 & -2 \\ 6 & -6 & 0 \end{bmatrix}$, $BA = \begin{bmatrix} -1 & 2 & 4 \\ 5 & 6 & -1 \\ 6 & 4 & -2 \end{bmatrix}$

19. Neither is defined.

21. $AB = A$, BA not defined

23. $\begin{bmatrix} -9 & 1 & 8 \\ -2 & -3 & -10 \end{bmatrix}$

25. $A^2 = \begin{bmatrix} 2 & -3 \\ -3 & 5 \end{bmatrix}$, $A^3 = \begin{bmatrix} 5 & -8 \\ -8 & 13 \end{bmatrix}$

37. $AB = \begin{bmatrix} 1 & 1 & 2 & 2 & 3 & 3 & 4 & 4 & 5 & 5 \\ 2 & 1 & 3 & 2 & 4 & 3 & 5 & 4 & 6 & 5 \\ 3 & 1 & 4 & 2 & 5 & 3 & 6 & 4 & 7 & 5 \\ 4 & 1 & 5 & 2 & 6 & 3 & 7 & 4 & 8 & 5 \\ 5 & 1 & 6 & 2 & 7 & 3 & 8 & 4 & 9 & 5 \\ 6 & 1 & 7 & 2 & 8 & 3 & 9 & 4 & 10 & 5 \\ 7 & 1 & 8 & 2 & 9 & 3 & 10 & 4 & 11 & 5 \\ 8 & 1 & 9 & 2 & 10 & 3 & 11 & 4 & 12 & 5 \\ 9 & 1 & 10 & 2 & 11 & 3 & 12 & 4 & 13 & 5 \\ 10 & 1 & 11 & 2 & 12 & 3 & 13 & 4 & 14 & 5 \end{bmatrix}$

Section 2.4 (page 93)

1. $\begin{bmatrix} 1 & -1 & 3 \\ 1 & 0 & -1 \\ -2 & 1 & 0 \end{bmatrix} \begin{bmatrix} x \\ y \\ z \end{bmatrix} = \begin{bmatrix} 1 \\ 0 \\ -1 \end{bmatrix}$

3. $\begin{bmatrix} 1 & -1 & 1 \\ -1 & 1 & 1 \end{bmatrix} \begin{bmatrix} x \\ y \\ z \end{bmatrix} = \begin{bmatrix} 0 \\ 0 \end{bmatrix}$

5. $AB = BA = I$

7. $\begin{bmatrix} 1/2 & 0 \\ 3/2 & 1 \end{bmatrix}$

9. $\begin{bmatrix} 1 & -2 & 5 \\ 0 & 1 & -2 \\ 0 & 0 & 1 \end{bmatrix}$

11. not invertible

13. -6

15. $(7/2, -3/2, -3/2)$

17. $(12, 7)$

19. $(4, -4, -3)$

21. no inverse

25. $\begin{bmatrix} 1 & 1/4 \\ 0 & 1/4 \end{bmatrix}$

27. $(1, 0, 0, 0, 0)$

Section 2.5 (page 102)

1. homogeneous

3. 2

5. 3

7. (a) $ad \neq bc$
(b) $ad = bc$ but $af \neq ce$
(c) $ad = bc$ and $af = ce$

9. infinitely many solutions

11. one solution or infinitely many solutions

15, 17, 19. answers vary

Section 3.1 (page 112)

	minor	cofactor
3, 1	-7	-7
3, 2	-1	1

1. (above table)

3. 3

5. -5

7. 15

9. 0

11. 120

13. a^3

15. $k^2(ad - bc)$

17. $t^2 - 3t + 3$

21. $\{-1, 3\}$

Section 3.2 (page 122)

1. 2

3. -24

5. $-1/720$

7. $-x(x + 1)(2x - 1)$

9. $-(\lambda - 1)(\lambda + 1)(\lambda - 2)(\lambda + 2)$

11. $x \neq -1, 0, 1/2;\ t \neq 3,\ (1 \pm \sqrt{33})/2;\ \lambda \neq \pm 1, \pm 2$

15. $(-1)^k \det A$, where $k = n/2$ if n is even, and $k = (n - 1)/2$ if n is odd.

25. -1.94

Section 3.3 (page 131)

1. $x = 1,$
$y = -1$

3. $x = 3/5,$
$y = 3/2,$
$z = 17/10$

5. $w = 0,$
$x = 2,$
$y = 1,$
$z = -1$

7. Adj $A = \begin{bmatrix} 4 & -2 \\ 3 & 1 \end{bmatrix}$, $A^{-1} = \begin{bmatrix} 2/5 & -1/5 \\ 3/10 & 1/10 \end{bmatrix}$

9. Adj $A = \begin{bmatrix} -27 & 9 & 9 \\ 18 & -6 & -6 \\ 9 & -3 & -3 \end{bmatrix}$, no inverse exists

15. $w = 26.60,$
$x = -2.16,$
$y = 38.15,$
$z = 13.27$

Section 4.1 (page 142)

1. $\mathbf{w}_1 = 3\mathbf{u} + 2\mathbf{v}$

3. $c = -2$

5. $(0, 1, 2) = (1/2)(1, 2, 3) + (1/2)(-1, 0, 1)$

7. independent

9. independent

11. $(1, 1) = 1(1, 0) + 1(0, 1)$

13. independent

15. dependent by theorem 2

17. independent by theorem 3

19. dependent by theorem 3

21. dependent from definition

29. independent

31. $\mathbf{v}_4 = \mathbf{v}_1 - (1/2)\mathbf{v}_2 + (1/2)\mathbf{v}_3$

Section 4.2 (page 148)

13. does not span

15. does not span

Section 4.3 (page 158)

1. not a basis

3. basis

5. basis

7. $\{(1, -1, 1), (2, 0, 1)\}$

9. $\{(-1, 1, 2, -1), (1, 0, -1, 1)\}$

11. $\{(1, 1, 0), (1, -1, 1), (1, 0, 0)\}$

13. $\{(1, 0, 0, 1), (0, 1, 1, 0), (1, 1, 1, 2), (0, 1, 0, 0)\}$
15. $4 > 3$, too many vectors
17. part iii of theorem 5 $(m = n = 4)$
19. 2, $\{(1, 1, 1, 0), (-2, -1, 0, 1)\}$
21. 2, $\{(1, 2, 0), (0, 0, 1)\}$
23. 2, $\{(1, -1, 3, 1), (-2, 0, -2, 0)\}$
25. 1

Section 4.4 (page 163)

1. row rank = column rank = 2
3. row rank = column rank = 3
5. $\{(1, 0, 0), (0, 1, 0)\}$
7. $\{(1, 0, 1, 2), (-1, 1, 1, -1), (3, 1, 0, 1)\}$
9. $\{(1, 2, 0), (0, 0, 1)\}$
11. $\{(1, -1, 3), (0, 1, 1), (1, 1, 0)\}$

Section 4.5 (page 170)

1. orthonormal basis   **3.** not orthogonal   **5.** orthonormal basis
7. $1(1/3, -2/3, 2/3) + 2(-2/3, 1/3, 2/3) + 3(2/3, 2/3, 1/3)$
9. $\{(4\sqrt{5}/15, \sqrt{5}/3, 0, -2\sqrt{5}/15), (-2/3, 2/3, 0, 1/3), (\sqrt{5}/5, 0, 0, 2\sqrt{5}/5)\}$
11. $\{(\sqrt{2}/2, -\sqrt{2}/2, 0), (\sqrt{6}/6, \sqrt{6}/6, -\sqrt{6}/3)\}$
13. $\{(-\sqrt{2}/2, \sqrt{2}/2, 0, 0), (0, 0, 1, 0), (0, 0, 0, 1)\}$
15. $\{(1, 0, 0, 0), (0, 1, 0, 0), (0, 0, 1, 0), (0, 0, 0, 1)\}$

Section 5.1 (page 180)

5. linear
7. not linear
9. $n = 3$, $m = 2$, $T(\mathbf{x}_0) = (16, -3)$
11. $n = 2$, $m = 3$, $T(\mathbf{x}_0) = (-4, -3, 2)$
17. $\begin{bmatrix} -\sqrt{2}/2 & -\sqrt{2}/2 \\ \sqrt{2}/2 & -\sqrt{2}/2 \end{bmatrix}$
19. $\begin{bmatrix} 0 & 1 \\ 1 & 0 \end{bmatrix}$

Section 5.2 (page 191)

1. $\begin{bmatrix} 1 & 0 & 0 \\ 1 & 1 & 0 \\ 1 & 1 & 1 \end{bmatrix}$
3. $\begin{bmatrix} 0 & 0 & 1 \\ 0 & 1 & 0 \\ 1 & 0 & 0 \end{bmatrix}$
5. $[0 \ldots 0 \ 1 \ 0 \ldots 0]$, the ith elementary row vector
7. $\begin{bmatrix} 5 & 0 & 0 \\ 3 & 1 & 0 \\ 1 & -3 & 0 \\ 0 & -5 & 0 \end{bmatrix}$
9. $\begin{bmatrix} 1 & 0 & 1 \\ 1 & 2 & 0 \\ 2 & 1 & 1 \end{bmatrix}$
11. $\begin{bmatrix} -5 & 0 & 5 \\ 0 & 0 & 0 \\ 5 & 0 & -5 \end{bmatrix}$

Section 5.3 (page 201)

1. v_1 is, v_2 is   **3.** v_1 is, v_2 is not

5. w_1 is not, w_2 is   **7.** w_1 is, w_2 is

9. nullity 2, basis for $\text{Ker}(T) - \{(-1, 1, 1, 0), (-1, 0, 0, 1)\}$

11. nullity 1, basis for $\text{Ker}(T) - \{(3/5, -4/5, 1)\}$

13. rank 2, basis for $\text{Im}(T) - \{(1, 1, 2), (2, 0, -4)\}$, not onto

15. rank 2, basis for $\text{Im}(T) - \{(3, 1), (1, 2)\}$, onto

21. $T^{-1}(x, y, z) = ((3/2)x - (1/2)y - (1/2)z, x - y, (-1/2)x + (1/2)y + (1/2)z)$

23. not invertible   **25.** $T_{-\theta}$

27. $T_{1/c}$ if $c \neq 0$, not invertible if $c = 0$

Section 5.4 (page 208)

1. $(1/4)(1, -5)_\mathscr{B}$

3. $(1/5)(-8, 3, -3, 6)_\mathscr{B}$

5. $\dfrac{1}{4}\begin{bmatrix} 2 & 1 \\ 2 & -1 \end{bmatrix}$, $(1/4)(1, -5)_\mathscr{B}$

7. $\dfrac{1}{5}\begin{bmatrix} 3 & -1 & 2 & -2 \\ -3 & 1 & 3 & 2 \\ 3 & -1 & 2 & 3 \\ -1 & 2 & -4 & -1 \end{bmatrix}$ $(1/5)(-8, 3, -3, 6)_\mathscr{B}$

9. $\begin{bmatrix} 1 & 0 \\ 2 & -1 \end{bmatrix}$

11. $\dfrac{1}{6}\begin{bmatrix} 17 & 16 & 33 \\ 8 & 16 & 12 \\ -9 & -18 & -21 \end{bmatrix}$

Section 6.1 (page 218)

1. eigenvalue: 1; eigenvectors: every $x \neq 0$ in R^n.

3. no eigenvalues

5. eigenvalue: 1; eigenvectors: $x = (0, c)$, $c \neq 0$
eigenvalue: -1; eigenvectors: $x = (c, 0)$, $c \neq 0$

7. $\lambda^2 - 9 = 0$; $\lambda = \pm 3$ **9.** $\lambda^2 - 2\lambda + 1 = 0$; $\lambda = 1$

11. $-\lambda^3 + 2\lambda = 0$; $\lambda = 0, \pm\sqrt{2}$ **13.** $-\lambda^3 + 3\lambda^2 = 0$; $\lambda = 0, 3$

15. $(1 - \lambda)^2(-2 - \lambda) = 0$; $\lambda = 1, -2$ **17.** $\lambda^4 - 4\lambda^3 = 0$; $\lambda = 0, 4$

19. $\lambda = 3$: all $(2c, c)$, c in R; basis $\{(2, 1)\}$
$\lambda = -3$: all $(-c, c)$, c in R; basis $\{(-1, 1)\}$

21. $\lambda = 1$: all (c, c), c in R; basis $\{(1, 1)\}$

23. $\lambda = 0$: all $(-c, 0, c)$, c in R; basis $\{(-1, 0, 1)\}$
$\lambda = \sqrt{2}$: all $(c, -c\sqrt{2}, c)$, c in R; basis $\{(1, -\sqrt{2}, 1)\}$
$\lambda = -\sqrt{2}$: all $(c, c\sqrt{2}, c)$, c in R; basis $\{(1, \sqrt{2}, 1)\}$

25. $\lambda = 0$: all $(-c - d, c, d)$, c, d in **R**; basis $\{(-1, 1, 0), (-1, 0, 1)\}$
$\lambda = 3$: all (c, c, c), c in **R**; basis $\{(1, 1, 1)\}$

27. $\lambda = 1$: all $(c, 0, 0)$, c in **R**; basis $\{(1, 0, 0)\}$
$\lambda = -2$: all $(0, c, 0)$, c in **R**; basis $\{(0, 1, 0)\}$

29. $\lambda = 0$: all $(-b - c - d, b, c, d)$, b, c, d in **R**; basis $\{(-1, 1, 0, 0), (-1, 0, 1, 0), (-1, 0, 0, 1)\}$
$\lambda = 4$: all (c, c, c, c), c in **R**; basis $\{(1, 1, 1, 1)\}$

31. 1, 3, 4   **33.** 1, 3, 6, 10

37. no eigenvalues except for $\theta = 0$, $\lambda = 1$ and $\theta = \pi$, $\lambda = -1$

Section 6.2 (page 229)

1. diagonalizable; $P = \begin{bmatrix} -1 & 2 \\ 1 & 1 \end{bmatrix}$; $D = \begin{bmatrix} -3 & 0 \\ 0 & 3 \end{bmatrix}$

3. not diagonalizable.

5. diagonalizable; $P = \begin{bmatrix} 1 & -1 & 1 \\ \sqrt{2} & 0 & -\sqrt{2} \\ 1 & 1 & 1 \end{bmatrix}$; $D = \begin{bmatrix} -\sqrt{2} & 0 & 0 \\ 0 & 0 & 0 \\ 0 & 0 & \sqrt{2} \end{bmatrix}$

7. diagonalizable; $P = \begin{bmatrix} -1 & -1 & 1 \\ 1 & 0 & 1 \\ 0 & 1 & 1 \end{bmatrix}$; $D = \begin{bmatrix} 0 & 0 & 0 \\ 0 & 0 & 0 \\ 0 & 0 & 3 \end{bmatrix}$

9. not diagonalizable.

11. diagonalizable; $P = \begin{bmatrix} -1 & -1 & -1 & 1 \\ 1 & 0 & 0 & 1 \\ 0 & 1 & 0 & 1 \\ 0 & 0 & 1 & 1 \end{bmatrix}$; $D = \begin{bmatrix} 0 & 0 & 0 & 0 \\ 0 & 0 & 0 & 0 \\ 0 & 0 & 0 & 0 \\ 0 & 0 & 0 & 4 \end{bmatrix}$

13. orthogonal   **15.** not orthogonal

17. $\begin{bmatrix} 1/\sqrt{2} & 1/\sqrt{2} \\ 1/\sqrt{2} & -1/\sqrt{2} \end{bmatrix}$

19. $\begin{bmatrix} 1/2 & -1/\sqrt{2} & 1/2 \\ 1/\sqrt{2} & 0 & -1/\sqrt{2} \\ 1/2 & 1/\sqrt{2} & 1/2 \end{bmatrix}$

21. $\begin{bmatrix} 1/\sqrt{2} & 1/\sqrt{6} & 1/\sqrt{3} \\ -1/\sqrt{2} & 1/\sqrt{6} & 1/\sqrt{3} \\ 0 & -2/\sqrt{6} & 1/\sqrt{3} \end{bmatrix}$

23. $\begin{bmatrix} 1/\sqrt{2} & 0 & 1/\sqrt{2} & 0 \\ -1/\sqrt{2} & 0 & 1/\sqrt{2} & 0 \\ 0 & 1/\sqrt{2} & 0 & 1/\sqrt{2} \\ 0 & -1/\sqrt{2} & 0 & 1/\sqrt{2} \end{bmatrix}$

27. A is diagonalizable if $(a - d)^2 + 4bc > 0$ or if $a = d$ and $b = c = 0$.
A is not diagonalizable if $(a - d)^2 + 4bc < 0$.

Section 6.3 (page 239)

1. -2 **3.** 2   **5.** $(i, 0, 0)$

7. $\begin{bmatrix} i & -2i & 1 \\ -1+i & i & 1+2i \\ -1-i & 3 & -2+2i \end{bmatrix}$

9. $\begin{bmatrix} 8-i & 2+i & 1+3i \\ 2+i & 3-i & -1+2i \\ -1-3i & 1-2i & 3-i \end{bmatrix}$

11. $1 + 2i$

13. Neither A nor B is hermitian.

15. yes

17. no

19. $\{(-1, 0, 1), (0, -1, 1)\}$

21. $\{(1, 1)\}$

23. $\lambda_1 = -5$, $\mathbf{x}_1 = (-(3+4i)s, 5s)$, basis $= \{(-(3+4i), 5)\}$
$\lambda_2 = 5$, $\mathbf{x}_2 = ((3+4i)t, 5t)$, basis $= \{(3+4i, 5)\}$

25. $\lambda_1 = 1$, $\mathbf{x}_1 = (0, r, 0)$, basis $= \{(0, 1, 0)\}$;
$\lambda_2 = -2i$, $\mathbf{x}_2 = ((1-i)s, 0, s)$, basis $= \{(1-i, 0, 1)\}$
$\lambda_3 = 2i$, $\mathbf{x}_3 = (-(1+i)t, 0, t)$, basis $= \{(-1-i, 0, 1)\}$

27. $\lambda_1 = i$, $\mathbf{x}_1 = (s, t, 0)$, basis $= \{(1, 0, 0), (0, 1, 0)\}$

29. $\lambda_1 = -1$, $\mathbf{x}_1 = (0, ir, r)$, basis $= \{(0, i, 1)\}$
$\lambda_2 = 1$, $\mathbf{x}_2 = (s, -it, t)$, basis $= \{(1, 0, 0), (0, -i, 1)\}$

31. $P = \begin{bmatrix} -(3+4i)/(5\sqrt{2}) & (3+4i)/(5\sqrt{2}) \\ 1/\sqrt{2} & 1/\sqrt{2} \end{bmatrix}$, $D = \begin{bmatrix} -5 & 0 \\ 0 & 5 \end{bmatrix}$ (hermitian)

33. $P = \begin{bmatrix} 0 & 1-i & 1+i \\ 1 & 0 & 0 \\ 0 & 1 & 1 \end{bmatrix}$, $D = \begin{bmatrix} 1 & 0 & 0 \\ 0 & -2i & 0 \\ 0 & 0 & 2i \end{bmatrix}$

35. not diagonalizable

37. $P = \begin{bmatrix} 0 & 1 & 0 \\ i/\sqrt{2} & 0 & -i/\sqrt{2} \\ 1/\sqrt{2} & 0 & 1/\sqrt{2} \end{bmatrix}$, $D = \begin{bmatrix} -1 & 0 & 0 \\ 0 & 1 & 0 \\ 0 & 0 & 1 \end{bmatrix}$ (hermitian)

39. Matrices of exercises 29 and 30 are unitary.

Section 7.1 (page 250)

1. vector space

3. is not; not closed under addition

5. vector space

7. is not; $1\mathbf{y} = \mathbf{x} \neq \mathbf{y}$

9. vector space

11. subspace

13. is not; not closed under addition or scalar multiplication

15. subspace

Section 7.2 (page 260)

1. independent

3. dependent; $1 = (1/2)(x+1) + (-1/2)(x-1)$

5. independent

7. dependent; $1 = (1) \sin^2 x + (1) \cos^2 x$

9. independent

11. generates

13. generates

15. generates

17. does not generate

19. generates

21. basis

23. dependent

25. basis

27. dependent, does not generate

29. basis

31. 12

33. 4

35. 2

37. 3

Section 7.3 (page 270)

1. $(2, -1, 0, 3)$

3. $(-3/2, -7/2)$

5. $(4, -5, -9/2, 0)$

7. independent

9. dependent; $(1)\begin{bmatrix} 1 & 2 \\ -1 & 0 \end{bmatrix} + (2)\begin{bmatrix} 0 & -1 \\ 1 & 1 \end{bmatrix} = \begin{bmatrix} 1 & 0 \\ 1 & 2 \end{bmatrix}$

11. $\begin{bmatrix} 0 & 1 & 0 \\ 0 & 0 & 1 \\ -1 & 0 & 0 \\ 0 & 0 & 0 \end{bmatrix}$

13. $\begin{bmatrix} 1 & 0 & 0 \\ 0 & 0 & -1 \\ 0 & 0 & -1 \\ 1 & -1 & 0 \end{bmatrix}$

17. $\text{Ker}(T) = \{\mathbf{0}\}$; one-to-one

19. $\text{Ker}(T) = \{\mathbf{0}\}$; one-to-one

21. $\begin{bmatrix} 0 & 1 & 0 & 0 & 0 \\ 0 & 0 & 2 & 0 & 0 \\ 0 & 0 & 0 & 3 & 0 \\ 0 & 0 & 0 & 0 & 4 \end{bmatrix}$

23. $\begin{bmatrix} 0 & 0 & 0 \\ 1 & 0 & 0 \\ 0 & 1/2 & 0 \\ 0 & 0 & 1/3 \end{bmatrix}$

Section 7.4 (page 280)

1. Yes

3. No, properties (b), (c), and (d) fail

5. No, property (d) fails

7. No, property (d) fails

9. Neither orthogonal nor orthonormal

11. Orthogonal and orthonormal

13. $\{1/\sqrt{2}, (\sqrt{3/2})x, (1/2)\sqrt{5/2}(3x^2 - 1), (1/2)\sqrt{7/2}(5x^3 - 3x)\}$

15. $\{1, \sqrt{2}(e^x - e + 1)/\sqrt{4e - e^2 - 3}\}$

Section 8.1 (page 291)

1. $p(x) = -3 + 2x + x^2$

3. $p(x) = -1 - x^2 + x^3$

5. $p(x) = 1 - 2x^2 + x^3$

7. $p(2.5) = 0.4045$

9. $p(x) = -(4/5) + (13/10)x$

11. $p(x) = (3.0) - (13.7)x + (7.5)x^2$

13.
$$7b_1 \quad\quad + 28b_3 \quad\quad\quad = -28$$
$$28b_2 \quad\quad + 196b_4 = 0$$
$$28b_1 \quad\quad + 196b_3 \quad\quad = 20$$
$$196b_2 \quad\quad + 1588b_4 = 0$$

15. $p(x) = 303 - 86x$

Section 8.2 (page 299)

1. Current through a: 5/6, b: 1/6, e: 1

3. Current through a: 1.08, b: 0.69, c: 0.39, d: 0.80, e: 0.41

5. A: $28s$, B: $10s$, C: $15s$ for $s > 0$

7. A: $2s$, B: s, C: s for $s > 0$

9. A: 12.51, B: 18.80, C: 9.46

11. A: 49.20, B: 39.28, C: 31.20

Section 8.3 (page 307)

1. $t_{32} = 0.4$

3. $T = \begin{bmatrix} 0.2 & 0.1 & 0.2 \\ 0.5 & 0.8 & 0.4 \\ 0.3 & 0.1 & 0.4 \end{bmatrix}$

5. (1/3, 1/3, 1/3)

7. (1/6, 17/30, 4/15)

9. 0.62

11.

	W	D	B
W	0	2/3	1/3
D	1/2	0	1/2
B	1/3	1/3	1/3

$T = \begin{bmatrix} 0 & 1/2 & 1/3 \\ 2/3 & 0 & 1/3 \\ 1/3 & 1/2 & 1/3 \end{bmatrix}$

13. (5/18, 1/3, 7/18)

15. 1/81

17. $(1/(2s + t))(s, s, t)$; $s, t \geq 0$, $s + t > 0$

21. (0.131, 0.689, 0.180)

Section 8.4 (page 319)

1. $(x')^2 + (y')^2 - \dfrac{(z')^2}{2} = 1$ (hyperboloid of one sheet)

3. $\dfrac{(x')^2}{4} + \dfrac{(y')^2}{24} - \dfrac{(z')^2}{6} = 0$ (cone)

5. $\dfrac{(x')^2}{18} + \dfrac{(y')^2}{36} + \dfrac{(z')^2}{36} = 1$ (ellipsoid)

7. $8(x'')^2 + 2(y'')^2 - 2(z'')^2 = 1$ (hyperboloid of one sheet)

9. $(x'')^2 - \dfrac{(z'')^2}{2} = 1$ (hyperbolic cylinder)

15. $3(x')^2 - (y')^2 + 6x' - 4y' = 0$

17. $-3(x'')^2 + (y'')^2 = 1$ (hyperbola)

Section 8.5 (page 328)

1. $\mathbf{p}(x) = 1/5$

3. $\mathbf{p}(x) = 2(14 \ln 2 - 9) + 6(2 - 3 \ln 2)x$

5. $\mathbf{p}(x) = -(3/35) + (6/7)x^2$

7. $\mathbf{p}(x) = (873 \ln 2 - 603) + 6(137 - 198 \ln 2)x + 30(13 \ln 2 - 9)x^2$

9. $\mathbf{p}(x) = -(3/35) + (6/7)x^2$ 　　　　　**11.** $\mathbf{p}(x) = 1 + 2 \sin x$

13. $\mathbf{p}(x) = 1 + 2(\pi^2 - 6) \sin x$ 　　　　**15.** $\mathbf{p}(x) = 1 + \cos x - \sin 2x$
$$-1 + 2 \sin x - \sin 2x$$

Section 9.1 (page 338)

1.

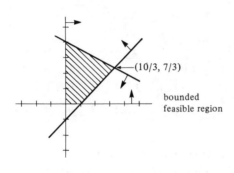

(10/3, 7/3)

bounded feasible region

3.

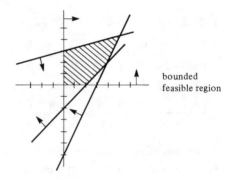

bounded feasible region

5.

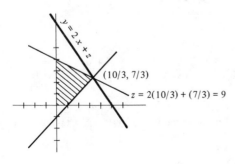

$y = 2x + z$

(10/3, 7/3)

$z = 2(10/3) + (7/3) = 9$

7.

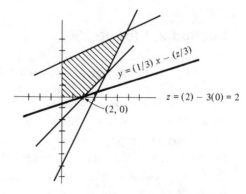

$y = (1/3)x - (z/3)$

(2, 0)

$z = (2) - 3(0) = 2$

9. $x - y + s_1 \qquad = 2$
$3x - y \qquad + s_2 = 6$
$x, y, s_1, s_2 \geq 0$

11. row-reduced
columns 2, 3, 5
$x_1 = 0, x_2 = 1, x_3 = 2, x_4 = 0, x_5 = 4$

13. row-reduced
columns 2, 3, 5
$x_1 = 0, x_2 = 6, x_3 = 2, x_4 = 0, x_5 = 1$
or columns 3, 4, 5
$x_1 = x_2 = 0, x_3 = 2, x_4 = 6, x_5 = 1$

15. Let x_1 be amount of savings, x_2 stocks, and x_3 property.

Standard form:

$$x_1 + x_2 + x_3 \leq 50{,}000$$
$$x_1 \qquad\qquad \leq 10{,}000$$
$$x_1 + x_2 \qquad \leq 20{,}000$$
$$x_1, x_2, x_3 \geq 0$$
$$z = 0.08x_1 + 0.15x_2 + 0.12x_3$$

Equation form:

$$x_1 + x_2 + x_3 + x_4 \qquad\qquad = 50{,}000$$
$$x_1 \qquad\qquad + x_5 \qquad = 10{,}000$$
$$x_1 + x_2 \qquad\qquad + x_6 = 20{,}000$$
$$x_1, x_2, x_3, x_4, x_5, x_6 \geq 0$$
$$z = 0.08x_1 + 0.15x_2 + 0.12x_3$$

Section 9.2 (page 350)

1. Bottom row:

-9	1	0	0	0	1

3.

	c*	x_1	x_2	x_3	x_4	x_5	b
x_2	-2	3	1	2	0	0	2
x_5	2	1	0	-3	0	1	5
x_4	0	2	0	-1	1	0	8
		-5	0	-11	0	0	6

5.

	c*	x_1	x_2	x_3	x_4	x_5	x_6	b
x_5	0	2	1	1	3	1	0	12
x_6	0	1	-3	-1	1	0	1	8
		-1	2	-1	-4	0	0	0

7. Solution of $(0, 2, 1, 0, 4)$ is maximal since no simplex indicators are negative.

9. Solution of $(11, 0, 8, 0, 0, 0)$ is minimal since no simplex indicators are positive.

11. $x_1 = x_2 = 0, x_3 = 1, x_4 = 9, x_5 = 8, z = 17$

13. $x_2 = 12, x_1 = x_3 = x_4 = 0, z = -24$

15. Savings $0, stocks $20,000, property $30,000, earnings $6600

Section 9.3 (page 356)

1. $2x - 3y - z + s_1 \qquad\qquad = 5$
$y + 2z \qquad - s_2 + a_1 \qquad = 8$
$x \qquad - 2z \qquad\qquad + a_2 = 3$
$x, y, z, s_1, s_2, a_1, a_2 \geq 0$

3. $2x - y \qquad\qquad + a_1 \qquad = 7$
$x + 2y + s_1 \qquad\qquad = 18$
$-3x + 5y \qquad - s_2 \qquad + a_2 = 3$
$x, y, s_1, s_2, a_1, a_2 \geq 0$

5. $x = 3, y = 8, z = 0, s_1 = 23, s_2 = 0$

7. $x_1 = 38/7, x_2 = 27/7, s_1 = 34/7, s_2 = 0$

9. No maximum exists.

11. $x = 32/5, y = 29/5$, minimum value: $-23/5$

13. $x_1 = 7, x_2 = x_3 = 0, x_4 = 12$, maximum value: 38

15. A: 50%, B: $16\frac{2}{3}$%, C: $33\frac{1}{3}$%, D: 0%; cost $0.26\frac{2}{3}$/lb.

Index